Information Systems Engineering and Management

Volume 85

The book series "Information Systems Engineering and Management" (ISEM) publishes innovative and original works in the various areas of planning, development, implementation, and management of information systems and technologies by enterprises, citizens, and society for the improvement of the socio-economic environment.

The series is multidisciplinary, focusing on technological, organizational, and social domains of information systems engineering and management. Manuscripts published in this book series focus on relevant problems and research in the planning, analysis, design, implementation, exploration, and management of all types of information systems and technologies. The series contains monographs, lecture notes, edited volumes, pedagogical and technical books as well as proceedings volumes.

Some topics/keywords to be considered in the ISEM book series are, but not limited to: Information Systems Planning; Information Systems Development; Exploration of Information Systems; Management of Information Systems; Blockchain Technology; Cloud Computing; Artificial Intelligence (AI) and Machine Learning; Big Data Analytics; Multimedia Systems; Computer Networks, Mobility and Pervasive Systems; IT Security, Ethics and Privacy; Cybersecurity; Digital Platforms and Services; Requirements Engineering; Software Engineering; Process and Knowledge Engineering; Security and Privacy Engineering, Autonomous Robotics; Human-Computer Interaction; Marketing and Information; Tourism and Information; Finance and Value; Decisions and Risk; Innovation and Projects; Strategy and People.

Indexed by Google Scholar. All books published in the series are submitted for consideration in the Web of Science.

For book or proceedings proposals please contact Alvaro Rocha (amrrocha@gmail.com).

Fadele Ayotunde Alaba · Ilésanmí Gabriel Ajibola

Digital Pedagogy: Bridging the Gap Between Software and the Humanities

Springer

Fadele Ayotunde Alaba
Department of Computer Science
Federal University of Education Zaria
Zaria, Nigeria

Ilésanmí Gabriel Ajibola
TETFund Centre of Excellence
Federal University of Education
Zaria, Nigeria

ISSN 3004-958X ISSN 3004-9598 (electronic)
Information Systems Engineering and Management
ISBN 978-3-032-20334-2 ISBN 978-3-032-20335-9 (eBook)
https://doi.org/10.1007/978-3-032-20335-9

This Springer imprint is published by the registered company Springer Nature Switzerland AG
The registered company address is: Gewerbestrasse 11, 6330 Cham, Switzerland

Competing Interests The authors have no competing interests to declare that are relevant to the content of this manuscript.

Introduction

In recent decades, education has undergone a dramatic transformation with the integration of digital technologies into pedagogy. From learning management systems to collaborative tools, digital platforms now play a vital role in shaping how knowledge is delivered, consumed, and assessed. The need for *digital pedagogy*—teaching practices that thoughtfully incorporate digital tools—has become especially critical in the humanities, where debates about relevance, innovation, and accessibility are at the forefront of academic discourse. This book offers a comprehensive exploration of how digital tools, theories, and practices are reshaping teaching and learning in the humanities.

Purpose and Scope of the Book

The purpose of this book is to provide educators, scholars, and institutional leaders with a *theoretical and practical framework* for engaging with digital pedagogy in the humanities. It bridges the gap between software and scholarship, theory and practice, providing both academic insight and actionable strategies.

The scope encompasses:

- The evolution of pedagogy from traditional to digital modes.
- Integration of tools like coding, digital storytelling, and online collaboration.
- Strategies for curriculum design, assessment, and interdisciplinary teaching.
- Ethical, cultural, and institutional dimensions of digital education.

This book is especially designed for instructors, early-career researchers, instructional designers, and digital humanists interested in *enhancing learning experiences* through technology while maintaining academic integrity and critical inquiry.

Structure of the Book

The book is organized into 12 chapters, progressing from foundational theories to applied practices:

- Introduction to Digital Pedagogy: Concepts, debates, and core principles.
- The Digital Turn in the Humanities: Historical context and discipline-specific transformations.
- Pedagogical Theories in the Digital Age: Theoretical grounding for digital education.
- Tools and Platforms: A survey of key technologies and applications.
- Coding for Humanists: The rationale and methods for teaching code in humanities contexts.
- Designing Digitally Enriched Curricula: Strategies for aligning pedagogy and technology.
- Interdisciplinary Collaboration: Models for co-teaching and cross-disciplinary innovation.
- Data, Archives, and Cultural Memory: Preservation, metadata, and access.
- Digital Storytelling: Narrative innovation across media formats.
- Assessment and Evaluation: Methods for grading and feedback in digital contexts.
- Barriers and Institutional Challenges: Equity, access, training, and policy.
- Future Directions: Emerging technologies and global perspectives.

Significance of Quantitative Research

While digital pedagogy often emphasizes qualitative insight and creativity, *quantitative research* plays a vital role in its development and evaluation. Learning analytics, survey data, and statistical measures help assess *student engagement, learning outcomes, and tool efficacy*. Quantitative methods allow institutions to make data-informed decisions, tailor instructional interventions, and justify the allocation of technological resources. By combining both qualitative and quantitative approaches, and digital pedagogy can be both deeply reflective and evidence-driven.

The Researcher's Role

The researcher in digital pedagogy occupies a *hybrid role*—as a scholar, practitioner, designer, and innovator. They must be comfortable navigating both theoretical discourse and technical experimentation. The digital pedagogue curates content, selects appropriate tools, designs inclusive learning environments, and evaluates impact. They are also tasked with staying current in rapidly evolving technological landscapes, advocating for ethical use, and mentoring others in adopting digital approaches with critical awareness.

Conclusion

As education continues to evolve, digital pedagogy stands at the intersection of tradition and innovation. This book serves as both a *compass and a toolkit*—offering direction, inspiration, and practical guidance for those committed to enhancing humanities education in a digital age. By embracing digital literacy, pedagogical theory, and technological creativity, educators can foster more inclusive, engaging, and future-ready learning environments.

Contents

Chapter 1
Introduction to Digital Pedagogy

1.1 Defining Digital Pedagogy

Digital pedagogy is the concept and critical practice of teaching and learning in the digital era, not only the use of digital instruments. Beyond incorporating technology into classrooms, it questions how digital technologies affect educational knowledge generation, access, engagement, and power structures [1] (Fig. 1.1).

1.2 What is Digital Pedagogy?

Digital pedagogy is the purposeful and critical use of digital tools, settings, and literacies in the teaching and learning processes. It is not limited to online learning but also includes mixed, hybrid, and face-to-face modalities in which digital engagement is part of the learning process. "Digital pedagogy is less about using tools and more about using them thoughtfully and reflectively."

Key elements include:

1. critical involvement with technology.
2. pedagogical intentionality vs technical novelty
3. learner-centred and interactive techniques.
4. flexible, adaptable, and responsive
5. focus on equality, access, and integrity (Fig. 1.2).

F. A. Alaba and I. G. Ajibola, *Digital Pedagogy: Bridging the Gap Between Software and the Humanities*, Information Systems Engineering and Management 85,
https://doi.org/10.1007/978-3-032-20335-9_1

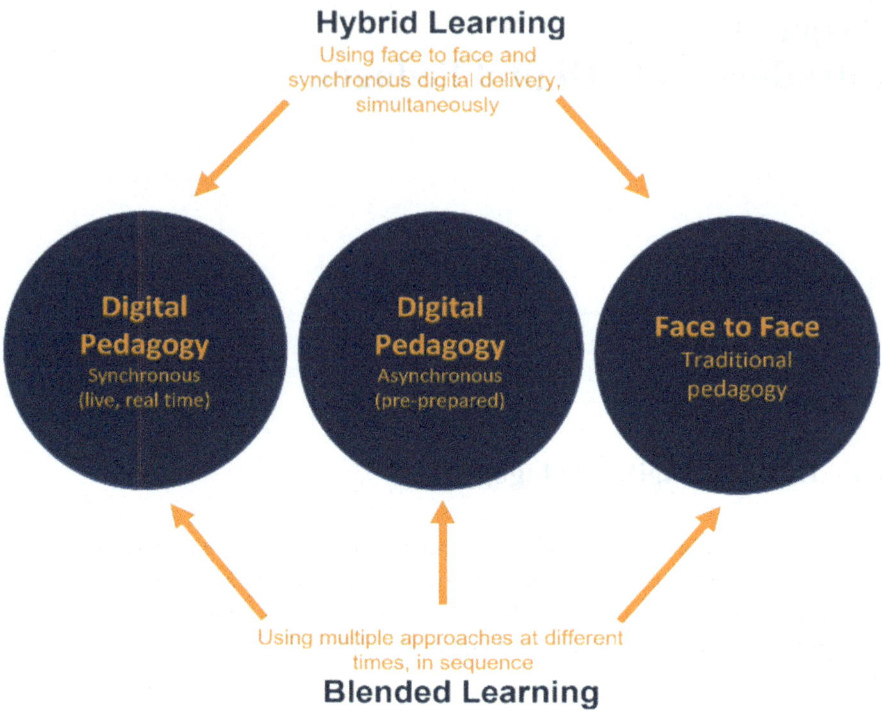

Fig. 1.1 Blended Learning [2]

1.3 Digital Pedagogy Versus Educational Technology

Digital pedagogy and educational technology are two closely related but distinct concepts in today's learning environments. Digital pedagogy is the concept and practice of teaching using digital resources, with an emphasis on how educators critically interact with technology to improve learning experiences. It focusses on educational methodologies, learner-centred approaches, and the deliberate integration of digital resources to provide meaningful education. Educational technology, on the other hand, refers to the tools and platforms that teachers employ, such as learning management systems, interactive apps, and AI-powered resources. It is more about the technology the software, hardware, and digital infrastructure that enable learning. While digital pedagogy defines how technology is utilised in education, educational technology offers the tools to successfully apply these techniques [4] (Table 1.1).

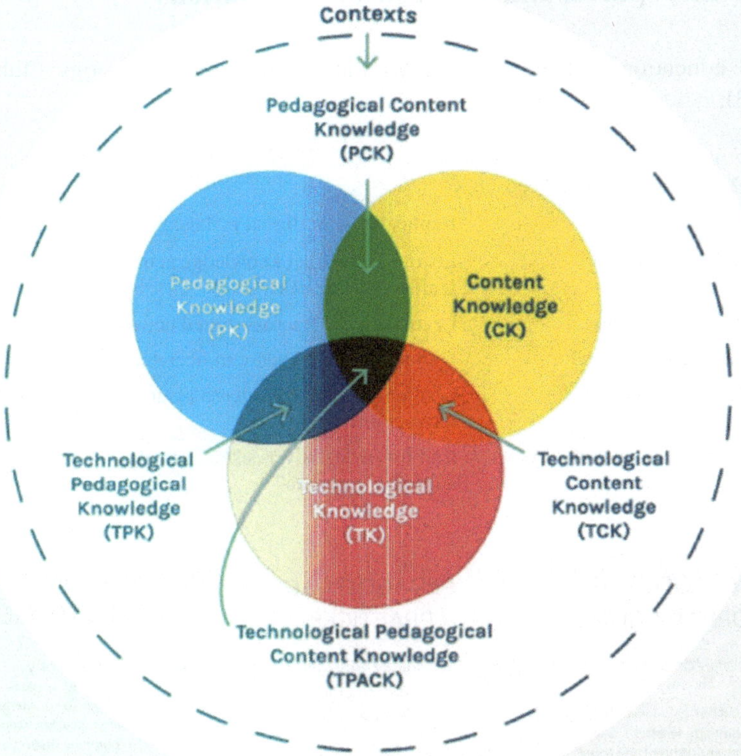

Fig. 1.2 The Basics of digital pedagogy [3]

Table 1.1 Digital pedagogy and educational technology

Aspect	Digital pedagogy	Educational technology
Focus	Teaching philosophy and critical practice	Tools, systems, and platforms
Goal	Empowerment, equity, and meaning-making	Efficiency, delivery, and automation
Application	Context-driven, experimental, reflective	Standardized, system-driven
Discourse	Humanistic, learner-oriented	Technical, product-oriented
Examples	Open pedagogy, critical digital literacy, participatory design	Learning Management Systems, MOOCs, proctoring tools

1.4 Philosophical and Theoretical Foundations

Several educational theories are incorporated into digital pedagogy (Table 1.2; Fig. 1.3).

Table 1.2 Theory and digital pedagogy

Theory	Implications for digital pedagogy
Constructivism (Piaget)	Learners construct knowledge actively; digital tools scaffold this process
Connectivism (Siemens)	Learning occurs across digital networks and platforms
Critical Pedagogy (Freire)	Digital tools must serve to liberate, not dominate
Posthumanism	Challenges human-centered assumptions in digital interaction
Feminist and decolonial theories	Call for inclusivity, resistance to algorithmic bias and epistemic injustice

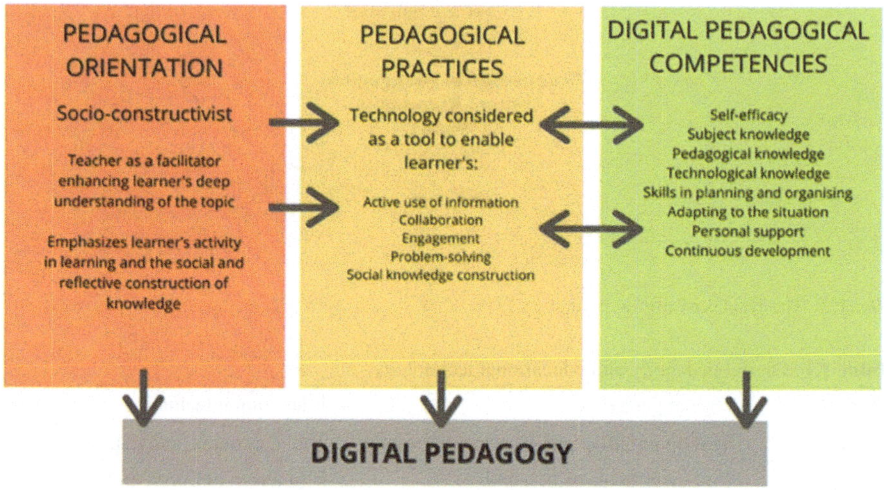

Fig. 1.3 Conceptualization of dimensions and a model for digital pedagogy

1.5 Core Practices in Digital Pedagogy

Core digital pedagogy approaches integrate technology into teaching meaningfully and effectively. Content knowledge ensures instructors know their topic, pedagogical knowledge entails knowing successful teaching tactics, and technical knowledge emphasises digital tool competency. Digital pedagogy also promotes learner-centeredness and cooperation over passive information intake. It encourages critical technology use to improve learning rather than dominate pedagogy (Fig. 1.4; Table 1.3).

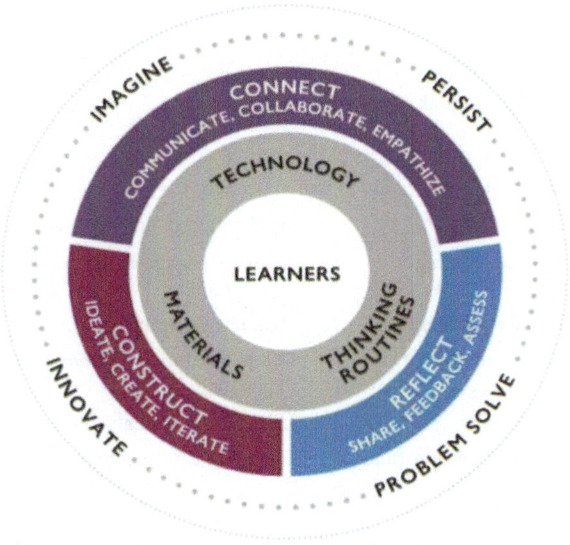

Fig. 1.4 Pedagogy for powerful learning

Table 1.3 Practice and description

Practice	Description
Open pedagogy	Students co-create content; learning materials are openly licensed (OER)
Critical digital literacy	Encouraging analysis of algorithms, data privacy, and media manipulation
Collaborative and networked learning	Using tools like wikis, blogs, forums for co-authorship and knowledge exchange
Inclusive design	Creating accessible content (e.g., captions, alt-text, low-bandwidth options)
Multimodal expression	Allowing students to submit work in diverse formats (video, podcast, art)

Fig. 1.5 Digital versus traditional

Table 1.4 Application in digital pedagogy

Model	Application in digital pedagogy
Flipped learning	Learners engage with materials asynchronously before class interaction
Blended learning	Strategic integration of face-to-face and online elements
HyFlex model	Offers flexibility to attend online or in-person with equivalent experience
Gamification	Uses game elements to increase engagement and feedback
Project-Based learning	Encourages authentic problem-solving with digital tools

1.6 Pedagogical Approaches in Digital Contexts

Digital pedagogy refers to the integration of technology into teaching and learning, transforming traditional educational methods into more interactive and flexible experiences (Fig. 1.5; Table 1.4).

1.7 Digital Pedagogy in Different Contexts

The digital transformation of education has redefined how learners engage with content, peers, and educators across different educational stages. From higher education to K-12 schooling and informal lifelong learning, the integration of technology presents both significant benefits and complex challenges. A nuanced understanding of these developments is essential to ensuring that educational practices remain equitable, effective, and ethically sound. In higher education, digital integration emphasizes student autonomy, research integration, and open scholarship. Online learning platforms and digital libraries allow university students to access vast academic resources independently, encouraging self-directed learning and critical thinking. Moreover, open-access journals and institutional repositories have made academic research more publicly accessible, aligning with the broader goals of open scholarship. These innovations support academic freedom and collaboration across disciplines and geographies, enhancing global research networks and interdisciplinary learning. However, digital learning environments in higher education are not without their drawbacks. Issues like academic monitoring, plagiarism detection, and digital weariness challenge the integrity and sustainability of digital engagement. The widespread use of plagiarism-detection software like Turnitin, while helpful in maintaining academic standards, also raises concerns about privacy and student trust. Simultaneously, constant screen exposure and the pressure to be perpetually online contribute to mental fatigue and decreased engagement—a phenomenon increasingly referred to as "Zoom fatigue" or digital burnout [5].

In K-12 education, the use of digital tools must be carefully aligned with age-appropriate learning goals and scaffolded digital skills development. Younger students require structured and supportive environments to develop responsible technology use. Digital tools should be introduced gradually, with guidance to build foundational skills in literacy, numeracy, and critical thinking. Scaffolded support ensures that learners are not overwhelmed and can gradually progress toward independence. Equally important in early education is the inclusion of play-based learning, the teaching of digital citizenship, and the active participation of parents. Play-based learning leverages digital games and interactive tools to make learning enjoyable and meaningful, while lessons in digital citizenship help children understand online safety, etiquette, and the consequences of digital actions. Parental involvement—through learning apps, digital communication platforms, and virtual meetings—enhances transparency and encourages a collaborative approach to learning. Beyond formal education, informal and lifelong learning has been transformed by platforms such as Coursera, edX, and YouTube, which have democratized access to high-quality educational content. These platforms allow individuals to learn new skills, explore personal interests, or advance their careers at their own pace, regardless of geographic or socioeconomic barriers. This flexibility supports a culture of continuous learning, crucial in a rapidly changing job market and knowledge economy [6, 7].

Despite these benefits, the digital education ecosystem also faces critical issues around quality assurance, accreditation, and data exploitation. Not all online courses meet rigorous academic standards, and the lack of unified accreditation can make it difficult for learners to demonstrate their qualifications. Moreover, concerns about the collection and commercial use of user data by edtech companies raise ethical questions about learner privacy and consent. In conclusion, the digitization of education offers transformative potential across learning stages but also demands thoughtful implementation. By addressing the unique needs of each educational phase and safeguarding ethical standards, digital education can serve as a powerful tool for inclusive, lifelong learning in the twenty-first century.

1.8 Ethical Considerations

As digital technologies become deeply embedded in educational systems, digital pedagogy—the approach to teaching and learning with technology—must evolve to critically examine not just what tools are used, but how, why, and with what consequences. While digital platforms offer opportunities for innovation, collaboration, and access, they also introduce a host of ethical concerns that must be consciously addressed to uphold the values of equity, autonomy, and human dignity in education. One of the most pressing ethical concerns is data privacy and surveillance. Learning management systems (LMS) and educational platforms collect vast amounts of student data, often without clear transparency or consent. This data is used for performance tracking, behavioural analytics, and sometimes even predictive modelling. While intended to improve learning outcomes, these practices can cross into invasive surveillance, potentially undermining trust between students and educators. Educators must advocate for transparent data practices, limit unnecessary monitoring, and educate students on their digital rights. Another growing concern is algorithmic bias in educational AI. Machine learning systems used for grading, assessment, or adaptive learning can replicate and amplify existing social inequalities. If trained on biased data, these systems may disproportionately disadvantage students from marginalised backgrounds. For example, AI tools may underperform for non-native English speakers or learners with disabilities. Ethical digital pedagogy requires that educators interrogate how algorithms are developed, audited, and deployed, ensuring fairness and accountability [8, 9].

Digital equity and access gaps are also central to ethical digital education. Not all students have reliable access to devices, stable internet, or digital literacy support—especially those in rural areas or from low-income families. When learning moves online, these disparities can widen, making equitable participation in education a challenge. Educators must design inclusive content, offer offline alternatives, and work with institutions to support underserved learners. A subtler but equally important issue is techno-solutionism—the belief that technology alone can solve complex educational problems. This mindset often side-lines the relational and human aspects of education, such as empathy, mentorship, and emotional support. Digital pedagogy must resist reducing education to metrics and automation, and instead preserve the social, dialogic nature of learning that fosters critical thinking, identity formation, and community engagement. Moreover, platform dependence and commercialisation pose risks to academic freedom and pedagogical autonomy. When educational institutions rely heavily on proprietary platforms, they risk becoming locked into commercial ecosystems that prioritise profit over pedagogy. This commodification of learning content, student data, and academic processes can erode the public mission of education [10].

To address these complex dynamics, educators must foster critical awareness and agency—both in themselves and in their students. This involves questioning the ethical implications of digital tools, advocating for inclusive and just technologies, and empowering learners to navigate digital systems with a critical eye. Digital pedagogy should not merely adapt to technology but challenge it to serve democratic, equitable, and human-centred goals. In conclusion, ethical digital pedagogy demands vigilance, reflection, and advocacy. As education continues to evolve in digital spaces, it is essential that we centre people, not platforms, and commit to technologies that uphold, rather than compromise, educational values [11].

1.9 Case Examples of Digital Pedagogy

When it comes to digital pedagogy, several programs and institutions are at the forefront of innovation. To encourage independence in virtual communities, the US-based University of Mary Washington launched the Domain of One's initiative, which gives students the tools they need to build and maintain their unique online personas. More diversity and inclusion in classrooms are goals of decolonial digital pedagogy projects spearheaded by South Africa's University of Cape Town. These projects include the development of multilingual Open Educational Resources (OER). Worldwide, FemTechNet is working to spread feminist pedagogy by creating online communities where people can work together to share and learn from one another's expertise. In Canada, a program called Indigenous Digital Storytelling is working to preserve cultural tales via the use of digital media technologies and oral traditions. The revolutionary power of digital education to promote diversity, inclusion, and critical thinking in the classroom is on full display in these initiatives.

1.10 Evolution from Traditional to Digital Education

Technology has transformed education from classroom-based to flexible digital platforms. Technology has radically transformed the landscape of education, shifting it from traditional, classroom-based instruction to a vast array of flexible digital platforms. This transformation has not occurred in isolation, but rather through a complex interplay of historical, cultural, and technical developments. From chalkboards and printed textbooks to learning management systems, AI-driven tutors, and real-time virtual classrooms, each phase of innovation has shaped not only the tools of education but also its pedagogical principles, accessibility, and cultural norms. Historically, education was a physical, synchronous experience—structured around schools, textbooks, and teacher-led instruction. The advent of mass printing enabled wider dissemination of knowledge, while later developments such as radio and television introduced distance learning. However, it was the digital revolution—marked by the rise of the internet, personal computers, and mobile devices—that fundamentally redefined the possibilities for teaching and learning [8].

Culturally, the shift reflects changing expectations around education as a lifelong, personalised, and globally connected process. Learners today demand greater flexibility, immediate access to resources, and more interactive, self-directed experiences. Platforms like MOOCs, educational apps, and AI-powered chatbots respond to this demand by offering scalable, on-demand, and often free learning experiences that transcend geographic and institutional boundaries. Technically, advancements in cloud computing, artificial intelligence, big data, and connectivity have underpinned the emergence of tools such as adaptive learning systems, virtual simulations, and intelligent tutoring agents. These tools not only deliver content but also assess, personalise, and adapt to individual learner needs in real time—far beyond the capabilities of traditional methods. The shift from analogue to digital education is not merely a matter of technology adoption. It is a multifaceted process, shaped by social, economic, and political factors. Policies promoting digital inclusion, public–private investment in edtech, and the global push for 21st-century skills have all contributed to this transformation. At the same time, the digital divide remains a critical concern, with access to technology still uneven across regions and populations.

Pedagogically, digital transformation challenges educators to rethink their roles—from knowledge transmitters to facilitators and co-learners. It calls for new approaches to assessment, collaboration, and learner engagement, as well as critical reflection on the ethical implications of digital surveillance, data privacy, and algorithmic bias. In conclusion, the evolution of education from chalkboards to chatbots represents more than a technological upgrade—it signals a paradigm shift in how society conceives of learning, teaching, and knowledge itself. This chapter will further explore these developments, offering a critical analysis of the opportunities and tensions that define the digital age of education [9, 12]. This section analyses the historical, cultural, and technical developments from chalkboards to chatbots, examining pedagogy, access, tools, and learning culture. It describes the shift as a multifaceted process driven by social, economic, and political factors (Fig. 1.6).

Fig. 1.6 Evolution from traditional to digital education

1.11 Traditional Education: Characteristics and Limitations

Traditional education refers to a long-established model of teaching and learning that is primarily in-person, teacher-centred, and classroom-based. Rooted in centuries of practice, this approach emphasises discipline, memorisation, and standardisation, often within highly structured environments such as schools and universities. While traditional education has provided foundational knowledge and social development for generations, it also presents several limitations in the context of contemporary learning needs. One of the core characteristics of traditional education is its hierarchical structure. Teachers are positioned as the primary source of knowledge, while students are passive recipients [13, 14]. Lessons are typically delivered through lectures, textbooks, and repetition, with assessment focused on examinations and the ability to recall information. Discipline and routine are seen as essential, fostering obedience and time management. Traditional education is also characterised by its uniformity and standardisation. Curricula are often fixed and regulated by national or institutional bodies, with limited flexibility to adapt to individual learning styles or interests. This model prioritises academic achievement over creativity, emotional intelligence, or collaborative skills. However, this approach comes with significant limitations. It tends to neglect learner diversity, failing to accommodate students with different cognitive, emotional, or social needs. The focus on rote learning may discourage critical thinking and problem-solving. Additionally, it often lacks real-world relevance and can demotivate learners who do not thrive under rigid structures [15].

Moreover, traditional education relies heavily on physical presence and synchronous learning, which can be a barrier for those in remote areas, those with disabilities, or working learners. The model's rigidity also struggles to keep pace with the demands of the 21st-century knowledge economy, where adaptability, digital literacy, and lifelong learning are essential. In the digital age, these limitations have become increasingly evident, prompting the rise of more flexible, learner-centred, and technology-enhanced approaches. While traditional education has played a vital role in shaping societies, there is a growing need to rethink and reform its structures to make education more inclusive, dynamic, and future-oriented [16] (Table 1.5).

Table 1.5 Traditional education model

Aspect	Traditional model
Learning environment	Physical classroom
Role of the teacher	Knowledge authority and transmitter
Assessment methods	Written exams, standardised tests
Access and equity	Limited by geography, socioeconomics, and institutional gatekeeping
Curriculum delivery	Linear, textbook-driven, time-bound
Learner role	Passive recipient of information

1.12 Emergence of Digital Education

The transition to digital education started with the incorporation of computers into schools in the 1980s, accelerated by the emergence of the Internet in the 1990s, and burst with the spread of smartphones and cloud computing in the 2010s. The emergence of digital education marks a transformative shift in how knowledge is delivered, accessed, and experienced. This evolution began gradually in the 1980s with the introduction of computers into classrooms, initially used for basic literacy programs and simple computer science lessons. At this stage, digital tools were supplementary and largely confined to specific subjects or lab settings [8, 9, 17].

The 1990s witnessed a significant acceleration with the rise of the internet, which allowed for the integration of online resources, educational software, and email communication between students and teachers. Schools began experimenting with digital libraries, interactive learning programs, and early forms of distance education. The internet enabled access to a vast array of information beyond the classroom walls, laying the foundation for globalised learning.

In the 2010s, the spread of smartphones, tablets, and cloud computing revolutionised digital education. These technologies facilitated real-time access to learning platforms, mobile applications, and collaborative tools such as Google Classroom and Microsoft Teams. Cloud-based systems allowed students and educators to store, share, and interact with content from anywhere, supporting the development of blended and remote learning models.

Digital education also embraced multimedia, gamification, and adaptive learning systems, offering personalised pathways based on student performance and preferences. The popularity of MOOCs (Massive Open Online Courses) through platforms like Coursera and edX further demonstrated the scalability of digital education, making high-quality content available to millions globally.

The COVID-19 pandemic served as a catalyst, pushing institutions at all levels to adopt digital solutions rapidly. This shift highlighted the potential of technology to sustain education during disruptions, but also exposed gaps in digital access, literacy, and infrastructure. In essence, the emergence of digital education reflects a continuous process of innovation, shaped by technological advances, social demand, and pedagogical rethinking. What began as a supplementary tool has become a core component of contemporary education, redefining the roles of teachers, the nature of classrooms, and the meaning of access and inclusion (Fig. 1.7; Table 1.6).

1.12.1 From Instruction to Interaction

Digital education has shifted the pedagogical needle from monologic to dialogic, changing the classroom into a dynamic environment of interaction, co-creation, and participation (Fig. 1.8; Table 1.7).

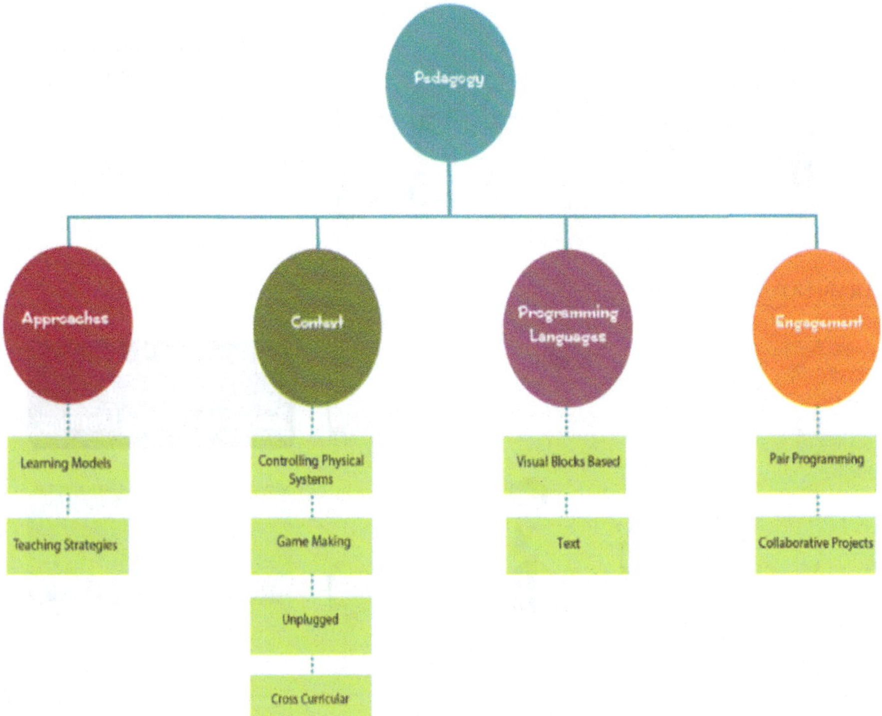

Fig. 1.7 Computing pedagogy

Table 1.6 Technological epoch

Technological epoch	Signature tools	Pedagogical impact
Early digital	PowerPoint, email, projectors	Supplemented teaching
Web 1.0	Static websites, forums, basic LMS	Enabled online delivery
Web 2.0	Blogs, wikis, social media	Fostered user participation and creation
Web 3.0 & AI	Adaptive systems, blockchain, virtual agents	Personalized, decentralised learning

1.13 Institutional Shifts

As the world becomes increasingly shaped by digital technologies, educational systems must adapt both structurally and strategically to remain relevant and effective. This adaptation is not simply about integrating new tools, but about fundamentally rethinking how education is designed, delivered, and governed in the digital age. One of the key shifts involves redesigning the curriculum to embed digital

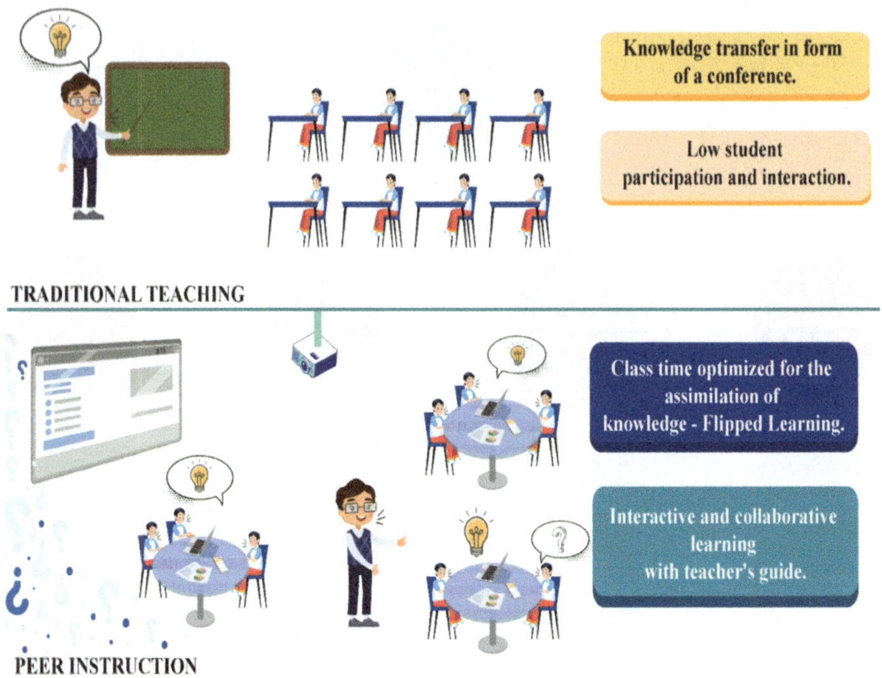

Fig. 1.8 From instruction to interaction [18]

Table 1.7 Instruction to interaction

Dimension	Traditional	Digital
Content delivery	One-way (lecturing)	Two-way/multimodal (video, text, gamified)
Learning mode	Passive reception	Active construction
Feedback	Delayed	Instant and automated
Collaboration	In-class group work	Global, real-time digital collaboration
Assessment	High-stakes exams	Continuous, formative, data-driven

literacy, computational thinking, and data fluency as core competencies—just as reading and arithmetic once were. Students must not only know how to use digital tools but understand the systems and ethics behind them. Equally important is teacher training. Educators must be equipped not only with technical skills but with pedagogical strategies for using digital platforms effectively. Continuous professional development is essential for adapting to fast-evolving digital environments.

Infrastructure investment is also crucial. This includes reliable internet access, up-to-date hardware, and secure learning platforms. Without such foundational resources, digital education cannot be inclusive or effective. Finally, policy changes are needed to support open education, protect learner data, and promote equity.

Governments must establish frameworks for digital inclusivity and responsible data governance, ensuring all learners have access to quality digital learning. In sum, adapting education to digital realities requires a comprehensive, future-focused transformation across curriculum, teaching, infrastructure, and policy.

1.14 Cultural and Social Transformation in the Digital Education Era

The digital transformation of education is not purely technical—it is deeply cultural and social. It alters how knowledge is created, accessed, and valued. One of the most significant shifts is the decentralisation of authority in learning. Students no longer rely solely on teachers or textbooks; instead, they access diverse sources such as YouTube tutorials, educational podcasts, blogs, and social media, empowering them to learn anytime, anywhere. This shift has redefined learner expectations. Today's students increasingly seek flexibility, personalisation, and relevance, expecting education to align with their interests, pace, and life contexts. Simultaneously, globalisation in education—enabled by digital platforms—has created borderless classrooms, where learners collaborate with peers across cultures and continents, enriching the educational experience with global perspectives [19].

1.15 Continuities and Hybridity in Education

Despite the rapid adoption of digital tools, many aspects of traditional education remain intact. The role of the teacher as mentor and guide, the value of structured learning, and face-to-face interaction in classrooms are still central to many learning environments. Instead of replacing traditional systems, digital education often leads to hybrid models, blending online content with in-person engagement. This hybridity preserves the strengths of both models: the rigour and community of classroom learning alongside the flexibility and scalability of digital tools. As a result, the future of education is not purely digital or traditional but a continuum of approaches adapted to diverse learners and contexts [20] (Table 1.8).

1.16 Challenges in the Transition

Digital pedagogy confronts various obstacles that reduce its efficacy and accessibility. The digital gap restricts online learning due to unequal access to devices, internet, and digital literacy. Students and instructors experience technostress, which causes anxiety and cognitive overload from continual digital involvement, resulting

Table 1.8 Continuities and hybridity

Traditional feature	Digital integration
Lectures	Flipped classroom videos, webinars
Group work	Online forums, collaborative documents, virtual labs
Exams	Online quizzes, proctoring software, simulations
Libraries	Digital repositories, open access archives
Office hours	Video conferencing, chatbots, asynchronous help forums

Table 1.9 Case study comparison

Parameter	Traditional university (1990)	Digitally-Transformed university (2025)
Enrollment model	In-person admission	Hybrid/remote/global admission portals
Learning materials	Printed textbooks	eBooks, OER, video tutorials
Instruction	Fixed schedule lectures	On-demand modules, synchronous webinars
Assessment	Pen-and-paper exams	Interactive platforms, peer review, portfolios
Library	Physical stacks	Digital databases, remote access
Support services	Office-based	AI chatbots, 24/7 online portals

in burnout and lower productivity. Loss of social presence online may limit human contact and engagement, making meaningful partnerships and collaborative learning tougher. Data ethics and surveillance raise challenges about permission, data security, and fairness in digital education due to student privacy and algorithmic prejudice. Additionally, pedagogical resistance happens when instructors lack training or trust in digital technologies, hindering the move to technology-integrated education. These issues demand deliberate solutions that prioritise accessibility, well-being, ethics, and educator support [21, 22] (Table 1.9).

1.17 Core Principles of Pedagogical Innovation

Pedagogical innovation involves rethinking teaching and learning, not just adding new tools. Innovative pedagogy must be learner-centred, inclusive, collaborative, and context-responsive in an age of fast technological, cultural, and epistemological change [23] (Fig. 1.9).

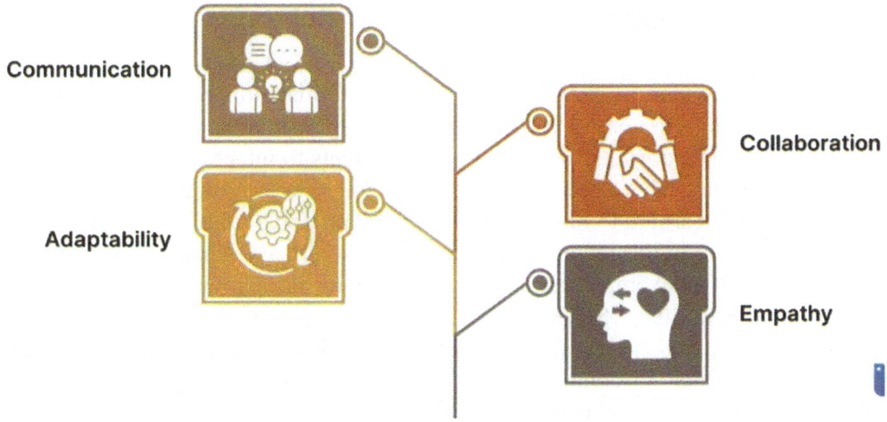

Fig. 1.9 Features of pedagogy principles [23]

1.18 Learner-Centeredness

Learner-centeredness is a foundational principle of modern education, especially in digital and blended learning environments. Unlike traditional education models, which prioritise the teacher as the primary source of knowledge, learner-centered education shifts the focus to the needs, interests, and agency of the student. It recognises that each learner is unique, with different backgrounds, abilities, learning styles, and goals. Digital tools play a crucial role in supporting this shift. Platforms that offer adaptive learning, personalised feedback, and flexible pacing allow students to take ownership of their learning. Online courses, mobile apps, and open educational resources give learners the freedom to explore subjects beyond standard curricula, promoting self-directed and lifelong learning.

At its core, learner-centeredness fosters engagement, motivation, and autonomy. It encourages collaboration, critical thinking, and real-world problem-solving, making learning more meaningful and relevant. Educators become facilitators and co-learners, creating environments where learners feel empowered to explore, question, and grow. In a rapidly changing world, learner-centered approaches are essential for equipping individuals with the resilience, adaptability, and curiosity needed to thrive in diverse and dynamic contexts (Table 1.10).

Table 1.10 Learner-Centredness

Traditional approach	Innovative learner-centered approach
Curriculum-driven	Needs-driven and interest-driven
Standardized outcomes	Personalised learning paths
Passive reception	Active exploration, inquiry, and construction
Teacher as authority	Teacher as facilitator and co-learner

1.19 Interdisciplinary and Real-World Relevance

Contemporary education increasingly values interdisciplinary as a means of connecting learning to complex, real-world challenges. Innovative teaching breaks down conventional subject silos, encouraging students to integrate knowledge from various disciplines. This approach fosters deeper understanding and transferable skills, preparing learners to navigate uncertainty and solve multifaceted problems. One key principle is the incorporation of multidisciplinary perspectives to tackle real-life issues. For example, aligning curricula with the United Nations Sustainable Development Goals (SDGs) enables learners to engage with global concerns, such as climate change, inequality, and sustainable cities, while also addressing local and cultural contexts. This enhances relevance and encourages active civic participation. Skills such as critical thinking, collaboration, creativity, and communication are prioritised over rote memorisation. A strong example is a digital humanities course that blends history, data visualisation, and narrative storytelling to analyse colonial archives. Students learn not only about historical content but also develop technical and interpretive skills by examining these sources from multiple perspectives, encouraging ethical reflection and analytical depth [24, 25].

1.20 Collaborative Learning and Co-Creation

In the twenty-first century, collaboration is not just a skill—it is a mode of learning. Collaborative learning encourages students to construct knowledge together, rather than absorbing it passively. This shift transforms classrooms into active, dialogic spaces where learners explore ideas, negotiate meaning, and build collective understanding. Co-creation further deepens this model by positioning students as partners in the learning process. In innovative educational settings, students may contribute to curriculum design, assess each other's work, or co-develop projects with their peers and instructors. This not only boosts engagement but also cultivates ownership and agency in learning. Digital platforms enhance collaborative learning by offering tools for shared writing, group research, discussion forums, and peer feedback across distances and time zones. Whether through group presentations, team coding challenges, or collaborative policy simulations, learners engage more deeply when they participate in creating knowledge, not just consuming it [24, 26] (Fig. 1.10; Table 1.11).

In an era of rapid educational transformation, inclusivity and accessibility must remain central to innovative pedagogy. All students—regardless of their background, identity, or ability—should be able to participate meaningfully in the learning process. This requires more than physical access; it demands intentional design grounded in Universal Design for Learning (UDL) and principles of equity and justice. Strategies to support inclusive education include multimodal content delivery, which ensures that learners can engage through text, audio, video, and visual aids.

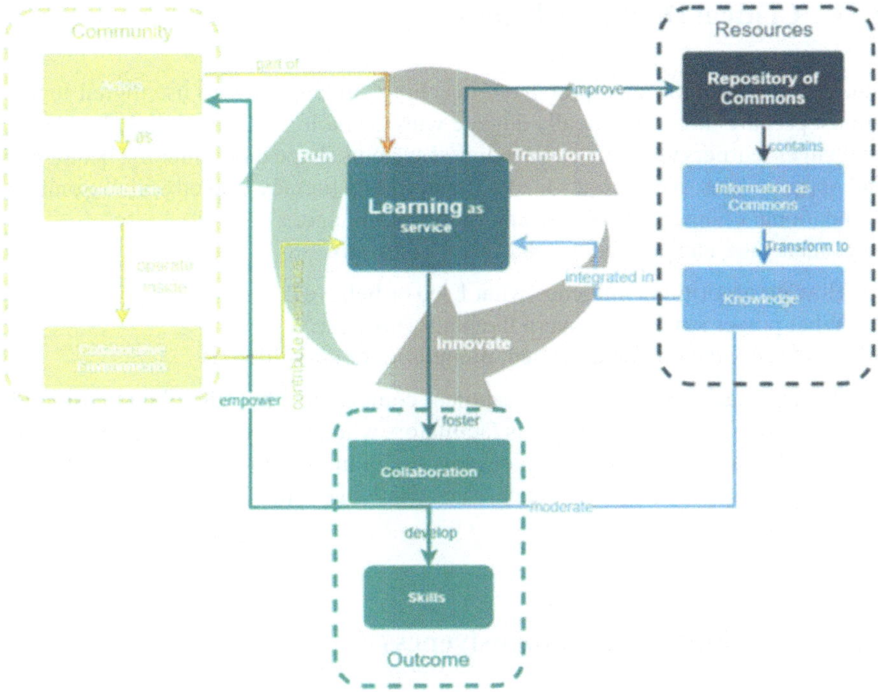

Fig. 1.10 A conceptual framework for skills building in collaborative online TVET learning communities [27]

Table 1.11 Collaborative mechanisms

Collaborative mechanisms	Tools and platforms
Group discussions	Padlet, slack, microsoft teams, discussion boards
Peer review	Google docs, peergrade, learning management systems
Co-created content	Wikis, blogs, shared databases
International collaboration	eTwinning, iEARN, global virtual exchange programs

Translation tools and captioning break down language barriers and support learners with hearing impairments. Adaptive technologies, such as screen readers or text-to-speech applications, empower students with disabilities to interact with digital materials. Beyond technical accessibility, curriculum content must reflect diverse identities, cultures, and experiences, enabling all students to see themselves represented in their learning. Inclusivity, when embedded intentionally, transforms digital pedagogy into a tool for empowerment and social justice.

1.21 Critical Digital Pedagogy

Innovative education must go beyond teaching students how to use digital tools—it must prepare them to critically engage with technology. Critical digital pedagogy challenges learners to reflect on how digital systems shape their thinking, behaviours, and opportunities. It integrates media literacy, data literacy, and ethical reasoning as foundational components of learning in the digital age.

Students are encouraged to ask critical questions:

1. How do algorithms influence what I see or believe?
2. What power structures govern the platforms I use?
3. How is my data being collected, monetised, or surveilled?

This approach fosters not just technical competence, but digital citizenship and ethical awareness. Instructors act as facilitators who guide learners in unpacking the social, political, and economic dimensions of digital life. In doing so, critical digital pedagogy equips students to navigate—and challenge—the systems that increasingly shape modern society.

1.22 Flexibility and Responsiveness

In today's dynamic world, flexibility and responsiveness are essential features of educational design. Institutions must be prepared to adapt to emerging technologies, shifting learner needs, and unexpected global events, as dramatically highlighted by the COVID-19 pandemic. Flexible pedagogies offer a blend of asynchronous and synchronous delivery, giving learners control over time and pace. They allow for multiple assessment formats, accommodating diverse learning styles and access situations. Modular learning pathways, including micro-credentials and stackable courses, enable learners to build knowledge incrementally, based on personal or professional goals. The integration of Open Educational Resources (OER) makes quality learning materials accessible to all, while feedback loops between students and instructors ensure that teaching evolves based on real-time input. Institutions that adopted flexible, cloud-based learning infrastructures during the pandemic were able to transition seamlessly, demonstrating that adaptability is not a luxury but a necessity in modern education.

1.23 Evidence-Based Practice and Reflective Teaching

Education innovation must be based on research and modified over time via feedback and reflection, in line with the notion of teachers as reflective practitioners and action researchers. Learning analytics are critical for measuring student engagement because they enable instructors to alter material and instructional practices based on real-time observations. input loops offer an organised process for gathering continuing learner input, ensuring that teaching methods are responsive to student requirements. Reflective journaling allows educators to track obstacles, experiments, and professional development, promoting a culture of self-evaluation and iterative progress. Furthermore, communities of practice enable educators to exchange ideas and techniques with colleagues, supporting collective learning and collaborative creativity. These processes work together to promote an adaptable, research-informed approach to education that focusses on continual progress and effectiveness.

1.24 The Need for Integration in the Humanities

In an era defined by rapid technological advancement, the Humanities stand at a crossroads—facing both an existential threat and an unprecedented opportunity. Traditionally grounded in disciplines such as literature, philosophy, history, languages, and cultural studies, the Humanities have long focused on introspection, critique, and the human experience. However, their perceived distance from technological development has contributed to widespread misconceptions about their continued relevance in the digital age [28].

Historically, the Humanities have maintained a degree of epistemic and institutional separation from the technical sciences. This divide has been reinforced by:

1. A belief in an inherent difference between qualitative and quantitative knowledge,
2. Institutional boundaries separating departments, budgets, and research priorities,
3. And a cultural reluctance within the Humanities to embrace standardisation and automation, due to concerns about losing nuance, subjectivity, and interpretive depth.

However, this isolation has had real consequences. Funding has diminished, enrolments have declined, and the Humanities have been increasingly marginalised in education policy and innovation discussions. The artificial "STEM versus Humanities" divide has created a competitive, rather than collaborative, academic climate—ultimately undermining the broader goals of education and research [29].

1.25 The Imperative of Relevance in a Digital Age

To reclaim their relevance, the Humanities must embrace integration with digital technologies and demonstrate how they contribute to understanding and shaping the modern world. In today's context of AI, big data, and pervasive digital platforms, the Humanities can offer not only content knowledge but critical tools and ethical perspectives to navigate digital transformation responsibly.

This integration involves multiple dimensions:

1. Equipping students with digital literacies—not just technical skills, but the ability to critically evaluate digital information, algorithms, and media.
2. Engaging in innovative research methods, such as text mining, digital archiving, and computational linguistics, which can open new pathways of inquiry in historical and literary analysis.
3. Preserving and disseminating cultural heritage in formats that are accessible to digital audiences, ensuring that valuable materials are not lost to obsolescence.
4. Providing ethical, philosophical, and historical context to discussions on digital disruption, artificial intelligence, surveillance, and online behaviour.

For instance, a literary historian might use machine learning algorithms to analyse recurring themes in literature over several centuries, uncovering patterns invisible through traditional close reading. Similarly, philosophers can contribute to debates around AI ethics, privacy, and human autonomy—issues that technical fields alone cannot fully address. The Humanities do not need to surrender their identity to remain relevant—they need to evolve and engage meaningfully with the digital landscape. By integrating digital tools, rethinking pedagogies, and embracing interdisciplinary, the Humanities can reaffirm their value as essential to understanding not only our past but the human future in a technologically saturated world.

1.26 Interdisciplinary and Transdisciplinary Synergies

Interdisciplinary and transdisciplinary techniques help solve complicated problems by collaborating across fields. Interdisciplinarity lets scholars mix viewpoints and methodologies from other domains to better understand a topic. It helps solve problems by connecting disciplines while preserving their methods. However, transdisciplinary incorporates practitioner, community, and stakeholder knowledge outside academic bounds. It promotes knowledge co-creation for comprehensive, revolutionary answers to real-world problems. Sustainability, health, and social innovation benefit from these synergies because different skills and lived experiences produce meaningful results [30] (Table 1.12).

Table 1.12 Potential synergy with humanities

Discipline	Potential Synergy with humanities
Computer science	Natural language processing, AI ethics, data visualization
Environmental science	Eco-criticism, climate narratives, indigenous ecological knowledge
Design & media	Digital storytelling, interactive exhibits, augmented reality in history education
Health sciences	Narrative medicine, medical humanities, historical epidemiology

Table 1.13 Traditional versus integrated humanities curriculum

Traditional curriculum	Integrated curriculum
Focus on canonical texts	Inclusion of digital texts, multimedia, and global voices
Emphasis on essays and seminars	Use of podcasts, blogs, digital exhibitions, and data work
Limited engagement with technology	Active use of software for annotation, archiving, mapping
Isolated learning modules	Interdisciplinary, project-based approaches

1.27 Addressing Real-World Challenges

The merging of the Humanities with digital activities is also critical in solving modern global challenges [31]:

1. Climate change: Humanities integrate scientific evidence with historical, ethical, and cultural narratives to motivate action.
2. Global inequality: Digital Humanities initiatives may amplify marginalized perspectives and democratize access to information.
3. Misinformation: Philosophical and rhetorical training prepares people to critically evaluate digital material and algorithms.
4. Ethical and historical viewpoints are critical for the appropriate design and implementation of AI systems.

1.28 Educational Innovation and Curriculum Reform

The need for integration is more acute in educational contexts. Curriculum in the humanities must change to include:

1. Digital tools and approaches (such as GIS, text encoding, and digital curation)
2. Collaboration initiatives integrating both Humanities and STEM students.
3. Courses on ethics of technology, digital culture, and media criticism (Table 1.13).

1.29 Challenges and Resistance to Integration

Despite the advantages, a number of difficulties remain [10]:

1. Many institutions lack adequate digital infrastructure.
2. Faculty skills in digital technologies and approaches vary.
3. Concerns with academic integrity and rigor in digital ventures.
4. Budget limits favor STEM over Humanities.

Overcoming these obstacles requires institutional support, investment in professional development, and a cultural change towards seeing the Humanities as dynamic and growing areas.

1.30 The Role of Digital Humanities

In the evolving educational landscape, Digital Humanities (DH) serves as a powerful paradigm for integration between the Humanities and digital technology. It bridges the gap between traditional humanistic inquiry and emerging digital tools, offering new ways to research, teach, and engage with knowledge. DH challenges the notion that the Humanities are static, elitist, or resistant to change, reimagining them as inclusive, interactive, and socially responsive disciplines. At its core, DH provides platforms that enhance access and engagement. These include digital archives, interactive maps, and open-source repositories, which allow scholars and the public to explore cultural materials in dynamic, participatory ways. For instance, historical documents once limited to physical archives can now be viewed, annotated, and analysed by anyone with internet access [32].

Moreover, DH promotes collaborative and interdisciplinary scholarship. It brings together historians, artists, computer scientists, linguists, and data analysts to co-create research that is richer and more multifaceted than any single discipline could achieve alone. Importantly, DH also opens doors for public participation. Through digital exhibitions, websites, podcasts, and social media, DH projects invite communities to engage with history, literature, and culture in accessible and often co-creative ways. This public-facing approach repositions the Humanities not as a closed academic domain but as a vibrant, democratic space for collective exploration. In doing so, Digital Humanities not only modernises humanistic scholarship but also demonstrates its vital relevance in addressing contemporary cultural, ethical, and social challenges in the digital age [32].

1.31 The Role of Technology in Education

Technology has become an indispensable element of modern education, fundamentally reshaping how knowledge is delivered, accessed, and evaluated. No longer confined to textbooks and chalkboards, today's educational environments are enriched by a wide array of digital tools and platforms that enhance both teaching and learning. The integration of technology into classrooms—through interactive whiteboards, learning management systems, video conferencing tools, and educational apps—has transformed traditional pedagogical methods. Teachers now use multimedia content, simulations, and gamified learning to present information in more engaging and diverse ways, catering to different learning styles [33].

Table 1.14 Timeline: major milestones in educational technology

Year/period	Technological milestone	Impact on education
1440 s	Printing press (Gutenberg)	Mass production of books, democratisation of knowledge
1920s–1950s	Radio and TV	Distance learning for remote communities
1980s–1990s	Personal computers	Introduction of computer-based learning in schools
2000s	Internet and E-learning platforms	Online courses, digital libraries, and virtual classes
2010s–present	Mobile Tech, AI, AR/VR	Personalised learning, gamification, immersive education

Technology has also significantly expanded access to education. Online courses, open educational resources (OER), and mobile learning platforms enable students from various backgrounds and locations to pursue education without the constraints of geography or rigid schedules. This has opened opportunities for lifelong learning, upskilling, and inclusive education. Moreover, digital tools support formative and personalised assessment, allowing educators to monitor student progress in real time and adjust instruction accordingly. Students can receive immediate feedback and access resources tailored to their pace and level of understanding. By promoting interactivity, accessibility, and flexibility, technology not only enhances student engagement but also fosters critical 21st-century skills such as digital literacy, collaboration, and independent learning. However, its use must be balanced with ethical awareness, inclusivity, and sound pedagogical design to ensure equitable and meaningful learning experiences for all [33, 34].

1.32 Historical Evolution of Educational Technology

There has been a long history of technological advancements in the classroom. The educational environment has changed with each new wave of innovation, from the printing press that made reading possible to the advent of radio and television that allowed for distant study. The introduction of desktop computers, laptops, the web, and mobile phones in the latter half of the twentieth and early twenty-first century was a watershed moment [35] (Table 1.14).

1.33 Enhancing Access and Equity

The possibility for technology to democratise learning is one of its most significant contributions to the field of education [35]:

1. Students in underprivileged or rural locations may now access top-notch educational materials online via remote learning.
2. Users with impairments are given more agency via the use of assistive technologies, such as screen readers, speech-to-text software, and adapted equipment.
3. Open Educational Resources (OERs): These freely accessible resources allow for curriculum customisation and lessen reliance on costly textbooks.
4. The term "Massive Open Online Courses" (MOOCs) describes the free or low-cost classes offered by prestigious universities via websites such as FutureLearn, Coursera, and edX.
5. To illustrate the point, in Kenya, rural areas are being connected to the internet via the use of solar-powered tablets that have educational materials already installed.

1.34 Facilitating Personalized Learning

Technology allows educators to personalise training to specific learners' needs:

1. Learning Management Systems (LMS) such as Moodle and Google Classroom provide organised yet flexible classrooms.
2. Adaptive Learning Platforms (like Khan Academy and DreamBox) employ algorithms to alter material difficulty depending on student performance.
3. Gamified Learning motivates and engages learners via the use of points, badges, and challenges.
4. Analytics and dashboards enable instructors to monitor students' progress in real time and intervene as needed.
5. These technologies encourage learner agency by promoting autonomy, self-pacing, and mastery learning paradigms.

1.35 Transforming Pedagogical Approaches

Technology has greatly transformed education, transforming conventional teaching approaches into more dynamic and interactive learning experiences. Instead of teacher-centered lectures, educators are increasingly using flipped classrooms, in which students interact with pre-recorded information prior to class, allowing for deeper discussions and hands-on activities. Rote memorisation has given way to problem-based and inquiry-driven learning, which promotes critical thinking and practical application of information. One-size-fits-all training has given way to differentiated and adaptive methods, which use personalised learning aids to customise material to specific student requirements.

Technological instruments give rapid feedback, which improves the evaluation process. Formative assessments, such as Kahoot, Quizizz, and Socrative, enable teachers to monitor student comprehension during sessions. E-portfolios demonstrate continuing student learning and reflection. Auto-graded quizzes in LMS systems save time while providing consistent grading. Technology facilitates global communication and collaboration, including video conferencing tools like Zoom and Microsoft Teams, collaborative documents and project management platforms, and social media and discussion boards.

1.36 Challenges in Implementation

Despite its promise, technology integration in education confronts several challenges:

1. The digital divide means that not all pupils have access to dependable internet or gadgets.
2. Teacher Training: Many instructors lack the digital proficiency required for successful integration.
3. Infrastructure gaps: Inadequate connection, old equipment, and insufficient technical assistance impede implementation.
4. Concerns about equity: If not handled properly, overreliance on technology may worsen inequities.
5. Privacy and Security: The growing usage of digital technologies creates worries about data security and cyber dangers.

Addressing these difficulties requires systemic investment, policy change, and capacity development at all levels.

1.37 Emerging Trends in Educational Technology

As technology continues to evolve, so too does its impact on education. The twenty-first century has seen a wave of emerging trends in educational technology that are reshaping how teaching and learning occur across all levels and contexts. These trends reflect a shift toward more personalised, data-driven, and globally connected educational experiences.

1. One significant trend is the rise of AI and machine learning in education. AI powers adaptive learning platforms that tailor content to individual learners' needs, offering real-time feedback and personalised pathways. Tools like chatbots and AI tutors also provide round-the-clock support, making learning more accessible and responsive.
2. Gamification and game-based learning continue to grow, using elements of play to increase engagement, motivation, and problem-solving skills. Educational games and simulations help students learn by doing, fostering deeper understanding in subjects like science, history, and coding.

3. Augmented Reality (AR) and Virtual Reality (VR) are also transforming the classroom, offering immersive experiences that allow students to explore historical events, conduct virtual science experiments, or practice real-world skills in safe environments.
4. Data analytics and learning dashboards enable educators to track student performance with greater precision, identifying learning gaps and adapting instruction accordingly. This trend supports more informed decision-making and timely interventions.
5. The shift toward microlearning and modular content, especially in corporate and lifelong learning settings, is another key trend. Learners can now engage in short, focused lessons through mobile apps, video platforms, or online modules that fit into busy schedules.
6. Finally, cloud-based collaboration tools like Google Workspace, Microsoft Teams, and Canvas have become standard in both in-person and remote learning environments, supporting communication, group work, and resource sharing across distances.

These trends illustrate that educational technology is no longer just a support tool—it is a driving force in reimagining education. While these innovations offer exciting possibilities, they also require thoughtful integration to ensure accessibility, equity, and pedagogical soundness.

1.38 Debates Around Technology in the Classroom

The use of technology in the classroom has provoked heated discussions among educators, legislators, scholars, and parents. While many people believe that digital technologies improve learning results, others warn against over-reliance on technology, citing worries about screen time, data privacy, distraction, and pedagogical dilution.

1.39 Technology as a Catalyst for Engagement Versus A Source of Distraction

One of the most prevalent arguments involves student participation.

Proponents argue that technology, including gamification, interactive simulations, and multimedia, may boost student motivation and involvement. Platforms such as Kahoot!, Edpuzzle, and virtual labs provide interactive alternatives to passive learning. Critics contend that technology like cellphones and computers may be sources of distraction. Notifications, social media, and multitasking may shorten attention spans and impede cognitive function. According to a 2019 research by the London School of Economics, prohibiting phones in schools boosted academic

Table 1.15 Personalisation versus standardisation tension

Argument for	Argument against
Adaptive systems improve differentiation	Algorithms may reinforce inequity
Learners progress at individual pace	May reduce opportunities for collaborative learning
Data-driven insights inform pedagogy	Risk of over-monitoring and data misuse

achievement among low-achieving children, illustrating the complex influence of technology usage.

1.40 Personalised Learning Versus Standardisation Concerns

Advocates: Technology provides personalisation, enabling students to study at their own speed and concentrate on their areas of weakness. AI-powered systems can change information in real time, which improves learning efficiency.

Sceptics: Some argue that algorithm-driven learning systems may entrench prejudices while ignoring the social and emotional aspects of education. There is also fear that an overemphasis on technology-based examinations would result in standardisation and the loss of instructional variation (Table 1.15).

1.41 Equity and Access: Bridging or Widening the Gap?

Equity and access are fundamental concepts in addressing disparities in education, healthcare, technology, and social services. Ideally, promoting equity and access should bridge existing gaps, enabling marginalized or underserved populations to achieve comparable outcomes to more privileged groups. However, the reality can be more complex. Efforts to enhance access—such as introducing digital learning platforms or expanding healthcare coverage—may unintentionally widen disparities if underlying structural inequities are not addressed. For example, providing online education helps many, but students without reliable internet or devices remain excluded, thus deepening the digital divide. Similarly, healthcare access improves outcomes only if accompanied by culturally sensitive services and affordability. True equity requires more than equal access; it demands tailored support that accounts for different needs and barriers. Without this, well-intentioned initiatives risk benefiting those already advantaged while leaving others further behind. Therefore, to bridge rather than widen gaps, policies must be holistic, inclusive, and responsive to the complexities of social inequality [21, 36].

1.42 Teacher Empowerment Versus De-Skilling

Teacher empowerment refers to giving educators greater autonomy, professional development opportunities, and a voice in decision-making processes. Empowered teachers can adapt teaching methods to meet their students' diverse needs, innovate in the classroom, and feel motivated and valued. This fosters better educational outcomes and job satisfaction. Conversely, de-skilling occurs when teachers' professional judgment and expertise are undermined by overly standardized curricula, rigid testing regimes, or excessive administrative control. When teachers are forced to follow scripts or focus solely on exam results, their creativity and critical thinking are stifled. This reduces teaching to a mechanical task, weakening their professional identity and potentially harming student learning.

The balance between empowerment and de-skilling is delicate. Policies and practices should aim to support teacher autonomy while providing clear frameworks and resources, ensuring that educators remain skilled professionals capable of meeting evolving educational challenges.

1. Optimistic Perspective: Technology empowers teachers by automating administrative work, allowing for novel pedagogies, and facilitating professional development via online platforms.
2. Critical Perspective: There is worry that internet businesses are altering education without sufficient participation from educators, resulting in a feeling of de-skilling. Teachers may feel compelled to utilise instruments that they have not been trained to use successfully, causing overuse or reliance.

For example, the use of learning analytics dashboards without sufficient training has resulted in some instructors misinterpreting data, hurting student support.

1.43 Authentic Learning Versus Technocentrism

Authentic learning emphasizes real-world relevance, critical thinking, and deep understanding by engaging students in meaningful, contextualized tasks. It encourages collaboration, problem-solving, and connecting knowledge to students' lives and communities, fostering lasting skills beyond rote memorization. Technocentrism, on the other hand, focuses heavily on the use and integration of technology as the primary driver of educational improvement. It often assumes that new gadgets, software, or digital tools inherently enhance learning, sometimes prioritizing technology over pedagogy [37].

While technology can support authentic learning by providing rich resources and interactive opportunities, an overemphasis on Technocentrism risks sidelining the core principles of meaningful engagement. Without thoughtful integration, technology may become a distraction or a superficial addition rather than a tool for deeper understanding. Balancing authentic learning and technology use involves centering pedagogy first, ensuring technology serves clear educational goals rather than driving them. For example, using AR applications in literature classrooms without context or meaningful engagement may simplify complicated tales to dazzling visualisations, resulting in a loss of interpretive depth [38].

1.44 Sustainability and Environmental Concerns

Sustainability focuses on meeting present needs without compromising the ability of future generations to meet theirs. It integrates environmental protection, social equity, and economic development, aiming for a balanced approach to resource use and human well-being. Environmental concerns are central to sustainability, addressing issues such as climate change, pollution, biodiversity loss, and resource depletion. These challenges threaten ecosystems and human societies worldwide, calling for urgent action to reduce carbon emissions, promote renewable energy, conserve natural habitats, and manage waste responsibly. Sustainable development requires systemic changes in how societies produce, consume, and interact with the environment. It involves governments, businesses, communities, and individuals working together to adopt green technologies, sustainable agriculture, and circular economies. Educating people about environmental stewardship and encouraging sustainable lifestyles are vital to fostering a culture that values and protects the planet for the long term [39].

The environmental cost of technology is also emerging as a critical debate [40]:

1. Green Tech Advocates: Sustainable gadgets, cloud computing, and digital tools may help to minimise paper consumption and encourage environmentally friendly activities.
2. Environmentalists argue that the creation and disposal of digital gadgets contribute to e-waste and resource depletion. Constant improvements create unsustainable consumerist cycles.
3. Fact: According to the United Nations, worldwide e-waste reached 53.6 million metric tonnes in 2019, with education technology accounting for a considerable portion.

1.45 Balancing Human Connection and Technological Mediation

In today's digital age, technology mediates much of our communication, transforming how we connect in personal, professional, and educational contexts. While technological tools—like video calls, social media, and messaging apps—enable instant and wide-reaching interaction, they can also diminish the depth and authenticity of human connection. Balancing human connection with technological mediation involves recognising technology as a facilitator, not a replacement, for meaningful relationships. Face-to-face interactions offer rich emotional cues—body language, tone, presence—that technology often struggles to fully replicate. Over-reliance on mediated communication may lead to feelings of isolation, misunderstanding, or superficial bonds. However, technology can also enhance connection when used thoughtfully, bridging physical distances and providing access for those otherwise isolated. The key is mindful integration—leveraging technology to complement, rather than substitute, genuine human interaction [41, 42].

Striking this balance requires intentional choices about when and how to use technology to maintain empathy, trust, and emotional engagement in our increasingly connected world.

1. Education is fundamentally a human endeavour. While technology may improve communication, it may also undermine genuine human relationships:
2. Technology links students and teachers across geographical barriers, increasing global understanding and cooperation.
3. Alienation Concern: Excessive screen usage may diminish face-to-face engagement, empathy, and socio-emotional development, particularly among younger pupils.
4. Reflection Point: How can educators utilise technology to enhance, rather than replace, human connections?

1.46 Digital Fluency Versus Digital Literacy

As digital technologies become more prevalent in education, the words digital literacy and digital fluency are sometimes used interchangeably. They do, however, represent separate ideas with distinctive consequences for teaching and learning. Understanding the distinction between literacy and fluency is critical for developing curriculum, assessments, and professional development to prepare students for the complexity of the digital world.

Digital literacy refers to the ability to effectively and critically navigate, evaluate, create, and communicate information using digital technologies. It goes beyond basic computer skills, encompassing a broad set of competencies including understanding online safety, discerning credible sources, managing digital identities, and using digital tools for problem-solving and collaboration. Being digitally literate means not only knowing how to operate devices and software but also having the critical thinking skills to interpret digital content, understand its impact, and use technology responsibly and ethically. It is essential for participation in modern society, enabling individuals to access information, engage in education and work, and communicate across diverse platforms. As technology rapidly evolves, digital literacy also involves continuous learning to adapt to new tools and environments, ensuring that people remain competent and confident in the digital world [43–45].

In contrast, digital fluency is the advanced ability to not only use digital tools effectively but to creatively and critically apply them to solve complex problems, communicate ideas, and innovate. It goes beyond digital literacy, which focuses on basic skills and understanding, by emphasizing adaptability, strategic thinking, and the confident integration of technology into various contexts. A digitally fluent individual can seamlessly navigate multiple digital platforms, troubleshoot challenges, and leverage technology to enhance productivity and collaboration. They think critically about the implications of digital actions, ethics, and data privacy, making informed decisions that reflect both technical skills and social awareness. Digital fluency is increasingly vital in education, work, and everyday life, where success depends on dynamic and flexible use of technology rather than simple operation [46].

1.47 Key Differences: Literacy Versus Fluency

There are two distinct degrees of competence when it comes to using technology: digital literacy and digital fluency. To guarantee that people can make effective use of digital tools, digital literacy centres on developing both basic and advanced skills. Digital content consumption and platform navigation are two of the fundamental competencies emphasised. Conversely, digital fluency necessitates a more sophisticated and strategic understanding of the when, why, and how use digital technologies in different circumstances. Because of its flexibility, users may easily adapt their abilities to new technology. People who are more fluent are more likely to produce, remix, and invent their own material instead of just consuming it. This is because fluency encourages creativity. From an educational standpoint, teaching student's fundamental computer skills is digital literacy, whereas teaching them digital fluency prepares them for complex digital settings by cultivating critical thinking and problem-solving. Individuals are certain to be proficient users and considerate producers when these talents are combined to form well-rounded digital interaction.

1.48 Educational Implications

1. **Curriculum Design:** Many educational systems emphasise digital literacy, particularly at the primary and secondary levels. However, as the digital landscape changes, curricula must adapt to prioritise proficiency. Coding, for example, should be taught as both a technical talent and a language for creative problem solution.

2. **Assessment Practices:** Traditional exams often measure procedural knowledge (for example, preparing a document or utilising search keywords). Fluency necessitates performance-based evaluations, such as digital storytelling, coding challenges, and collaborative projects, which demonstrate students' ability to use technologies in meaningful ways.
3. **Professional Development:** Teachers also need assistance in transitioning from literacy to fluency. Professional development must extend beyond tool orientation training to include chances for experimentation, critical discourse, and reflective practice with technology in teaching.

1.49 Case Study: New Zealand's Digital Technologies Curriculum

New Zealand's education system is an excellent example of incorporating digital fluency into national standards. From early childhood through high school, the "Digital Technologies Hangarau Matihiko" curriculum encourages computational thinking, design thinking, and the production of digital results. Students not only utilise tools, but also comprehend how they function and use them in culturally appropriate ways.

1.50 Challenges in Achieving Fluency

1. Equity of Access: Students from underprivileged neighbourhoods may not have had enough exposure to various technologies, restricting their ability to gain fluency.
2. Teacher Preparedness: Many educators function at a literacy level and may lack the confidence or skills to promote fluent learning.
3. Policy gaps: Educational policies often emphasise infrastructure and access while ignoring the pedagogical changes required for fluency development.
4. Over-standardisation: Systems that depend primarily on strict curriculum and high-stakes testing may unintentionally inhibit the innovation needed for proficiency.

1.51 The Role of Critical Thinking

At the core of digital fluency is critical thinking. A digitally fluent person challenges sources, recognises algorithmic bias, assesses data veracity, and considers the ethical consequences of digital acts. This critical mindset is essential in an age of disinformation, surveillance capitalism, and fast automation.

1.52 Preparing Students for Future Work and Civic Life

Employers in the post-digital economy place a higher priority on digital adaptation. Roles in artificial intelligence, remote collaboration, creative industries, and e-governance need more than just fundamental skills; they also require the capacity to re-skill, synthesise, and problem-solve in a variety of digital contexts. Furthermore, digital citizenship is essential for democratic involvement. Fluency enables people to participate responsibly in digital communities, advocate for rights online, and make important contributions to digital public debate.

1.53 Role of Educators in the Digital Age

The digital revolution of education has fundamentally changed the role of educators. In the digital era, educators are more than just information transmitters; they are also facilitators, mentors, curators, co-learners, and change agents. They are leading the way in developing 21st-century abilities such as critical thinking, teamwork, creativity, and digital citizenship, all while navigating new ethical, technological, and pedagogical difficulties.

1.54 Shifting Paradigms: From Instructor to Facilitator

Traditional educational approaches positioned the teacher as the major source of information. In contrast, digital pedagogy promotes learner autonomy, inquiry-based learning, and self-directed discovery. Educators today:

1. Encourage active learning using digital tools and platforms.
2. Design blended or flipped classes that promote peer cooperation.
3. Collect materials from OERs.
4. Provide personalised learning paths depending on student requirements and choices.
5. The transition to constructivist and connectivist pedagogies focusses on how learners acquire, analyse, and generate information, rather than merely what they learn.

1.55 Essential Digital Competencies for Educators

Educators in the digital era are expected to be both technologically adept and inventive in their teaching methods. The European Framework for the Digital Competence of Educators (DigCompEdu) (Table 1.16).

Table 1.16 Competency area

Competency area	Key skills
Digital resources	Selecting, creating, and modifying digital content
Teaching & learning	Integrating digital tools to support active and collaborative learning
Assessment	Using technology for formative and summative evaluation
Empowering learners	Adapting teaching strategies to individual needs and accessibility
Facilitating digital literacy	Helping learners critically and responsibly engage with digital technologies

1.56 Educators as Designers of Learning Experiences

Digital pedagogy requires educators to become learning designers, creating adaptable and multimodal learning environments. This involves:

1. Storyboarding Interactive Lessons
2. Incorporating multimedia (films, podcasts, simulations).
3. Designing gamified material and online challenges.
4. Blending synchronous and asynchronous activities.
5. Developing digital exams to evaluate higher-order thinking
6. Educators must think like architects, engineers, and artists to create learning environments that are engaging, inclusive, and contextually relevant.

1.57 Addressing Equity and Inclusion

In digitally mediated learning settings, educators play an important role in ensuring fair access and participation. They must:

1. Identify and reduce digital inequalities (for example, device or internet access inequities).
2. Use Universal Design for Learning (UDL) to cater to varied learning demands.
3. Implement culturally sensitive instruction that reflects students' identities.
4. Advocate for accessible material, such as subtitles, transcripts, and screen reader compatibility.
5. Promote digital citizenship and responsible online behaviour.
6. The educator's ethical role includes pushing for systemic reform and inclusive practices at all levels of education.

1.58 Emotional and Social Roles of Educators

Despite the availability of AI instructors and virtual learning environments, the human aspect in education is irreplaceable. Educators continue:

1. Develop a feeling of belonging and psychological safety.
2. Serve as mentors and role models in both real and virtual classrooms.
3. Develop trust and empathy via social-emotional learning (SEL).
4. Identify and help pupils with mental health or socio-emotional difficulties.

The pandemic highlighted the emotional labor of teaching, as educators often served as first responders in times of panic and uncertainty.

1.59 Continuous Professional Development (CPD)

To stay relevant, educators must pursue continual professional development that extends beyond technical training. This includes:

1. Participate in online professional learning communities (PLCs).
2. Attending webinars, MOOCs, and digital pedagogy bootcamps.
3. Read research and reflective practice literature.
4. Collaborating with colleagues in action research and curriculum design
5. Receiving institutional support and time to explore and upskill
6. CPD must be contextual, adaptable, and sensitive to the changing needs of digital pedagogy.

1.60 Benefits of Digital Pedagogy

1. Improved accessibility and flexibility.
2. Digital education enables students to study at any time and from any location, erasing geographical, temporal, and physical restrictions.
3. Students in rural or underserved locations have access to high-quality educational materials.
4. Asynchronous learning works well with a variety of schedules, particularly for adult and working learners.
5. Recorded lectures and digital notes enable students to review at their own speed.
6. Personalised Learning Paths.
7. Digital platforms can tailor to learners' requirements by providing personalised feedback, materials, and pace.
8. Adaptive learning systems alter difficulty in response to student feedback.
9. Students may choose customised pathways based on their interests and ability levels.
10. Learning analytics provide early intervention and customised help.
11. Interactive and Multimodal Content.
12. Digital pedagogy enables the use of a variety of media text, audio, video, simulations, and games to increase participation.
13. Alternative representations of material aid both visual and aural learners.

14. Simulations and virtual laboratories provide safe and reproducible testing.
15. Gamification promotes motivation and active involvement.
16. Real-Time Communication and Collaboration.
17. Digital technologies enable interactivity across place and time.
18. Video conferencing, chat applications, and discussion boards offer both synchronous and asynchronous communication.
19. Students may participate in global teams and gain intercultural competency.
20. Collaborative writing tools (e.g., Google Docs, Notion) support peer feedback and co-creation.
21. Scalability and Cost Efficiency
22. Especially in higher education and MOOCs, digital pedagogy allows for broad content delivery at relatively low cost.
23. Thousands of students may take a single course at the same time.
24. Open Educational Resources (OERs) lower textbook and material expenses.
25. Automated grading solutions save time while providing rapid feedback.

1.61 Limitations and Challenges of Digital Pedagogy

1. Digital Divide and Inequity.
2. Access to digital learning is unevenly distributed.
3. Students from low-income neighbourhoods may not have access to dependable internet or gadgets.
4. Rural locations often have weak connection and infrastructural restrictions.
5. Digital disparity may worsen existing educational gaps.
6. Lack of human touch and social presence.
7. Even with interactive gadgets, the lack of physical presence might have an influence on social-emotional development.
8. Students may feel alone, disinterested, or unsupported.
9. Teachers struggle to interpret body language and emotional signs online.
10. Building rapport and community might be more challenging in virtual environments.
11. Technology Dependence and Fatigue.
12. Overreliance on digital technologies might result in cognitive and physical strain.
13. Screen fatigue, eye strain, and sedentary behaviour are all rising issues.
14. Technical difficulties, such as program breakdowns or inadequate connection, might impede learning.
15. Not all students or teachers are comfortable using digital media.
16. Pedagogical limitations.
17. Digital pedagogy often necessitates a reconsideration of teaching methodologies and learning goals.
18. Not all information or talents work well in digital media (for example, hands-on skills and some arts).

Table 1.17 Comparative pros and cons

Aspect	Benefits	Limitations
Access	Global, flexible, inclusive	Digital divide and tech inequity
Learning design	Personalized, multimodal, adaptive	Limited effectiveness for some subjects and learner types
Interaction	Enables remote collaboration	Reduced emotional connection, possible isolation
Assessment	Automated, instant feedback	Can overlook critical thinking, creativity, and nuance
Cost and reach	Scalable, reduces some traditional expenses	Requires investment in infrastructure and support
Control and autonomy	Empowers learner choice and pacing	Demands high self-motivation and digital literacy
Ethics and privacy	Learning analytics can inform instruction	Surveillance, data misuse, and ethical ambiguity

19. Poorly constructed online courses might lead to passive or superficial learning.
20. Automated evaluations may not capture creativity or critical thinking.
21. Data privacy and surveillance concerns
22. The collecting and analysis of massive volumes of student data is central to digital education.
23. Concerns over who controls student data and how it is utilised are growing.
24. Surveillance capabilities on certain systems, such as proctoring software, create ethical concerns.
25. Data leaks and security flaws pose serious threats.

1.62 Comparative Pros and Cons

In the context of education and learning, the term "digital pedagogy" refers to the use of digital technology. It has revolutionized education by providing students with new methods to connect with the material, but it also comes with a number of challenges (Table 1.17).

References

1. Zhao, S., Pan, Q., Gao, D., Cheng, J.: Integrating internet of things and mixed reality to teach performance-based architectural design: a case study of shading devices. Educ. Inf. Technol., (2022). https://doi.org/10.1007/s10639-022-10998-6
2. Ghavifekr, S., Kunjappan, T., Ramasamy, L.: Teaching and learning with ICT tools: issues and challenges from teachers' perceptions. Malaysian Online J. Educ. Technol. **4**(2), 38–57 (2016)
3. Assar, S.: Information and communications technology in education. Int. Encycl. Soc. Behav. Sci. Second Ed. **12**, 66–71 (2015). https://doi.org/10.1016/B978-0-08-097086-8.92104-4

4. Korpi, H., Sjögren, T., Mikkonen, K., Piirainen, A.: A systematic review and metasynthesis of qualitative studies on the competencies of health and rehabilitation science educators. NERP, **64**(2), pp. 270–282 (2020). Available: https://www.academia.edu/download/88209414/cr_201tec_202019_20paper_2021.pdf
5. Lou, C.: Social media influencers and followers: theorization of a trans-parasocial relation and explication of its implications for influencer advertising. J. Advert. (2021). https://doi.org/10.1080/00913367.2021.1880345
6. Miorandi, D., Sicari, S., De Pellegrini, F., Chlamtac, I.: Internet of things: Vision, applications and research challenges. Ad Hoc Netw. **10**(7), 1497–1516 (2012). https://doi.org/10.1016/j.adhoc.2012.02.016
7. Lou, C.: Social media influencers and followers: theorization of a trans-parasocial relation and explication of its implications for influencer advertising. J. Advert. **51**(1), 4–21 (2022). https://doi.org/10.1080/00913367.2021.1880345
8. Cynthia Chizoba Ekechi, Excel G Chukwurah, Lawrence Damilare Oyeniyi, Chukwuekem David Okeke.: A review of small business growth strategies in african economies. Int. J. Adv. Econ., **6**(4), pp. 76–94 (2024), https://doi.org/10.51594/ijae.v6i4.1071
9. Joshi, S., Sharma, M.: Impact of sustainable supply chain management on performance of SMEs amidst COVID-19 pandemic: an Indian perspective. Int. J. Logist. Econ. Glob. **9**(3), 248 (2022). https://doi.org/10.1504/ijleg.2022.10044917
10. Abdul-Ganiy Kayode Ahmad: E-Learning: an implication of Covid-19 pandemic for the teaching and learning of Arabic and Islamic Studies in Ogun State's Tertiary Institutions. Int. J. Soc. Learn. **2**(2), 217–234 (2022). https://doi.org/10.47134/ijsl.v2i2.115
11. Tiron-Tudor, A., Deliu, D.: Reflections on the human-algorithm complex duality perspectives in the auditing process, **19**(3) (2022). https://doi.org/10.1108/QRAM-04-2021-0059
12. Arias Velásquez, R.M., Mejía Lara, J.V.: Knowledge management in two universities before and during the COVID-19 effect in Peru. Technol. Soc., **64**(2020) (2021), https://doi.org/10.1016/j.techsoc.2020.101479
13. Obi, K.C., Iwuanyanwu, O.: Innovation in the assessment of primary school mathematics: an approach to revamping nigeria education. J. Teach. Perspect., **18**(1), pp. 51–66 (2023). Available: http://repositorio.unan.edu.ni/2986/1/5624.pdf%0Ahttp://fiskal.kemenkeu.go.id/ejournal%0Ahttps://doi.org/10.1016/j.cirp.2016.06.001%0A10.1016/j.powtec.2016.12.055%0A10.1016/j.ijfatigue.2019.02.006%0A10.1
14. Kanmodi, K.K. et al.: Strategies adopted by oral physicians, oral and maxillofacial surgeons, and oral pathologists in patient education on oral cancer: A Nigerian study. Cancer Rep., no. September, pp. 1–10, (2023). https://doi.org/10.1002/cnr2.1929
15. Goh, E., Sigala, M.: Integrating Information & Communication Technologies (ICT) into classroom instruction: teaching tips for hospitality educators from a diffusion of innovation approach. J. Teach. Travel Tour. **20**(2), 156–165 (2020). https://doi.org/10.1080/15313220.2020.1740636
16. Evans, G.: A new dawn or false hope? Exploring the early implementation of curriculum for Wales. Educ. Inq. **00**(00), 1–15 (2023). https://doi.org/10.1080/20004508.2023.2297506
17. Idris, A., Yusof, R.: Adoption of E-voting system in nigeria: a prospect for poverty alleviation. Proc. Int. Conf. E-Commerce, pp. 92–96 (2015). Available: http://www.icoec.my/index.php/proceedings/5-icoec-2015-proceedings/97-adoption-of-e-voting-system-in-nigeria-a-prospect-for-poverty-alleviation
18. Rahman, R.A., Halid, S., Zamri, N.: Academic dishonesty and financial reporting fraud intention. pp. 2650–2655 (2022)
19. Aljowder, T., Ali, M., Kurnia, S.: Development of a maturity model for assessing smart cities: a focus area maturity model. Smart Cities **6**(4), 2150–2175 (2023). https://doi.org/10.3390/smartcities6040099
20. Deivanayagam, T.A., Osborne, R.E.: Breaking free from tunnel vision for climate change and health. PLOS Glob. Public Heal. **3**(3), e0001684 (2023). https://doi.org/10.1371/journal.pgph.0001684

21. Offenheiser, R.C., Holcombe, S.H.: Challenges and opportunities in implementing a rights-based approach to development: An Oxfam America perspective. Nonprofit Volunt. Sect. Q. **32**(2), 268–301 (2003). https://doi.org/10.1177/0899764003032002006
22. Hofer, E., Musakwa, W., van Lanen, S., Gumbo, T., Netsch, S., Gugerell, K.: Inclusivity insights: two urban development projects in Johannesburg. J. Hous. Built Environ., 11, (2021) https://doi.org/10.1007/s10901-021-09916-y.
23. Canagarajah, S.: Decolonizing academic writing pedagogies for multilingual students. TESOL Q. **58**(1), 280–306 (2024). https://doi.org/10.1002/tesq.3231
24. Wilson, L., Hamwi, S., Zanni, F., Lomazzi, M.: Global public health policies: gathering public health associations' perspectives. Glob. Health Action, **16**(1) (2023). https://doi.org/10.1080/16549716.2023.2183596
25. Heydari, J., et al.: Towards a circular economy for packaging waste by using new technologies: The case of large multinationals in emerging economies. J. Clean. Prod., **11**(1), pp. 438–450 (2019). Available: https://doi.org/10.1016/j.resconrec.2017.08.017
26. Erin, O.A., Bamigboye, O.A.: Evaluation and analysis of SDG reporting: evidence from Africa. J. Account. Organ. Chang. **18**(3), 369–396 (2022). https://doi.org/10.1108/JAOC-02-2020-0025
27. Winkelmann, R., et al.: Social tipping processes towards climate action: A conceptual framework. Ecol. Econ., **192** (2022). https://doi.org/10.1016/j.ecolecon.2021.107242
28. Kehayov, M., Holder, L., Koch, V.: Application of artificial intelligence technology in the manufacturing process and purchasing and supply management. Procedia Comput. Sci. **200**(2019), 1209–1217 (2022). https://doi.org/10.1016/j.procs.2022.01.321
29. Olubusola Odeyemi, Kehinde Feranmi Awonuga, Noluthando Zamanjomane Mhlongo, Ndubuisi Leonard Ndubuisi, Funmilola Olatundun Olatoye, Andrew Ifesinachi Daraojimba.: The role of AI in transforming auditing practices: A global perspective review. World J. Adv. Res. Rev., **21**(2), pp. 359–370 (2023). https://doi.org/10.30574/wjarr.2024.21.2.0460
30. Chehelgerdi, M., Chehelgerdi, M.: The use of RNA-based treatments in the field of cancer immunotherapy. BioMed Central, **22**(1) (2023). https://doi.org/10.1186/s12943-023-01807-w
31. Arufe-Giráldez, V., Sanmiguel-Rodríguez, A., Ramos-Álvarez, O., Navarro-Patón, R.: News of the Pedagogical models in physical education—a quick review. Int. J. Environ. Res. Public Health, **20**(3) (2023). https://doi.org/10.3390/ijerph20032586
32. Nurudeen Olalekan Orunbon, Ganiyu Abiona Ibikunle, Adewale Abiodun Badmus.: Principals' digital transformational leadership, teachers' organisational commitment and job satisfaction During Covid-19 in lagos state education district V, Nigeria. Int. J. Humanit. Technol. Civiliz., **8**(1), pp. 68–73 (2023), https://doi.org/10.15282/ijhtc.v8i1.9424
33. Olowa, T., Witt, E., Morganti, C., Teittinen, T., Lill, I.: Defining a BIM-Enabled learning environment—an adaptive structuration theory perspective. Buildings, **12**(3) (2022). https://doi.org/10.3390/buildings12030292
34. Okai, S., Uddin, M., Arshad, A., Alsaqour, R., Shah, A.: Cloud computing adoption model for universities to increase ICT proficiency. SAGE Open **4**(3), 1–10 (2014). https://doi.org/10.1177/2158244014546461
35. Kozma, R.B., Andersen, R.E.: Qualitative case studies of innovative pedagogical practices using ICT. J. Comput. Assist. Learn. **18**(4), 387–394 (2002). https://doi.org/10.1046/j.0266-4909.2002.00250.doc.x
36. Uwuigbe, jafaru.: Dividend policy and firm performance. J. Real Estate Portf. Manag., **7**(1), pp. 79–88 (2012). Available: http://www.jstor.org/stable/24882251
37. Tondeur, J.: Enhancing future teachers' competencies for technology integration in education: Turning theory into practice. Seminar.net, **14**(2), pp. 216–224 (2018)
38. Kasturi, G.S., Jain, A., Singh, J.: Detection and classification of radio frequency Jamming attacks using machine learning. J. Wirel. Mob. Networks, Ubiquitous Comput. Dependable Appl., **11**(4), pp. 49–62 (2020),https://doi.org/10.22667/JOWUA.2020.12.31.049
39. Oderinde, F.O., Akano, O.I., Adesina, F.A., Omotayo, A.O.:"Trends in climate, socioeconomic indices and food security in Nigeria: Current realities and challenges ahead. Front. Sustain. Food Syst., **6**(2022) https://doi.org/10.3389/fsufs.2022.940858

40. Ning, X., Qi, J., Wu, C., Wang, W.: Reducing noise pollution by planning construction site layout via a multi-objective optimization model. J. Clean. Prod. **222**, 218–230 (2019). https://doi.org/10.1016/j.jclepro.2019.03.018
41. Almajali, D., Al-Radaideh, A.T., Nussir, N.A., Eid, A.A., Al-Fakeh, F.A., Masad, F.: Antecedents of mobile banking app adoption during COVID19: A perspective of Jordanian consumer. Int. J. Data Netw. Sci. **7**(1), 477–488 (2023). https://doi.org/10.5267/j.ijdns.2022.8.011
42. Khan, M.T., Idrees, M.D., Rauf, M., Sami, A., Ansari, A., Jamil, A.: Green supply chain management practices' impact on operational performance with the mediation of technological innovation. Sustain. **14**(6), 1–22 (2022). https://doi.org/10.3390/su14063362
43. Alaba, F.A., Amina, M., Ibrahim, U., Sulaimon, H.A., Zainab Aliyu, A.: Exploring the role of the internet of things (IoT) on economic growth in Nigeria: challenges and opportunities. Eng. Technol. Trends, **2**(6), (2024)
44. Relations, E., Repository, P.F.: An exploration of the practices of locational flexibility in developing economies: insights from the nigerian higher education sector abstract. **42**, pp. 166–179 (2021)
45. Fadele Ayotunde Alaba, Hakeem Adewale Sulaimon, Madu Ifeyinwa Marisa, Owamoyo Najeem.: Smart contracts security application and challenges: a review. Cloud Comput. Data Sci., **5**(1), pp. 15–41 (2023), https://doi.org/10.37256/ccds.5120233271
46. Fadele, A.A., Othman, M., Hashem, I.A.T., Alotaibi, F.: Internet of things Security: A Survey. J. Netw. Comput. Appl. **88**(April), 10–28 (2017). https://doi.org/10.1016/j.jnca.2017.04.002

Chapter 2
The Digital Turn in the Humanities

2.1 Introduction to Digital Humanities

DH is an interdisciplinary discipline that combines the humanities and digital technology. It combines classical humanistic research (literature, history, philosophy, art, and cultural studies) with computer tools and digital approaches to investigate novel ways of understanding, analysing, and sharing information. While the phrase acquired popularity in the early 2000s, its intellectual origins date back to the mid-twentieth century, when computers were first applied to ancient literature and archives. Digital Humanities is more than merely digitising books and incorporating software into the classroom; it signifies a paradigm change in how humanists approach research, teaching, and knowledge production. It raises important problems about representation, access, labour, infrastructure, and the politics of technology itself [1].

2.1.1 Historical Background and Evolution

The beginnings of DH may be traced back to the 1940s and 1950s, with early computational text analysis, particularly Father Roberto Busa's Index Thomisticus, a computerised concordance of Thomas Aquinas' writings. Over time, DH expanded beyond computational linguistics to include digital archives, spatial humanities, text encoding (e.g., TEI), data visualisation, network analysis, and interactive storytelling. The emergence of digital humanities centres, academic publications (such as Digital Humanities Quarterly), and degree programs marked an institutional acceptance of DH in the twenty-first century. Its rise has coincided with arguments over the role of the humanities in a digital era, eliciting both excitement and scepticism among academics [2].

F. A. Alaba and I. G. Ajibola, *Digital Pedagogy: Bridging the Gap Between Software and the Humanities*, Information Systems Engineering and Management 85,
https://doi.org/10.1007/978-3-032-20335-9_2

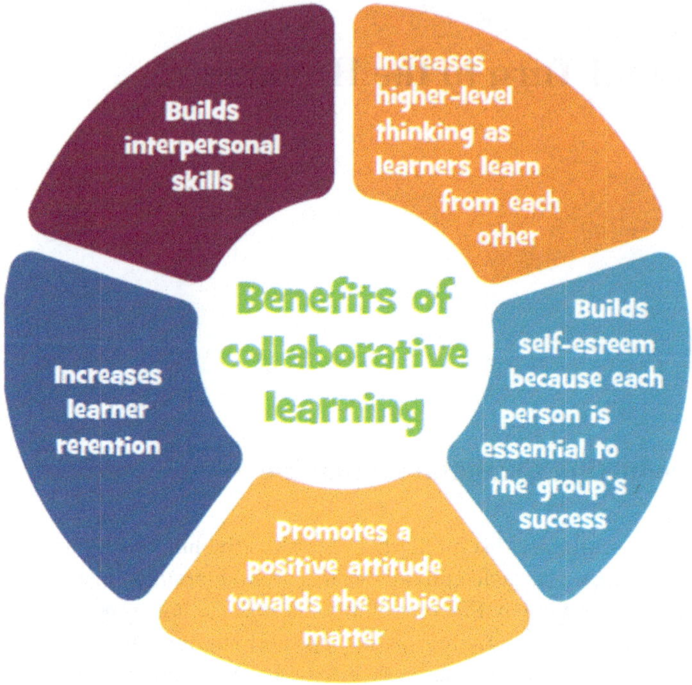

Fig. 2.1 Benefits of collaborative learning [3]

2.1.1.1 Key Areas and Tools in Digital Humanities

Digital Humanities comprises a diverse set of subfields and techniques, which often intersect with computer science, data science, library studies, and design (Fig. 2.1; Table 2.1).

2.1.2 Philosophical and Methodological Considerations

DH is not only about using technologies; it also sparks crucial methodological and philosophical debates [4]:

1. Epistemology: What constitutes knowledge in a digital context? Is data-driven analysis challenging or reinforcing established hermeneutics?
2. Ethics: Who has authority over the digital archive? How are biases represented in algorithms and data structures?
3. Access and Inclusion: Are digital resources available to everyone? What about locations that lack digital infrastructure?

Table 2.1 Areas and tools in digital humanities

Area	Description
Text mining and analysis	Use of algorithms to analyse large textual corpora for patterns and meaning
Digital archives	Creating searchable, annotated digital versions of physical texts and objects
Geographic information systems (GIS)	Spatial analysis and mapping of historical or cultural data
Network analysis	Studying relationships between people, texts, or events using graph theory
Digital publishing	New models for scholarly communication and interactive media
Computational linguistics	Syntax, semantics, and discourse analysis using machine learning
Augmented reality (AR)/virtual reality (VR)	Immersive experiences to reconstruct historical contexts

4. Authorship and cooperation: DH often necessitates multidisciplinary cooperation, blurring the distinctions between author, developer, and curator.

2.1.3 *Digital Humanities and Interdisciplinarity*

Interdisciplinarity is essential to the DH movement. Collaborations among humanists, data scientists, librarians, designers, and community partners lead to better insights and greater impact. For example, a project on colonial literature may utilise digital mapping to identify imperial expansion routes, as well as text mining to analyse thematic material. This interdisciplinary approach also allows students and researchers to develop hybrid skill sets that combine critical thinking with technological literacy [5] (Fig. 2.2).

2.1.4 *Global Perspectives and the Challenge of Representation*

Much of DH's growth has occurred at universities located in the Global North. Scholars, on the other hand, are increasingly emphasising the relevance of global digital humanities particularly the necessity for locally anchored initiatives, indigenous data sovereignty, multilingual resources, and inclusive infrastructure. African, Latin American, South Asian, and Indigenous people are working to challenge Western-centric digital paradigms and propose alternative epistemologies that decenter colonial legacies and power systems [7].

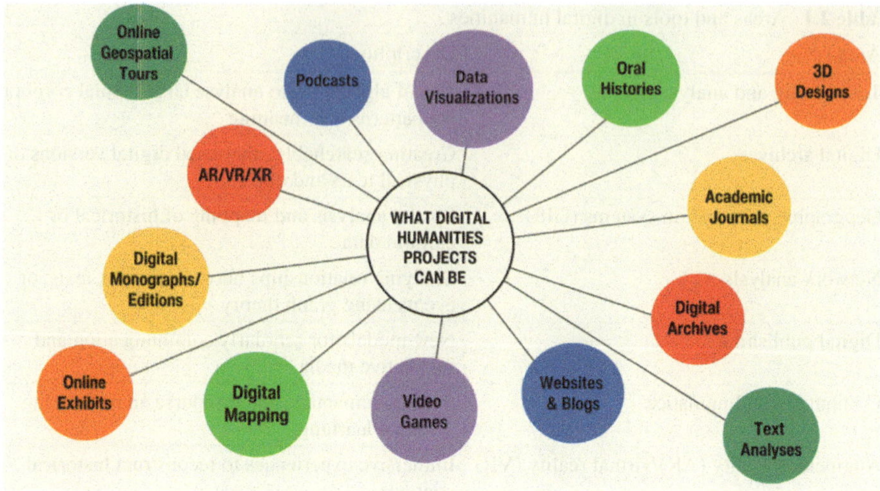

Fig. 2.2 Digital humanities [6]

2.1.5 Relevance to Education and Pedagogy

In the classroom, DH provides novel teaching techniques [8].

1. Interactive Learning: Students interact with historical sources via digital maps, archives, and games.
2. Skill development includes critical analysis, coding, digital curation, and project management.
3. Collaborative Pedagogy emphasises co-creation, collaborative collaboration, and multidisciplinary methods.
4. Public Humanities promotes connection with non-academic audiences via online displays, blogs, and open data.

Educators may use DH to make humanities instruction more relevant, applicable, and engaging for today's learners.

2.1.6 Challenges and Critiques

Despite its potential, DH confronts a number of challenges:

1. Funding and Sustainability: Many initiatives do not have long-term institutional backing.
2. The Digital Divide: Infrastructure constraints limit participation from the Global South and disadvantaged areas.

3. Labour Inequities: It often depends on insecure, unpaid labour, particularly from early-career researchers or graduate students.
4. Tool Fetishism: The risk of valuing technology above critical thought or intellectual depth.

Critics advocate for a balanced approach that considers both technical competence and humanistic research.

2.1.7 Looking Ahead

The future of DH will be determined by its capacity to remain critical, inclusive, and transformational. As AI, machine learning, and immersive technologies grow more common, DH must continue to raise challenging concerns about power, representation, and meaning. Emerging themes including feminist DH, Black DH, Indigenous DH, and decolonial DH are reshaping the discipline and broadening its ethical and political scope.

2.2 Historical Overview of Digital Transformations

Understanding the present and future of digital pedagogy in the humanities demands a critical examination of the past. This chapter presents a historical review of the important technological developments that influenced educational practices, academic communication, and knowledge production and consumption. From early computers in the postwar period to the widespread use of the internet and subsequent advances in artificial intelligence, digital changes have constantly transformed both the techniques and metaphors of education and research [9].

2.2.1 Pre-Digital Foundations and Analog Systems

Before digital technology, knowledge systems were mostly analogue. Manuscripts, printed books, microfilm, and analogue audio-visual media were the principal mediums for education and academic study. Libraries served as the major centres of knowledge, with cataloguing methods such as Dewey Decimal and Library of Congress Classification facilitating access to information [10] (Table 2.2).

Table 2.2 Key Characteristics of the Pre-digital era

Aspect	Pre-Digital Era
Storage	Physical media: books, film reels, microfilm
Dissemination	Slow, often localized or institution-bound
Communication	Primarily oral or print-based
Accessibility	Often limited to those within academic institutions

2.2.2 *Emergence of Educational Computing (1950s–1980s)*

The post-World War II era saw the creation of early computer systems, which were quickly applied to academic applications. In the 1940s, scholars such as Father Roberto Busa pioneered computational humanities by digitising Latin literature using punch cards and mainframe computers. In the 1960s and 1970s, universities started experimenting with computers in education, particularly in:

- Computer-Assisted Instruction (CAI): Programs such as PLATO (Programmed Logic for Automatic Teaching Operations) pioneered interactive screen-based learning.
- Word Processing and Text Analysis: Tools such as TACT (Textual Analysis Computing Tools) facilitated early literary analysis using digital methods.

These advancements started to move educational approaches away from instructor-centered paradigms and towards learner-centred ones.

2.2.3 *Rise of Personal Computing and the Internet (1980s–1990s)*

The 1980s marked a significant shift with the introduction of personal computers (PCs) into homes and schools. The Apple II, IBM PC, and other comparable devices made computers more accessible to educators and students. By the 1990s, the emergence of the internet had converted accessibility into connectivity [11]:

1. Email and bulletin boards enabled new scholarly interactions.
2. Hypertext and the World Wide Web enabled nonlinear learning and interactive content.
3. Online databases and digital archives, such as JSTOR, transformed academic access.

This time also witnessed the development of early e-learning platforms and the digitisation of significant humanities books, which aided in the democratisation of information.

Table 2.3 Web 2.0 and the social turn

Development	Impact on education and humanities
Learning management systems (e.g., Moodle, Blackboard)	Enabled blended and online learning
Wikipedia and open content	Changed how knowledge is curated and challenged traditional authority
Blogging and podcasts	Offered alternative scholarly and pedagogical platforms
YouTube and MOOCs	Massive Open Online Courses opened learning to global audiences

2.2.4 Web 2.0 and the Social Turn (2000s)

With the rise of Web 2.0, which is defined by user-generated content and participatory platforms (e.g., blogs, wikis, social media), digital learning became more collaborative, interactive, and social (Table 2.3).

2.2.5 Mobile Technology and the App Ecosystem (2010s)

Mobile devices, tablets, and smartphones were more popular in both official and casual learning throughout the 2010s. With the advent of mobile applications and cloud-based platforms, digital pedagogy became more accessible and portable. Key shifts included the following:

1. Students have access to materials whenever and wherever they need them.
2. Gamification with EdTech Apps: Increased engagement and personalised learning.
3. Cloud Collaboration Tools: Google Docs, Padlet, and Trello all enabled real-time, group-based projects.
4. Data Analytics in Education: Platforms started recording student behaviour to inform adaptive learning tactics.

The digital classroom has expanded beyond physical bounds, making education more hybrid and adaptable.

2.2.6 The Pandemic Pivot (2020–2022)

The COVID-19 epidemic was a watershed moment in the digital revolution. Institutions throughout the world were pushed to go online, hastening digital adoption in ways previously inconceivable [12] (Table 2.4).

Table 2.4 Pandemic pivot

Change	Outcome
Emergency remote teaching	Highlighted infrastructural and pedagogical inequalities
Mass adoption of zoom/teams	Normalized video conferencing for education
Digital fatigue	Exposed limits and psychological toll of digital immersion
Rise of hybrid models	Created lasting structural changes in pedagogy

Table 2.5 Key phases of digital transformation in humanities education

Era	Technological milestone	Impact on humanities
1950s–70 s	Mainframe Computing	Early text analysis, foundational digital tools
1980s–90 s	PC & Internet	Digitisation of archives, online learning
2000s	Web 2.0 & Social Media	Participatory platforms, new pedagogies
2010s	Mobile & Cloud Technologies	Flexible, ubiquitous learning
2020–2022	Pandemic & Hybrid Learning	Structural shifts, increased digital dependence
2023–Present	AI, XR, Blockchain	Intelligent, immersive, decentralised education

2.2.7 The Contemporary Landscape: AI, XR, and Blockchain

In the 2020s, AI, Extended Reality (XR), and Blockchain are altering the future of education and digital humanities.

1. AI and Machine Learning: Enable intelligent tutoring, grading automation, and content recommendation engines.
2. XR (AR/VR/XR): Improves immersive historical simulations and virtual museums.
3. Blockchain technology enables verified, decentralised credentialing and data security.

These tools are transforming digital teaching, shifting basic epistemologies and institutional frameworks (Table 2.5).

2.2.8 Affected Disciplines: Literature, History, Philosophy

The digital revolution of education has not had an equal impact on all academic subjects. While STEM areas often lead in technology adoption, the humanities particularly literature, history, and philosophy have interacted with and adapted to digital technologies in novel ways, reshaping their techniques, forms of expression, and teaching tactics.

2.3 Literature in the Digital Age

Digital Literary Studies

Digitisation and computer methodologies have transformed literature, spawning new areas such as digital literary studies and computational critique. Scholars increasingly utilise methods for remote reading, corpus analysis, and visualising literary networks [13] (Table 2.6).

These methods enable researchers to uncover theme trends, authorial styles, and intertextual links in big datasets.

Pedagogical Applications

Digital platforms improve literary education through:

1. Interactive Reading Platforms: Tools like Perusall allow for collaborative annotation and debate.
2. Audiobooks with Multimedia Storytelling: Increase access and engagement among different learners.
3. Fan fiction communities and online archives provide insights into reader response theory and story progression.

Despite these advances, critics caution that algorithmic reading may flatten literary complexity, arguing for a balanced mix of near and distant reading approaches.

2.3.1 History and the Digital Turn

The digitisation of archive resources is one of the most important effects of digital change in history, since institutions throughout the globe now allow online access to manuscripts, letters, pictures, and government data, democratising historical study. Platforms such as Europeana, which collects European cultural heritage objects; Chronicling America, which provides historic newspapers from the repository of Congress; and British History Online, a digital repository of essential historical documents, have transformed accessibility. This shift allows for distant research and comparative historical studies, extending both geographical and thematic views. Furthermore, digital history employs Geographic Information Systems (GIS) and

Table 2.6 Digital literary

Tool/method	Function
Voyant tools	Text mining and word frequency analysis
MALLET (Machine Learning for Language Toolkit)	Topic modeling and classification
Digital editions	Annotated hypertext versions of classic and modern texts

Table 2.7 Emerging topics in digital philosophy

Topic	Key concerns
AI and ethics	Moral responsibility, agency, machine learning bias
Surveillance and privacy	Limits of state and corporate oversight
Digital identity	Selfhood in the age of social media
Algorithmic decision-making	Transparency, fairness, and accountability

data visualisation tools to map spatial and temporal patterns over time. Examples include Mapping the Republic of Letters, which visualises correspondence among Enlightenment intellectuals, and the Trans-Atlantic Slave Trade Database, which provides insights into the movement of enslaved people. However, digital history has several obstacles, including issues regarding authenticity and source verification, since digital surrogates may not always adequate for thorough study. Furthermore, data gaps often reveal institutional biases in what is digitised and shared, whilst ethical curation raises concerns about handling sensitive or contentious histories online. As a result, connecting with digital history requires both technological ability and critical digital literacy [14].

2.3.2 Philosophical Inquiry into Technology

In addition to making use of digital technologies, philosophy also engages in critical reflection on such tools. As AI, surveillance, digital identity, and algorithmic prejudice become more important, the disciplines of philosophy of technology and digital ethics have arisen as essential areas of study [15] (Table 2.7).

2.3.3 Reimagining Pedagogy and Dialogue

Philosophers have the ability to communicate their views beyond the confines of academic circles via the use of online platforms such as PhilPapers, Academia.edu, and public philosophy blogs. The way philosophical discourse is developed is being reshaped by the proliferation of podcasts, online discussions, and YouTube lectures [16].

2.4 Tools Shaping Humanities Research

The digital era has brought up a new set of instruments that are changing humanities research techniques. Rather from being simply assistance, these technologies profoundly alter how researchers' access, analyse, interpret, and present information. From textual analysis software and digital archives to visualisation tools and collaboration platforms, the digital toolkit is diverse and ever-changing. Text remains central to much of humanities research. Digital tools for textual analysis allow scholars to process and interpret large volumes of text quickly and in novel ways [17, 18].

a. Natural Language Processing (NLP): NLP approaches allow for automatic interpretation of human language. In the humanities, they are employed to examine themes, emotion, linguistic style, and authorial voice. Voyant Tools, an open-source platform, offers word frequency analysis, collocation graphs, and topic modelling for big text corpora.
b. Stylometry and Authorship Attribution: Stylometric methods analyse writing style to identify writers, compare literary styles, and investigate textual development. For example, signature stylometric analysis aided in the identification of numerous contested Shakespearean plays and the examination of authorial consistency (Table 2.8).

2.4.1 Data Visualisation and Mapping Tools

In the field of humanities studies, visualisation is becoming an increasingly significant tool since it enables researchers to show data in ways that are both understandable and interpretable.

a. Data Visualisation Platforms: Researchers can present complicated information via graphs, networks, and charts with the use of platforms such as Tableau, Gephi, and Palladio. These platforms enable researchers to identify patterns that are not clearly obvious through conventional techniques. For instance, Gephi has been used to map links within medieval manuscripts, which has resulted in the identification of influential and transmitting clusters.
b. Geographic Information Systems (GIS): GIS helps scholars to investigate the geographical aspects of historical, literary, and philosophical material. Mapping

Table 2.8 Common textual analysis tools

Tool	Functionality	Application example
Voyant Tools	Word clouds, frequency, co-occurrence	Analysing recurring themes in novels
AntConc	Concordancing, keyword search	Linguistic research and corpora study
MALLET	Topic modeling using LDA	Identifying thematic patterns in texts
Stylo (R)	Stylometry and clustering	Authorship attribution

Table 2.9 Visualisation tools for the humanities

Tool	Purpose	Sample use case
Gephi	Network analysis and visualization	Mapping philosopher correspondence
Tableau	Interactive charts and dashboards	Visualising publication trends
Palladio	Time and network visualisation	Charting character interactions in novels
ArcGIS/QGIS	Spatial and temporal mapping	Historical mapping of trade routes

events, narratives, or intellectual networks offers a novel geographical and temporal viewpoint. For example, the Slave Voyages database depicts the transatlantic slave trade, including routes, ports, and demographic information (Table 2.9).

2.4.2 Digital Archives and Repositories

Scholarly study has been revolutionised as a result of the digitisation of primary sources, which has made it possible to get access to manuscripts, letters, pictures, and audio-visual materials from institutions all over the world without the need of physically travelling to different locations. Europeana, which houses over 50 million items from European museums and libraries, HathiTrust, which provides access to more than 17 million digitised volumes, and the British Library Digital Collections, which includes manuscripts, newspapers, and rare books, are examples of platforms that have been created as a result of initiatives that have been spearheaded by universities, libraries, and cultural organisations to digitise and open their collections. Furthermore, open-access repositories such as JSTOR Open Content, Project MUSE, and the Internet Archive have posed a threat to the conventional publishing methods. These repositories have brought about the democratisation of information and ensured that academic materials are made accessible to scholars and the general public in their entirety [19].

2.4.3 Annotation, Collaboration, and Writing Platforms

As a result of the proliferation of digital technologies that allow shared annotation, writing, and feedback, the progress of research has grown more collaborative and participatory throughout its history. Hypotheses.is is one example of a platform that enables social reading and collaborative annotation of online texts and PDFs. These platforms provide educational and research advantages by providing layered comments and conversation from several perspectives. On the other hand, applications such as Google Docs, overleaf (which is used for LaTeX papers), and GitHub allow for collaborative writing and revision history that are easily accessible. While Google Docs makes it possible to write and get comments in real time, which helps to encourage distant collaboration, overleaf simplifies the process of academic publication by handling the sophisticated formatting required for research. The breadth of digital humanities study has been expanded thanks to the use of GitHub, which is mostly recognised for its version control capabilities. GitHub has gained popularity in the field of humanities research for collaborative digital editions, as well as the preservation of academic resources and code repositories [20].

2.4.4 Audiovisual and Multimedia Tools

Tools that allow for the creation, editing, and analysis of audio-visual material are required since study in the humanities is increasingly dealing with form of media that does not include text. For the purpose of video and audio editing, software applications such as Audacity, Adobe Premiere Pro, and Final Cut Pro are extensively utilised in digital storytelling, oral history projects, and video essays. For instance, universities are now relying on these tools to produce high-quality documentation of personal narratives in oral history projects. 3D modelling and virtual reality applications are made possible by technologies such as SketchUp, Unity, and Omeka + Neatline. These technologies allow for the reconstruction of historical settings and the creation of interactive museum exhibitions, among other uses. An example that is particularly noteworthy is the Rome Reborn project, which use three-dimensional modelling to digitally reproduce Ancient Rome in order to provide educational experiences that are immersive. The growing significance of multimedia in the field of humanities studies is brought to light by these instruments [20].

2.4.5 Artificial Intelligence and Machine Learning

In the field of humanities study, the incorporation of artificial intelligence and machine learning is causing a transformation in the way that academics analyse and interpret cultural artefacts. These technologies make it possible to do automatic picture tagging, handwriting recognition, and predictive modelling, which opens up new study possibilities. Tesseract and Transkribus are two examples of OCR systems that are very important in the process of digitising archive materials. These programs turn scanned documents into text that can be searched. For example, machine learning algorithms have assisted in the analysis of the Dead Sea Scrolls, distinguishing various scribes based on handwriting patterns. Deep learning is also being used to the classification of artworks, the detection of frauds, and the deciphering of ancient manuscripts. Both of these applications go beyond text. By demonstrating how artificial intelligence is boosting the study of historical and cultural resources, these breakthroughs highlight how AI is making research activities that were previously inaccessible or time-consuming more efficient and accurate [21].

2.5 Humanities in the Age of Algorithms

Algorithms have emerged as important drivers of decision-making, communication, and cultural output in the twenty-first century. In this environment, the humanities have both an opportunity and a challenge: understand, criticise, and adapt to an algorithmic-driven society. Rather than being passive spectators, humanities academics are actively involved with algorithms, employing them to supplement research while questioning its social ramifications. An algorithm is a series of step-by-step instructions or rules used to solve a problem or complete a job. In digital systems, algorithms often govern [22]:

1. Data processing and retrieval (such as search engines)
2. Recommendation systems (such as social media feeds)
3. Classification and prediction (such as AI-based analysis)

In the humanities, these tasks overlap with fundamental questions concerning meaning, interpretation, context, and prejudice.

2.5.1 Algorithmic Reading and Interpretation

The advent of algorithmic tools in the humanities has allowed researchers to "read" texts and pictures on a massive scale, revealing new patterns and modes of interpretation. One popular method is distant reading, a phrase originated by Franco Moretti, which uses computer tools to analyse large textual corpora, focussing on broad trends, structures, and patterns rather than detailed readings. For example, topic modelling applied to nineteenth-century literature showed changing themes of gender, class, and morality throughout time. Another approach, sentiment and emotion analysis, use computers to identify emotional tone in text, enabling scholars to investigate affective qualities in literature, diaries, speeches, and even social media. However, opponents warn that these computer techniques may oversimplify sophisticated displays of sarcasm, ambiguity, or cultural complexity, thus reducing the richness of human communication. Despite their limits, these methods provide compelling new approaches to investigating large-scale cultural transitions and hidden patterns in historical and literary sources [23].

2.5.2 Cultural Production and Algorithmic Mediation

Algorithms increasingly play a crucial role in determining how culture is created, circulated, and consumed, creating important concerns for humanities study. On digital platforms such as YouTube, TikTok, and Spotify, recommendation algorithms filter material in ways that often highlight popular trends while ignoring niche or opposing viewpoints. This has prompted humanities academics to study how these systems affect diversity, representation, and cultural visibility, specifically if they reinforce existing prejudices or produce new kinds of cultural homogenisation. Generative AI technologies, such as ChatGPT, DALL·E, and MusicLM, are challenging conventional concepts of creativity and authorship by producing text, pictures, and music with minimum human involvement. This development raises philosophical discussions about originality, creative agency, and intellectual property, requiring a rethinking of what it means to produce and understand culture in the age of AI. As algorithms increasingly influence cultural experience, humanities studies must address the ethical, aesthetic, and social ramifications [24].

2.5.3 Ethics and Bias in Algorithmic Systems

The humanities play an important role in the digital era by critically examining algorithms' ethical elements, notably their link to power, surveillance, and social justice. Algorithmic bias shows how automated systems may perpetuate and exacerbate social disparities when trained on inaccurate or unrepresentative data. For example, face recognition algorithms have been proven to have much greater mistake rates for persons with darker skin tones, revealing racial biases in otherwise neutral systems. Meanwhile, the rise of algorithmic surveillance, from predictive policing to social credit scoring, jeopardises individual liberty, privacy, and basic rights. Humanities academics provide critical frameworks for analysing these challenges, using philosophy, cultural theory, and ethics to explore the larger implications of algorithmic governance. By asking who benefits from these systems and who is marginalised, the humanities give critical tools for fighting technological determinism and advocating for more fair and equitable digital futures [25] (Table 2.10).

Table 2.10 Humanities responses to algorithmic bias

Issue	Algorithmic example	Humanities response
Racial bias	Facial recognition tech	Critical race theory, postcolonial analysis
Gender representation	Image search results	Feminist media studies
Surveillance and autonomy	Predictive policing, spyware	Ethics, political philosophy
Algorithmic censorship	Social media moderation	Freedom of speech, digital rights

2.5.4 Algorithmic Thinking and Epistemology

New epistemological frameworks have emerged as a result of the proliferation of algorithmic systems, which often come into conflict with the conventional humanistic approaches to knowledge. Whereas the field of humanities studies has traditionally placed an emphasis on contextual and qualitative interpretation, algorithmic techniques have a tendency to place a greater emphasis on quantitative analysis, pattern identification, and reductive metrics. A fundamental conflict is created as a result of this: how can humanists make use of computational tools without oversimplifying complex cultural phenomena, such as works of literature, works of art, or historical events, into data-driven models that are flattened? One of the concerns that is connected to this is the "myth of algorithmic objectivity." Although these systems are presented in a manner that is impartial and unbiased, it is inevitable that they embody the beliefs, assumptions, and limits of the individuals who designed them. The academics of the humanities, drawing from the fields of hermeneutics, critical theory, and philosophy of technology, dispute this appearance of impartiality by revealing the ways in which power structures impact the decision-making process used by algorithms. Through the reintroduction of subjectivity, ambiguity, and historical contingency into discussions of computer systems, the humanities play a crucial role in preventing the unthinking acceptance of technologies that have the potential to obfuscate rather than enlighten human experience. By ensuring that technology serves as a tool for deeper knowledge rather than a substitute for human judgement, this critical engagement guarantees that technology fulfils its intended purpose [26].

2.5.5 Teaching and Learning with Algorithms

The ever-increasing impact of algorithmic systems is causing a transformation in the field of humanities education, necessitating the development of new literacy strategies and instructional methods. Students are now required to have "algorithmic literacy," which is an awareness of how algorithms organise information, affect cultural narratives, and even impact what constitutes knowledge. This is in addition to the usual activities of critical reading, historiography, or textual analysis that students have historically been taught. individuals are able to question the prejudices and preconceptions that are ingrained in digital systems thanks to this skill set, which ensures that individuals interact with technology as educated critics rather than as passive consumers. At the same time, educators are embracing pedagogical innovation by incorporating algorithmic technologies into the classroom. An effort is being made to revitalise the study of ancient texts and ideas via the use of data visualisation, artificial intelligence-assisted text analysis, and computational methodologies. For example, the course "Digital Shakespeare" employs network analysis to map character interactions, while the course "Philosophy of AI" leverages modern arguments about machine learning to examine age-old problems about consciousness and ethics.

Both of these courses are examples of those that apply network analysis. Students are able to get a better understanding of the linkages between humanistic research and the technological forces that are influencing their society via the use of these experiments, which not only make historical and literary studies more active. For the purpose of teaching students to navigate and confront the algorithmic elements of culture, power, and knowledge creation, educators are combining conventional instruction in the humanities with computational thinking [27].

2.5.6 The Algorithmic Public Sphere

The widespread role that algorithms play in mediating public discourse has significant repercussions for democratic processes, the preservation of collective memory, and the formation of cultural identities. These systems actively impact political discussions, public memory, and civic involvement, and they do so in ways that are sometimes completely unseen. They do this by managing what material is amplified or repressed. The growth of algorithmically reinforced echo chambers is a significant concern. These echo chambers are characterised by recommendation algorithms that give preference to material that is politically aligned or sensational, which in turn accelerates the spread of disinformation and polarisation. These processes are being dissected by humanities researchers in the fields of media studies, rhetoric, and communication, who are reacting by building tools for digital literacy and critical engagement. The work that they do helps the general public understand how the effect of algorithmic curation may be seen in perception and argument.

At the same time, the transitory character of information that is driven by algorithms, such as viral memes and popular social media postings, presents new obstacles for the preservation of cultural heritage. Digital archivists are increasingly seeing these transient artefacts as crucial archives of our algorithmic moment, and they are inventing strategies to preserve the context and growth of these artefacts. The purpose of this endeavour is to guarantee that future researchers will be able to study how platforms and their algorithms affected discourse in the twenty-first century, much as we already study newspapers or broadcast media from past periods. The humanities provide vital frameworks for comprehending and navigating our increasingly mediated public sphere by means of these dual approaches, which include criticising the power of algorithms while yet maintaining their products [28].

2.6 Digital Reading Versus Deep Reading

Reading has always been fundamental to the humanities, laying the groundwork for literary critique, historical analysis, philosophical research, and cultural knowledge. However, the digital revolution has fundamentally affected the way people read. Screens have replaced pages, hyperlinks provide a challenge to linearity, and scrolling conflicts with immersion. Digital reading is the process of reading texts on digital devices including smartphones, tablets, e-readers, and computers [29]. It covers a variety of formats:

1. Web articles and blogs.
2. eBooks and PDFs
3. Social media postings.
4. Online academic journals.
5. Hypertext fiction and interactive tales.

This kind of reading is generally distinguished by skimming, scanning, and bouncing between hyperlinks, which are regularly interrupted by alerts and multitasking.

2.6.1 Deep Reading: A Cognitive and Cultural Practice

Literary historian Maryanne Wolf coined the term "deep reading," which refers to an intensive, attentive, and critical interaction with a work. It's crucial for:

1. Critical Thinking and Inference
2. Empathy and Perspective-Taking
3. Memory consolidation and the synthesis of concepts
4. Interpretation of Metaphor, Irony, and Ambiguity

Deep reading takes time and is aided by consistent attention and steady, deliberate pace.

2.6.2 Cognitive Science Perspectives

Cognitive science is an interdisciplinary field that explores how the mind works—examining mental processes like perception, memory, learning, language, and decision-making. Drawing from psychology, neuroscience, linguistics, computer science, philosophy, and anthropology, cognitive science seeks to understand how people acquire, process, and use information. From a cognitive science perspective, learning is not just about absorbing facts but involves active mental engagement, prior knowledge, and context. Concepts such as cognitive load, metacognition (thinking about one's thinking), and schema theory (how knowledge is organized) inform educational strategies that align with how the brain functions.

For example, cognitive science suggests that chunking information, using retrieval practice, and integrating multisensory inputs can enhance memory and understanding. It also highlights the importance of attention and motivation in effective learning. By applying insights from cognitive science, educators, technologists, and

Table 2.11 Cognitive differences between digital and deep reading

Feature	Digital reading	Deep reading
Attention span	Shortened due to screen fatigue, alerts	Extended, focused
Comprehension	Lower for complex texts	Higher, supports inferential reasoning
Memory retention	Episodic, fragmented	Durable, interconnected
Emotional engagement	Reduced empathy	Heightened empathy and identification
Reading speed	Faster (skimming and scanning)	Slower (intentional and deliberate)

policymakers can design environments and tools that support deeper, more efficient learning and decision-making.

Numerous research in neuroscience and cognitive psychology have shown that digital reading and deep reading activate the brain differently (Table 2.11).

2.6.3 The Digital Reading Environment

Digital reading environments provide major distraction and multitasking issues, as hyperlinks, adverts, pop-ups, and alerts compete for attention, making sustained concentration difficult to achieve. Readers commonly multitask by jumping between tabs and programs, which may cause cognitive overload and reduced understanding depth. Furthermore, interface design and ergonomics play an important influence in reading efficiency and enjoyment. Font choice, layout, screen brightness, and preference for scrolling vs pagination all have an impact on reading fatigue, while extended screen exposure may cause eye strain and postural concerns, altering the length and efficacy of digital reading experiences [30].

2.6.4 Reading Behaviors in the Digital Age

Eye-tracking studies show that digital readers tend to follow a F-shaped pattern, reading the first few lines thoroughly before scanning vertically for keywords or headings. This nonlinear approach is supported by digital navigation habits, which include users jumping between parts using hyperlinks or search features rather than going linearly via text. Furthermore, digital platforms promote visual literacy by including infographics, memes, and emoji into communication, broadening interpretative options while sometimes obscuring the subtle linguistic complexity of conventional texts. Together, these shifts demonstrate how digital environments are reshaping reading practices by emphasising fragmentation, visual cues, and user-directed pathways over the sustained, sequential engagement characteristic of print [31].

Furthermore, the digital age has raised significant worries about the loss of profound reading abilities, especially among younger generations habituated to fast-paced, fragmented digital consumption. [32] a cognitive neuroscientist, warns that we are becoming simply information decoders. Our capacity to analyse text, to build the complex mental connections that occur when we read carefully and reflectively, is waning." This shortening attention span is most visible in educational settings, where pupils are increasingly unable to focus on long-form materials, and instructors see a decline in analytical writing and close reading competence. A surprising case study from a university literature course discovered that, although students enjoyed reading digitally, they consistently performed better on comprehension exams after interacting with paper texts, indicating that digital habits may jeopardise critical engagement even when users believe differently. These changes pose critical challenges about how to maintain deep literacy in an age of constant digital distraction.

2.6.5 Reclaiming Deep Reading in a Digital World

The argument between print and digital reading does not have to be binary; many educators now argue for a synergistic strategy that takes use of the benefits of both mediums. Digital tools excel at accessibility and interactive features (e.g., search functionality, annotation applications), although print is still better for prolonged, distraction-free engagement. To bridge this gap, teachers are implementing metacognitive strategies that teach students how to self-regulate their digital consumption, such as turning off notifications, using focus apps like Freedom or Forest, scheduling dedicated reading time, and combining digital texts with handwritten notes to improve retention. Meanwhile, creative platforms like Perusall and Hypothesis redefine digital reading as a communal, in-depth process, allowing for shared annotations and layered comments. These solutions reframe the issue not as a conflict between media, but as a chance to create conscious reading habits whether on screen or paper that prioritise critical thinking in an age of distraction [33].

2.6.6 Literary and Philosophical Reflections

The ethics and aesthetics of attention have been studies by humanists for a very long time. A number of philosophers, including [34, 35], have emphasised the importance of attentive reading in terms of its moral and creative potency. At this point in time, their observations take on a newfound sense of urgency. Because reading is increasingly being done on screens, humanists need to consider what it means to "read" in the truest sense that is, to not only digest information but to profoundly connect with the human experience that is communicated via language.

2.7 Digital Textual Analysis

Text has always been the foundation of humanities studies. Close reading, critical theory, and historical contextualisation have long been used by humanists to understand Shakespearean sonnets and decode political manifestos. However, in the digital era, textual analysis has grown tremendously as a result of the introduction of digital tools and computational approaches. Digital textual analysis is the use of computer tools and methodologies to analyse, interpret, and visualise vast amounts of textual data. It includes:

1. Text Mining
2. Corpus Linguistics
3. Stylometry
4. Topic Modelling
5. Sentiment analysis.
6. Network study of personalities or themes.
7. Keyword Frequency and Concordance

These tools enable researchers to identify patterns and trends that are difficult or impossible to notice via manual reading alone.

2.7.1 Historical Context: From Concordances to Corpora

Earlier attempts to measure and categorise textual information paved the way for digital textual analysis.

The evolution from concordances to corpora marks a significant shift in how language is studied, moving from manual text analysis to large-scale, computer-assisted linguistic research. Concordances, dating back to medieval times, were originally hand-compiled indexes of key words in texts—often religious ones like the Bible—listing where each word appeared and in what context. These tools helped scholars analyze patterns of word usage, themes, and theological concepts. With the advent of computers in the twentieth century, linguistic research expanded dramatically. Corpora (plural of corpus) emerged as large, structured collections of real-world texts—written or spoken—stored digitally for analysis. Unlike static concordances, corpora support dynamic querying, statistical analysis, and pattern recognition across millions of words. Pioneering examples include the Brown Corpus (1960s) and the British National Corpus (1990s).

Table 2.12 Common tools for digital textual analysis

Tool	Key function	Use case example
Voyant Tools	Word frequency, trends, concordance	Exploring keyword use in Victorian novels
AntConc	Concordancing and keyword analysis	Corpus linguistics for academic writing
MALLET	Topic modeling, clustering	Discovering themes in historical newspapers
NLTK (Python)	Tokenization, parsing, sentiment analysis	Analyzing tone in political speeches
Lexos	Text cleaning, comparison, and clustering	Stylometry in early modern poetry
Sketch Engine	Advanced corpus query and collocation	Studying semantic shifts in news articles

This shift revolutionized fields like computational linguistics, lexicography, and language teaching, enabling data-driven insights into authentic language use, variation, and change. Today, massive digital corpora like Project Gutenberg, Google Books Ngram Viewer, and HathiTrust provide unparalleled access to textual archives [36].

2.7.2 *Tools for Digital Textual Analysis*

Many digital tools, many of them web-based or open-source, facilitate textual analysis (Table 2.12).

2.7.3 *Case Studies of Digital Textual Analysis in the Humanities*

The introduction of computational approaches to literary studies is revealing new insights on classic literature. When researchers used Voyant Tools to analyse Shakespeare's entire works, they discovered distinct linguistic fingerprints between genres high-frequency words diverged sharply between comedies and tragedies, while concordance analysis revealed thematic clusters such as "death," "love," and "king." Network research revealed underlying relationship dynamics by mapping character interactions across plays.

Similarly, sentiment analysis using NLTK on Romantic poetry (including Blake, Wordsworth, and Shelley) revealed strong historical-literary relationships. The statistics showed that positive sentiment and dramatic emotional contrasts in Blake's Songs of Innocence and Experience decreased during the French Revolution. Most

Table 2.13 From close reading to distant reading

Aspect		Close reading	Distant reading
Scale		Individual text	Hundreds or thousands of texts
Focus		Interpretation, nuance, ambiguity	Patterns, trends, generalisations
Tools		Human intellect, critical theory	Algorithms, visualisation tools
Purpose		Deep understanding	Broad contextualization

compellingly, algorithmic mood monitoring corresponded to important historical events, providing measurable proof of how social upheavals influenced poetry language [37]. These case studies show that digital tools can:

1. Expose structural patterns that are undetectable to close reading alone.
2. Data-driven contextualisation bridges the gap between textual and historical study.
3. Enhance conventional interpretation without replacing it.

By combining computational rigour with humanistic inquiry, such methodologies broaden rather than limit the possibilities of literary studies.

2.7.4 From Close Reading to Distant Reading

Close reading is paying careful attention to individual passages, language usage, and literary methods. Franco Moretti popularised the term "distant reading," which encourages macro-level literary study across enormous corpora (Table 2.13).

2.7.5 Techniques in Digital Textual Analysis

DH researchers scour massive amounts of text for meaning using a toolbox of computer methods. Prompt frequency counts and machine learning applications are made possible by standardising words to their base forms (e.g., "running" → "run") and tokenisation, which separate texts into analyzeable units (words, sentences), which is the first step in the process.

By analysing the term "liberty" in 18th-century pamphlets, for instance, we can see how its meaning shifted from religious to political implications, but we may also learn about theme emphasis and contextual use via concordance and frequency analysis. "Family," "suffering," and "patriotism" are frequent themes in Civil War letters, with variations across time and locations. Topic modelling, for example, uses LDA algorithms, to discover latent themes in big corpora. This allows for wider pattern recognition.

Stylometry solves attribution disputes (like the one involving the Federalist Papers) or detects literary influence at the level of style by quantifying authorship fingerprints using metrics like sentence length or function-word frequency. At the same time, researchers may use sentiment and emotion analysis to chart literary affect, from simple polarity (positive/negative) to complex emotions (angry, joy), which helps them to understand the emotional journeys of books and how sentiment relates to past events [38].

2.7.6 Challenges and Critiques

While computational tools provide strong new ways to analyse texts, they also raise important epistemological and ethical concerns. One important worry is the loss of context: quantitative methods such as word-frequency analysis often flatten meaning by removing phrases from their cultural, historical, and rhetorical contexts. For example, counting uses of the word "freedom" without considering its many implications in abolitionist and libertarian rhetoric risks extreme oversimplification. These technologies can carry and reinforce algorithmic biases. When trained on corpora that over-represent Western, male-authored, or canonical literature, computational analyses may systematically marginalise minority voices and viewpoints, reinforcing existing disparities in the academic record. Perhaps most importantly, an overreliance on metrics risks limiting humanistic inquiry to what can be measured. While data visualisations and statistical outputs give captivating glimpses, they often fail to convey ambiguity, irony, and intertextuality the same characteristics that make literature and historical records valuable research subjects. Humanities researchers advise against substituting measurable surface patterns for profound, culturally based interpretation. This critical viewpoint assures that digital techniques broaden, rather than limit, humanistic modalities of knowing.

2.7.7 Ethics and Data Governance

Digital textual analysis introduces critical ethical considerations, particularly concerning privacy, consent, and representation. Social media platforms, for example, generate user content not originally intended for scholarly scrutiny. Mining this data without proper consent raises questions about surveillance and data misuse. Historical texts present additional challenges, often containing sensitive or problematic content requiring careful, context-aware interpretation. Furthermore, machine learning algorithms trained on biased datasets risk reinforcing societal prejudices, inadvertently perpetuating discrimination under the guise of objectivity. As a result, digital humanists must adopt robust data ethics practices, including obtaining informed consent where possible, maintaining transparency in data handling and analysis, and clearly acknowledging the limitations of the datasets and tools used.

Ethical data governance is not only a technical responsibility but also a reflection of the interpretive care central to humanistic inquiry [39].

2.7.8 *Pedagogical Applications*

The incorporation of digital textual analysis into humanities education is reshaping both curriculum and pedagogy. Tools like Voyant Tools enable students to explore recurring themes, keywords, and patterns in prescribed texts, fostering deeper engagement through data visualization. Sentiment analysis assignments allow learners to trace rhetorical shifts in political discourse over time, promoting historical awareness and methodological rigor. Collaborative digital annotation platforms, such as Hypothesis and Lumiere Reader, encourage communal reading practices and the layering of multiple interpretive voices onto a single text. These practices cultivate essential skills in digital literacy, critical thinking, and methodological transparency, equipping students to navigate and critique digital culture as informed participants and scholars [40].

2.7.9 *The Future of Textual Scholarship*

Emerging technologies—including AI, NLP, and large language models—are expanding the frontiers of textual scholarship. AI-assisted tools can detect stylistic patterns or authorial fingerprints across corpora, offering new insights into literary style and genre. Cross-lingual analysis, supported by automated translation and semantic modelling, enables comparative studies across linguistic boundaries. In addition, integrated text-image analysis is opening new possibilities for the study of graphic novels, illuminated manuscripts, and hybrid media forms. Multimodal analysis, which synthesizes text, audio, video, and metadata, supports holistic cultural interpretations. Despite these innovations, human interpretation remains indispensable. Algorithms can extend and augment scholarly analysis, but they cannot replicate the contextual understanding, cultural sensitivity, and interpretive nuance that define humanities research [40].

2.8 Preservation and Access of Digital Knowledge

The digital age has fundamentally transformed how knowledge is created, analysed, preserved, and accessed. The transition from physical to digital archives offers significant opportunities for the broader dissemination and durability of intellectual and cultural artefacts. Digitisation enhances the accessibility of rare or fragile materials, enabling global scholarly collaboration and public engagement. It also facilitates new forms of analysis through computational tools and methodologies. However, this shift also introduces substantial challenges. Digital knowledge—encompassing digitised historical documents, born-digital content (such as blogs, social media, and e-books), digital humanities projects, audiovisual media, and software repositories—is inherently vulnerable. Its preservation depends on the continued functionality of specific hardware, software, and data standards, many of which rapidly become obsolete. Questions of authenticity also arise; as digital reproductions may lack the material context of their physical counterparts. Additionally, digital infrastructure is unevenly distributed, raising concerns about equity of access across global and socio-economic divides.

Ensuring long-term access to digital knowledge demands coordinated strategies in digital preservation, including the use of open standards, metadata, and robust archival systems. Equally important are ethical considerations around ownership, consent, and cultural sensitivity. As digital collections continue to grow, libraries, archives, and scholars must remain vigilant in developing sustainable, inclusive, and forward-looking approaches to digital stewardship [41, 42].

2.8.1 Challenges of Digital Preservation

The process of preserving digital material entails a great deal more than merely conserving data; it necessitates the implementation of proactive techniques to counteract the vulnerability of technology and to guarantee accessibility over the long term. One of the most basic problems is the phenomenon of technology obsolescence. As software advances, some file formats, such as Flash or WordStar, fail to be usable, while physical media, such as CDs, deteriorate with time. The use of open, standardised formats (such as PDF/A, XML, and TIFF) and the maintenance of access via emulation or format migration are included as potential solutions.

The danger of bit rot and data corruption, which occurs when digital files steadily deteriorate, is also a particularly significant concern. In response to this, preservationists use error-correcting algorithms, checksums, and redundant storage systems in order to identify and restore any harm that may have occurred. Equally as important is the provision of information for the purpose of maintaining context; in the absence of characteristics such as authorship, date, or origin, a file runs the danger of becoming an indecipherable relic. There are frameworks available for incorporating this essential contextual scaffolding, and some examples of these frameworks are Dublin Core, METS/ALTO, and PREMIS [43].

The preservation process is further complicated by legal and ethical obstacles. It is possible that copyright limitations and privacy regulations would prohibit access to sensitive information or material that was created by users. In order to handle these limits while maintaining ethical stewardship, strategies such as Creative Commons license, data anonymisation, and permission processes are helpful [44].

Principles Essential to the Care of Sustainable Preservation:

1. The adoption of formats and systems that are resistant to technological change is a future-proofing strategy.
2. Redundancy: Store several copies that have been validated across archives that are located in different locations.
3. Combine the material with detailed, standardised metadata to ensure contextual integrity.
4. Compliance with the law: Strike a balance between the preservation aims, intellectual property rights, and privacy rights

Archivists have the ability to guarantee that digital assets continue to be accessible, interpretable, and valuable for future generations by addressing the myriad of issues that they confront. This is not just because they are data, but also because they are cultural legacy.

2.8.2 *Strategies for Sustainable Digital Preservation*

Strategies for the Long-Term Preservation of Digital Information.

The OAIS Model

An all-encompassing methodology for digital archiving is provided by the Open Archival Information System (OAIS) architecture (Table 2.14).

Redundancy and Geographic Distribution

It is common practice to refer to preservation plans as "LOCKSS," which stands for "Lots of Copies Keeps Stuff Safe.“ These plans must contain many copies that are kept in different geographical locations.

Periodic Migration

Migration is the process of moving data from outdated forms to supported ones, avoiding obsolescence. For instance, changing VHS tapes to digital MP4 formats provides access to video archives.

Table 2.14 OAIS functional components

Component	Description
Ingest	Accepts and prepares data for archiving
Archival storage	Stores data and ensures its integrity
Data management	Organises and maintains metadata
Administration	Oversees archive operations
Preservation planning	Plans for format migration, risk management
Access	Provides tools for discovery and retrieval

2.8.3 Access in the Digital Era

While conserving digital information is important, genuine stewardship entails reducing obstacles to access and encouraging inclusion. The Open Access (OA) movement has been groundbreaking in this respect, allowing academic work to be freely accessible online in order to democratise information, increase research impact, and speed collaboration. Similarly, Open Data regulations guarantee that datasets are available and reusable, allowing for verification and innovation as illustrated by programs such as the Digital Public Library of America (DPLA), which brings together open cultural heritage collections countrywide.

However, persistent inequities undermine these ideals. Paywalls and expensive journal subscriptions restrict academics from underfunded universities, while bandwidth inequities and technical limitations disproportionately affect scholars in the Global South. Even when the material is technically accessible, varying degrees of digital literacy might prevent meaningful participation. Pathways to Equitable Access:

1. Policy Advocacy: Expanding requirements for open-access publication and fighting exploitative subscription models
2. Infrastructure investment includes supporting low-bandwidth platforms and offline-access solutions.
3. Inclusive Design prioritises usability for varied audiences, incorporating bilingual interfaces and adaptable technology.

By tackling these structural challenges, the academic community may go beyond simple preservation and towards a genuinely participatory digital commons one in which information circulates as a public benefit rather than a commodity.

2.8.4 Case Studies in Digital Preservation and Access

Several innovative initiatives show how institutions may protect digital heritage while increasing access:

HathiTrust Digital Library.

1. A coordinated effort by academic institutions to preserve millions of digitised manuscripts.
2. Combats obsolescence via redundant storage and proactive format migration.
3. Tiering availability depending on permissions strikes a balance between access and copyright compliance.
4. Internet Archive
5. non-profit repository for books, music, software, and online history.
6. pioneered the Wayback Machine, storing billions of transitory online pages from "link rot" advocated for free universal access as a counterbalance to commercial platform control.

Europeana.

1. Europe's cultural heritage aggregator, linking over 4,000 organisations.
2. Focusses on multilingual discovery and metadata standardisation for cross-border research.
3. Engages audiences via virtual exhibits and crowdsourcing.

Shared Innovations:

1. Scalable Preservation: Fighting bit rot and format obsolescence institutionally.
2. Access philosophy varies from open (Internet Archive) to rights-managed (HathiTrust).
3. User-centred Design: Europeana focusses on multilingual, interactive participation.

These approaches demonstrate how technological preservation must be combined with appropriate access rules and participatory design in order to properly benefit global communities.

2.8.5 Role of Libraries, Museums, and Archives

Libraries, museums, and archives are leading the push to preserve cultural heritage for the digital age, with each using specialised tools and tactics to address specific issues.

1. Libraries have become centres of digital scholarship by:
2. Creating institutional repositories (with systems like DSpace, EPrints, and Fedora Commons) to preserve theses, publications, and datasets.
3. Collaboration with efforts such as HathiTrust to provide long-term access to digitised materials.
4. Advocate for open-access policies to democratise intellectual materials.

Museums are rethinking accessibility with technology:

1. Create virtual exhibits to share collections internationally.
2. Using 3D scanning and VR to highlight fragile or restricted artefacts (e.g., old texts, delicate sculptures)
3. Using augmented reality (AR) to provide contextual overlays for physical displays.
4. Archives address the issues of born-digital preservation.
5. Using digital forensics to retrieve data from outdated media (floppy discs, early websites).
6. Using TEI-XML and other markup languages to provide searchable and analyzeable metadata

Addressing ethical issues with storing emails, social media, and other ephemeral digital communications.

Shared Goals and Challenges:

1. Interoperability: Ensure that systems can interact across institutions (e.g., Europeana's aggregate model).
2. Engagement: Balancing preservation, public access, and educational value
3. Sustainability: Obtaining funds and technical expertise for long-term upkeep.

By combining conventional care with cutting-edge techniques, these organisations guarantee that cultural heritage is active, accessible, and resilient in a quickly changing digital context.

2.8.6 *Preservation and Access in the Global South*

Preservation efforts must be worldwide, recognising uneven access to technology and financing.

Challenge:

1. Underfunded institutions.
2. Lack of infrastructure for data storage and distribution.
3. Colonial Legacies in Knowledge Curation

Solution:

1. Community archiving projects.
2. Mobile-first digital platforms.
3. International collaborations for digitisation and training.

For example, the British Library's Endangered Archives Programme supports digitisation initiatives throughout Africa, Asia, and the Caribbean.

2.8.7 *Ethical Considerations*

There are significant ethical considerations that come into play when preserving and sharing digital knowledge, particularly with regard to ownership, privacy, and representation. When it comes to the management of records from Indigenous or minority groups, concerns of intellectual property and cultural sovereignty are particularly important. As a result, community-driven archives and free, prior, and informed consent (FPIC) are key best practices. Furthermore, the digitisation of personal data, such as letters or images, must be done in a way that respects the right to be forgotten while also minimising the possible hazards of abuse and monitoring. Because many digitisation projects favour Western canons, which creates gaps in cultural diversity, representation is another important concern that needs to be addressed. It is recommended that inclusive collection policies and diverse leadership in digital preservation be implemented in order to address this issue. This will ensure a wider and more equitable representation of the population [45].

2.8.8 Future Directions

The introduction of blockchain technology for tamper-proof provenance tracking, which ensures transparency in ownership and custody of digital artefacts; artificial intelligence-assisted curation, which automates classification, duplicate detection, and metadata generation while requiring safeguards against bias; and cloud-based Preservation-as-a-Service (PaaS) models, which enable smaller institutions to securely outsource storage and migration are all examples of emerging technologies that are revolutionising digital preservation. These advancements improve scalability, efficiency, and trust, but they must be balanced with ethical considerations, such as minimising algorithmic bias, ensuring equitable access, and maintaining interoperability, in order to cultivate a preservation ecosystem that is both technologically robust and socially responsible. This ecosystem will enable institutions to protect cultural heritage while simultaneously adapting to a digital landscape that is constantly evolving.

2.9 Critiques of the Digital Humanities Movement

The DH movement has grown significantly over the last two decades, transforming how researchers approach literature, history, philosophy, and cultural studies. By combining computational technologies with humanistic inquiry, DH has created new opportunities for study, cooperation, and public participation. However, this transition has not gone without criticism. The rise of DH has sparked critical debates about its impact on traditional humanistic scholarship, particularly concerning the loss of interpretive depth. This is because computational methods prioritise quantifiable patterns over the nuanced ambiguity, close reading, and subjective reflection that have been central to the humanities for a long time. This conflict is exemplified by the tensions surrounding Franco Moretti's "distant reading," which, while revealing macro literary trends, is accused of sidestepping the aesthetic and contextual richness of texts. In addition, critics warn against the phenomenon known as "technological determinism," which occurs when the tools themselves, such as text-mining algorithms, shape research agendas. This could potentially restrict inquiry to only what these methods can address, thereby posing a risk of reducing humanistic criticism to the limits of its digital instruments. These worries highlight a fundamental tension: while DH broadens the scope of analysis, it must continually engage with questions of epistemological integrity. This is necessary to ensure that computational approaches complement, rather than displace, the fundamental commitment of the humanities to interpretive complexity and critical autonomy [46].

2.9.1 Methodological Concerns

While digitising and quantifying cultural artefacts allows for large-scale research, it also creates substantial methodological and epistemological conflicts. Datafication threatens reducing the multidimensionality of literature, art, and philosophy to organised information, removing the subtleties (ambiguity, irony, historical contingency) that characterise their significance. For example, reducing poetry to word frequencies or political speeches to single phrases devalues context and symbolic resonance, but network visualisations of novels often neglect narrative temporality and aesthetic experience [47].

Similarly, an overreliance on quantitative tools (network analysis, sentiment analysis, topic modelling) risks marginalising traditional humanistic criticisms (rhetorical, psychoanalytic, feminist) that address power, subjectivity, and ideology. A major concern arises: Does counting substitute critical interpretation? While computational tools show macro patterns, they must be used carefully in conjunction with critical reading and theoretical frameworks to avoid turning the humanities to a positivist exercise. The difficulty is to strike a balance between scalability and depth, so that data acts as a lens rather than a constraint for cultural research.

Key Trade-offs:

1. Pattern versus Particularity: Data displays patterns while concealing singularities.
2. Efficiency versus Depth: Automated analysis accelerates research but may reduce criticism.
3. Objectivity versus Positionality: Metrics hide the subjectivity inherent in tool design.

The future of DH depends on critical data literacy using computational approaches while constantly challenging their limitations and biases.

2.9.2 Institutional and Economic Critiques

The advent of the DH has changed the academic scene, bringing both possibilities and institutional challenges. A major worry is the financial gap between DH and conventional humanities research, as institutions prioritise data-driven initiatives that correspond with STEM-oriented concepts of "innovation" and quantitative output. This approach risks marginalising subjects such as philosophy and classics, which depend on smaller-scale, interpretative work yet struggle to compete for resources in a period where massive DH laboratories get substantial funding and institutional commitment.

Compounding the problem is DH's vulnerability to neoliberal academic metrics, in which success is often limited to quantitative outputs grant money, tool development, and citation counts rather than the depth of critical inquiry. One professor noted sarcastically, "We are not building things to build knowledge; we are building things to build careers." This managerialist approach risks commodifying DH labour by rewarding programs that prioritise technological outcomes above transformational humanistic criticism. The drive to show "impact" might influence research objectives by prioritising flashy digital technologies over persistent engagement with theoretical or ethical issues [48].

Implications for the humanities ecosystem:

1. Resource Allocation: Does DH's financing advantage exacerbate divisions between the "technical" and "traditional" humanities?
2. Scholarly Values: How can the discipline avoid reduction to productivity measures while maintaining institutional support?
3. Critical Resistance: Can DH be used to criticise the structures (e.g., academic capitalism) that it is often required to serve?

Navigating these difficulties necessitates a concerted effort to ensure that DH reinforces, rather than undermines, the humanities' basic aim of cultivating critical thinking, historical reflection, and cultural debate outside the logic of the market.

2.9.3 Political and Ethical Critiques

Despite its revolutionary potential, the DH discipline has severe injustices that reflect wider academic inequality. An ongoing concern is the over-representation of whiteness and Western ideas in DH initiatives, leadership, and canon-building. While DH purports to democratise information, many programs, such as digital archives or text corpora, continue to prioritise Eurocentric materials while ignoring Indigenous, African, and Global South cultures. Scholars such as Roopika Risam argue for postcolonial DH, emphasising inclusive design and community-led archiving as ways to resist colonial paradigms.

Equally urgent are the labour hierarchies that support DH's collaborative approach. Though DH programs often rely on the knowledge of graduate students, librarians, and IT personnel, these participants suffer unstable employment and little credit compared to tenured teachers who lead the efforts. Lisa Spiro's assessment emphasises, "DH celebrates collaboration, but who gets tenure and who stays precarious?" This mismatch indicates a contradiction: DH's language of transparency often conceals oppressive labour practices, in which contingent workers shoulder the technical cost while permanent academics receive professional benefits.

Paths to Equity:

1. Representation: Actively diversify project leadership and content, emphasising marginalised perspectives in digital collections.
2. Labour Justice: Promote equitable salaries, credit, and career security for non-faculty DH contributions.
3. Critical Infrastructure: Create tools and standards that challenge colonial or neoliberal logics (for example, Mukurtu CMS for Indigenous cultural heritage).

Addressing these concerns requires more than simply inclusion; it also requires reframing DH as a location where technological innovation meets social justice, reshaping both approaches and institutional structures.

2.9.4 Disciplinary Resistance

The advent of DH has caused scepticism among conventional humanists, who often see it as a disruptive rather than complementing approach. Many academics are concerned that digital humanities marginalises non-digital study, enforces methodological uniformity, and prioritises flashy visualisations over substantial reasoning, comparing it to "putting neon lights on a Rembrandt." This contradiction echoes broader concerns regarding academic identity: is DH a discipline, a toolset, or an institutional rebranding of the humanities under technocratic pressures? While its methodological flexibility allows for multidisciplinary creativity, it also makes it subject to allegations of intellectual ambiguity, raising the issue of whether DH's expansiveness comes at the expense of academic depth.

However, these criticisms indicate opportunity. Instead of a binary conflict, the future may be critical integration, in which DH methodologies are used judiciously to address humanistic concerns rather than dictating them. For example, computational text analysis may reveal patterns in historical speech, but its conclusions need conventional hermeneutic evaluation. Similarly, data visualisations may reveal patterns, but they need contextual interpretation to avoid drawing sweeping conclusions [49].

Key Questions for the Field:

1. Can DH and conventional humanities coexist without one supplanting the other?
2. How can humanists convey the importance of both near and distant reading?
3. Does DH's institutional significance risk making old approaches seem antiquated, even when they are still necessary?

The route ahead is dependent on rejecting false dichotomies and encouraging dialogue ensuring that DH complements, rather than replaces, the humanities' essential commitment to critical thinking and cultural meaning.

2.9.5 *Inclusivity and Accessibility*

The potential of DH as a democratising force is jeopardised by continuing worldwide and accessibility gaps. Scholars in the Global South typically encounter infrastructural constraints, such as inconsistent internet, limited server resources, and low institutional finance, while Anglophone dominance in DH publications fosters linguistic and epistemic discrimination. While initiatives like Global Outlook Digital Humanities (GODH) promote multilingual cooperation and resource sharing, fundamental disparities exist.

Similarly, many digital humanities tools and platforms exclude handicapped researchers due to inadequate accessible design. Basic features like as screen reader compatibility, closed captioning, and keyboard navigation are sometimes considered as optional rather than essential. Scholars have emphasised that this negligence undermines DH's concept of openness, emphasising that universal design must be fundamental, not optional, in project development.

Paths to Equity:

1. Infrastructure Solidarity: Working with Global South institutions to co-develop low-bandwidth, offline technologies.
2. Accessibility by Default: Enforcing WCAG standards in DH tool development and ensuring disability representation on design teams.
3. Decentralising Knowledge: Encouraging non-English DH publications and conferences to broaden academic perspectives.

True inclusion in DH necessitates addressing not just what is examined, but also who may participate and under what circumstances.

2.9.6 *The Problem of Scale*

Scholars in the field of DH are challenging the idea that "bigger data equals better insights." They contend that humanistic investigations, like a psychoanalytic reading of Kafka's The Trial, necessitate careful examination of individual texts instead of conducting extensive computational analyses. This has led to significant methodological tensions within the DH field. This scepticism of size is accompanied with DH's struggle with "tool fetishism," in which the obsession with technical proficiency in software like as Python or Gephi trumps the intellectual merit of research issues, reducing scholarship to technical demonstrations instead of insightful interpretation. Digital technologies should be seen as means to a deeper knowledge rather than goals in themselves if DH is to tackle these difficulties. A balance between computational methodologies and conventional humanistic inquiry is necessary. In order for DH to complement, rather than replace, the interpretative depth inherent in humanistic research, there has to be a recommitment to question-driven research, methodological diversity, and critical engagement with technology outside of its application.

2.10 Rethinking Humanities Education

Humanities education has always emphasised critical thinking, interpretation, cultural awareness, and ethics. It has avoided instrumentalisation and vocational training by relying on literary study, debate, and philosophy. However, a fast-changing digital world, economic concerns, and a shifting student demography are redefining the humanities. In an era that prioritises economic utility as shown by shrinking departments, dwindling majors, and prevailing narratives that link education's worth to income potential there have been intense debates regarding the value of disciplines like literature, history, and philosophy in light of the worldwide decline in humanities enrolments and funding. This decline is often attributed to perceptions of these fields as being impractical compared to STEM fields. Nevertheless, there are numerous scholars who disagree with this declension narrative and instead see this time as a chance to rethink what the humanities are and how they should be taught. They argue that the humanities should continue to teach people to think critically, ethically, and culturally, but they also want to find new ways to engage with the public and innovate in education so that they can prove they are still relevant in solving problems like digital inequality and the erosion of democracy. We are at a critical juncture when we must decide how to preserve the humanities while also bringing them up to date with the demands of the modern world without sacrificing their essence to instrumentalist reasoning [50].

2.10.1 Pedagogical Innovation in the Digital Age

Blended learning environments, which combine online and in-person instruction with technologies like Learning Management Systems (LMS), video essays, and collaborative annotation platforms like Hypotheses, are among the innovative pedagogical approaches brought about by the digital transformation of humanities education. Is facilitating student-driven learning via digital storytelling and podcasting, increasing accessibility, and encouraging multimodal expression; nevertheless, educators must carefully prevent the watering down of meaningful involvement in virtual places. At the same time, flipped classrooms are becoming increasingly popular, which is shaking up the status quo of lecture-based learning. Instead of having students read and watch lectures beforehand, they can engage in more dynamic and active learning activities like group projects, debates, and textual analysis. This approach helps students develop their critical thinking and application skills while also expanding their humanistic knowledge base. These changes are the result of more than simply technology advancements; they are also part of a larger pedagogical rethinking that places an emphasis on adaptability, student agency, and collaborative learning in the face of the age-old problem of how to foster genuine intellectual community in both online and offline settings [50] (Table 2.15).

Table 2.15 Comparison of traditional versus flipped humanities classroom

Aspect	Traditional classroom	Flipped classroom
Content delivery	In-class lectures	Pre-recorded or assigned texts
In-Class activities	Passive listening	Collaborative analysis
Assessment	Exams and essays	Projects, presentations, peer review

2.10.2 Interdisciplinary and Curriculum Design

The humanities are experiencing a transformational transition, as they increasingly act as both multidisciplinary bridges and ethical anchors for higher education. Humanities programs are cultivating vital intersections with STEM and social science fields by breaking down traditional disciplinary silos through courses like "Philosophy of Artificial Intelligence" that examine algorithmic bias, "Climate Narratives and Environmental Justice" that merge literary analysis with ecological science, or "History of Medicine and Public Health" that contextualises pandemics through cultural studies. These hybrid offers not only broaden the humanities' intellectual scope, but also demonstrate their critical role in tackling complex global challenges ranging from AI governance to climate policy.

Simultaneously, the humanities are seen as the ethical foundation of multidisciplinary education. Medical humanities programs prepare prospective physicians to deal with patient narratives and bioethical quandaries, while technology ethics courses prepare engineers to examine the societal ramifications of innovation. This twin movement outward growth into other fields and inward reinforcement of humanistic values positions the humanities not as a dying field, but as a critical foundation for combining knowledge with moral thinking. While colleges focus on "wicked problem" solution, the humanities' emphasis on critical reflection, historical context, and nuanced interpretation provides skills for navigating an increasingly complicated world without sacrificing depth for usefulness [51, 52].

Key Implications:

1. Curriculum Innovation: Humanities departments are moving beyond conventional canonical topics to address social demands.
2. Institutional Influence: Professional schools (business, medical, and engineering) are implementing humanities-based ethical instruction.
3. Public Relevance: These developments challenge notions of humanities demise by proving their practical use.

The future of humanities education resides in this synergistic paradigm, which maintains analytical rigour while actively impacting how other disciplines interact with human values, narrative, and systemic criticism.

Table 2.16 Humanities skills and job market applications

Skill	Application in job market
Argumentation	Law, public policy, communications
Research and synthesis	Journalism, UX design, consulting
Cultural literacy	International relations, marketing
Ethics and reasoning	Nonprofit leadership, corporate governance

2.10.3 Skill Development and Employability

Contrary to persistent stereotypes, humanities graduates possess a range of highly transferable skills that are increasingly valued in the modern workforce. These include analytical and critical thinking, effective written and verbal communication, cross-cultural competence, ethical reasoning, and adaptability—all of which are essential in a global, diverse, and fast-changing employment landscape. Humanities education fosters the ability to interpret complex information, construct persuasive arguments, and navigate nuanced social and cultural contexts, preparing graduates for roles in sectors such as media, policy, education, technology, and beyond. However, a key challenge lies in translating these skills into career readiness. Innovative educational programs are addressing this gap by helping students articulate their capabilities and apply them in professional environments. Strategies include portfolio-based assessments, which showcase student work in real-world contexts; internships with structured reflective components that connect academic learning with practical experience; and integrated career development workshops embedded within the curriculum.

As noted by a UK university career services director: "The problem is not that humanities students lack skills; it's that they often don't know how to name them." Helping students recognise and communicate their competencies is essential to bridging the gap between education and employment, and to ensuring that humanities graduates thrive in diverse and dynamic career paths [51] (Table 2.16).

2.10.4 Equity, Inclusion, and Decolonisation in Curriculum

The humanities are experiencing major transformations as they embrace decolonisation and inclusion, changing curriculum and pedagogy to reflect a diverse worldview. Diversifying the canon today prioritises voices from the Global South, translated works, and critical interrogations of how the Western literary and philosophical traditions were produced, allowing students to interact with a genuinely global intellectual history. Along with this curriculum transition, inclusive pedagogies are reinventing classroom practice via Universal Design for Learning (UDL), culturally responsive teaching, and community-engaged research, ensuring that education adapts to varied learners rather than imposing old standards. Collectively, these trends are increasing the relevance of the humanities not by abandoning tradition, but by strengthening its responsibility to the whole range of human experience and knowledge creation.

Key outcomes:

1. Expanded Epistemologies: Students encounter knowledge systems that go beyond Eurocentric frameworks.
2. Accessible Excellence: Pedagogy prioritises equality in interaction and assessment.
3. Scholarly Legitimacy: Recognising that inclusion enhances rather than dilutes rigour.

This dual emphasis on what we teach and how we teach it situates the humanities as a location where critical tradition meets transformative practice, equipping students to negotiate and influence a complicated, interrelated world.

2.10.5 Public Humanities and Civic Engagement

The humanities are breaking down conventional academic barriers with public-facing work that connects college and community, converting students and teachers into active participants in civic debate and cultural preservation. Oral history projects documenting local narratives, public philosophy discussions in libraries, and student-curated museum exhibitions all demonstrate the importance of the humanities in fostering democratic engagement, collective memory, and social justice demonstrating how humanistic inquiry can address real-world issues while remaining intellectually rigorous. Simultaneously, digital channels ranging from podcasts and YouTube to social media and blogs are transforming how humanities information is disseminated, allowing researchers to reach a wider audience and democratise critical ideas. This trend towards digital publics calls into question established academic publication standards, supporting more dynamic, accessible forms of knowledge sharing but also raising concerns about sustainability, legitimacy, and the ethics of public involvement in an era of disinformation.

Key Implications:

1. Democratising Knowledge: Public and digital humanities make research available outside of academia.
2. Applied Learning: Students learn narrative, curation, and media literacy while contributing to community knowledge.
3. New challenges include balancing open-access goals with scientific rigour and ethical duties in digital contexts.

By adopting these outward-facing forms, the humanities reassert their importance not just as a discipline of study, but also as a practice of critical citizenship in both the physical and digital worlds.

2.10.6 *Institutional Support and Policy Implications*

The future of humanities education is dependent on institutional commitment to faculty empowerment and strategic advocacy, which ensures educators are prepared to navigate changing pedagogical settings while upholding the field's essential ideals. Universities must engage in targeted faculty development, disciplinary partnerships and emphasising the humanities' critical role in developing democratic discussion, ethical reasoning, and cultural literacy. This dual strategy, based on pedagogical adaptation but strengthened by institutional action, frames the humanities as a resilient force determining education's mission in an age of technological and social revolution.

Key strategies for sustainability:

1. Resource Allocation: Provide money for faculty development in digital, public, and inclusive pedagogies.
2. Narrative Shift: Explain the significance of humanities in terms of social effect (for example, misinformation resilience and empathy development).
3. Coalition Building: Collaborate with STEM and professional schools to demonstrate transdisciplinary interdependence.

The route ahead requires neither nostalgia nor capitulation, but rather reimagining stewardship honoring the humanities' heritage while expressing their modern relevance.

References

1. Nurudeen Olalekan Orunbon, Ganiyu Abiona Ibikunle, Adewale Abiodun Badmus.: Principals' digital transformational leadership, teachers' organisational commitment and job satisfaction during Covid-19 in lagos state education district V, Nigeria," Int. J. Humanit. Technol. Civiliz., **8**(1), pp. 68–73 (2023). https://doi.org/10.15282/ijhtc.v8i1.9424
2. Oyeronke, A., Fagbohun, M., Adebayo, O.: An assessment of computer and ICT skills among secondary school teachers in Ota Ogun State. (2013). Available: http://digitalcommons.unl.edu/libphilprac%5Cn, http://digitalcommons.unl.edu/libphilprac/846%5Cn, http://digitalcommons.unl.edu/libphilprac/
3. Omheni, N., Bouabidi, I., Gharsallah, A., Zarai, F., Obaidat, M.S.: Smart mobility management in 5G heterogeneous networks. IET Networks **7**(3), 119–128 (2018). https://doi.org/10.1049/iet-net.2017.0208
4. Pereira, C.S., Durao, N., Moreira, F., Veloso, B.: The importance of digital transformation in your business strategy. Mdpi, **14**(2), pp. 1–26 (2022). Available: https://blog.workana.com/en/entrepreneurship/the-importance-of-digital-transformation-in-your-business-strategy/
5. Tauseef, M., Kounte, M.R., Nalband, A.H., & Ahmed, M.R.: Exploring the joint potential of blockchain and AI for securing internet of things. Int. J. Adv. Comput. Sci. Appl. **14**(4), 885–895 (2023). https://doi.org/10.14569/IJACSA.2023.0140498
6. Reja, V.K., Varghese, K.: Impact of 5G technology on IoT applications in construction project management. In Proc. 36th Int. Symp. Autom. Robot. Constr. ISARC. **36**(5), pp. 209–217 (2019). https://doi.org/10.22260/isarc2019/0029.

7. Giantari, I.G.A.K., Yasa, N.N.K., Suprasto, H.B., Rahmayanti, P.L.D.: The role of digital marketing in mediating the effect of the COVID-19 pandemic and the intensity of competition on business performance. Int. J. Data Netw. Sci. **6**(1), 217–232 (2022). https://doi.org/10.5267/J.IJDNS.2021.9.006
8. Olika, C.D.: Risk management practices among selected state governments in southwestern Nigeria. Eur. J. Sci. Innov. Technol. **1**(2), pp. 24–31 (2024). Available: https://www.researchgate.net/profile/Hendra-Manurung/publication/356633046_From_Connectivity_to_Digital_Improving_Employee_Readiness_toward_Organizational_Change_in_Digital_Transformation/links/61a5b00671a23a0084c9b061/From-Connectivity-to-Digital-Improvi
9. New, T., Revolution, D.: Transformations through blockchain. Technology (2022). https://doi.org/10.1007/978-3-030-93344-9
10. Bhatija, V.P., Thomas, N., Dawood, N.: A preliminary approach towards integrating knowledge management with building information modeling (K BIM) for the construction industry. Int. J. Innov. Manag. Technol. **8**(1), 64–70 (2017)
11. Khan, J.: Impact of information communication technology on. Int. J. Res. **4**(9), 97–100 (2016). https://doi.org/10.5281/zenodo.158975
12. Syed, A.R.S. et al.: Mission of polio eradication in pakistan-a review on challenges and future recommendations. (2023). https://doi.org/10.20944/preprints202305.1206.v1
13. Kim, S., Ji, Y.: Positive ripple effects of corporate leaders' CSR donations amid COVID-19 on corporate and country reputations: Multi-level reputational benefits of CSR focusing on Bill Gates and Jack Ma. Public Relat. Rev. **47**(4), 102073 (2021). https://doi.org/10.1016/j.pubrev.2021.102073
14. Mothukuri, V., Khare, P., Parizi, R.M., Pouriyeh, S., Dehghantanha, A., Srivastava, G.: Federated learning-based anomaly detection for IoT security attacks IEEE Internet Things J., **4662**, pp. 1–10 (2021). https://doi.org/10.1109/JIOT.2021.3077803.
15. Nozari, H., Szmelter-jarosz, A., Ghahremani-nahr, J.: (AIoT) for the smart supply Chain (Case Study). (2022)
16. Zhang, Z., Shang, Y., Cheng, L., Hu, A.: Big Data capability and sustainable competitive advantage: the mediating role of ambidextrous innovation strategy. Sustain. **14**(14), 1–17 (2022). https://doi.org/10.3390/su14148249
17. Miglionico, A.: The use of technology in corporate management and reporting of climate-related risks. Eur. Bus. Organ. Law Rev. **23**(1), 125–141 (2022). https://doi.org/10.1007/s40804-021-00233-z
18. Di Maria, E., De Marchi, V., Galeazzo, A.: Industry 4.0 technologies and circular economy: The mediating role of supply chain integration. Bus. Strateg. Environ. **31**(2), 619–632 (2022). https://doi.org/10.1002/bse.2940
19. Brahmia, S., Brahmia, Z., Grandi, F., Bouaziz, R.: τJSchema: A framework for managing temporal JSON-based NoSQL databases. Lect. Notes Comput. Sci. (including Subser. Lect. Notes Artif. Intell. Lect. Notes Bioinformatics), **9828** LNCS, pp. 167–181 (2016). https://doi.org/10.1007/978-3-319-44406-2_13
20. Sykes, D., et al.: Comparison of rule-based and neural network models for negation detection in radiology reports. Nat. Lang. Eng. **27**(2), 203–224 (2021). https://doi.org/10.1017/S1351324920000509
21. Campillos-Llanos, L., Valverde-Mateos, A., Capllonch-Carrión, A., Moreno-Sandoval, A.: A clinical trials corpus annotated with UMLS entities to enhance the access to evidence-based medicine. BMC Med. Inform. Decis. Mak. **21**(1), 1–19 (2021). https://doi.org/10.1186/s12911-021-01395-z
22. Xia, P. et al.: Lome: Large ontology multilingual extraction. In: EACL 2021—16th Conf. Eur. Chapter Assoc. Comput. Linguist. Proc. Syst. Demonstr. **3**(6), pp. 149–159 (2021). https://doi.org/10.18653/v1/2021.eacl-demos.19
23. Savova, G.K. et al.: Mayo clinical text analysis and knowledge extraction system (cTAKES): Architecture, component evaluation and applications. J. Am. Med. Informatics Assoc., **17**(5), (2010). https://doi.org/10.1136/jamia.2009.001560

24. Lin, C.: Support vector machine solvers. Large-Scale Kernel Mach., pp. 1–27 (2019). https://doi.org/10.7551/mitpress/7496.003.0003
25. Leskovec, J., Rajaraman, A., Ullman, J.D.: Mining of massive datasets. (2014). https://doi.org/10.1017/CBO9781139924801
26. Catalano, C., Paiano, L., Calabrese, F., Cataldo, M., Mancarella, L., Tommasi, F.: Anomaly detection in smart agriculture systems. Comput. Ind. **143**, p. 103750 (2022). https://doi.org/10.1016/j.compind.2022.103750
27. Kumar, A., Starly, B.: 'FabNER': information extraction from manufacturing process science domain literature using named entity recognition. J. Intell. Manuf. **4**(3), 1–15 (2021). https://doi.org/10.1007/s10845-021-01807-x
28. Wang, C., Liu, S., Yang, H., Guo, J., Wu, Y., Liu, J.: Ethical considerations of using ChatGPT in health care. J. Med. Internet Res., **25**, (2023). https://doi.org/10.2196/48009
29. Katarya, R., Massoudi, M.: Recognizing fake news in social media with deep learning: a systematic review. In: 4th Int. Conf. Comput. Commun. Signal Process. ICCCSP **2020**, 2020–2023 (2020). https://doi.org/10.1109/ICCCSP49186.2020.9315255
30. Önder, M.: Contribution of plays and toys to children's value education. Asian J. Educ. Train. **4**(3), 146–149 (2018). https://doi.org/10.20448/journal.522.2018.42.146.149
31. Hakak, S., Kamsin, A., Idna Idris, M.Y., Gani, A., Herawan, T., Zerdoumi, S.: Preserving content integrity of digital holy quran: survey and open challenges. IEEE Access, pp. 1–1 (2017). https://doi.org/10.1109/ACCESS.2017.2682109
32. Liang, Z., et al.: EEGFuseNet: hybrid unsupervised deep feature characterization and fusion for high-dimensional EEG with an application to emotion recognition. IEEE Trans. Neural Syst. Rehabil. Eng. **29**(3), 1913–1925 (2021). https://doi.org/10.1109/TNSRE.2021.3111689
33. Emmert-Streib, F., Yang, Z., Feng, H., Tripathi, S., Dehmer, M.: An introductory review of deep learning for prediction models with big data. Front. Artif. Intell. **3**(February), 1–23 (2020). https://doi.org/10.3389/frai.2020.00004
34. Peters, M.E., Neumann, M., Zettlemoyer, L., Yih, W.T.: Dissecting contextual word embeddings: Architecture and representation. In: Proc. 2018 Conf. Empir. Methods Nat. Lang. Process. EMNLP 2018, pp. 1499–1509 (2020). https://doi.org/10.18653/v1/d18-1179
35. Zhang, Z., Zhang, Y., Zhou, T., Pang, Y.: Medical assertion classification in Chinese EMRs using attention enhanced neural network. Math. Biosci. Eng. **16**(4), 1966–1977 (2019). https://doi.org/10.3934/mbe.2019096
36. D'Souza, J., Ng, V.: Sieve-based entity linking for the biomedical domain. Assoc. Comput. Linguist. **2**(3), 297–302 (2020). https://doi.org/10.3115/v1/p15-2049
37. Li, F., Lin, Z., Zhang, M., Ji, D.: A span-based model for joint overlapped and discontinuous named entity recognition. ACL-IJCNLP 2021—59th Annu. Meet. Assoc. Comput. Linguist. 11th Int. Jt. Conf. Nat. Lang. Process. Proc. Conf., pp. 4814–4828, 2021, https://doi.org/10.18653/v1/2021.acl-long.372
38. Nguyen, T.H., Shirai, K.: Topic Modeling based Sentiment Analysis on Social Media for Stock Market Prediction. Proc. 53rd Annu. Meet. Assoc. Comput. Linguist. 7th Int. Jt. Conf. Nat. Lang. Process. (Volume 1 Long Pap., pp. 1354–1364 (2015). https://doi.org/10.1016/j.eswa.2015.07.052
39. Zupan, L.: 20 key risks to consider by Internal Audit before 2020. KPMG (2020)
40. Peters, G.T., Bagshaw, K.: Corporate governance mechanisms and financial performance of listed firms in Nigeria : A content analysis. Glob. J. Contemp. Res. Accounting, Audit. Bus. Ethics, **1**(2), pp. 103–128 (2014)
41. Lou, C.: Social media influencers and followers: theorization of a trans-parasocial relation and explication of its implications for influencer advertising. J. Advert. **51**(1), 4–21 (2022). https://doi.org/10.1080/00913367.2021.1880345
42. Obi and, K.C., Iwuanyanwu, O.: "Innovation in the assessment of primary school mathematics: an approach to revamping nigeria education. J. Teach. Perspect., **18**(1), pp. 51–66 (2023). Available: http://repositorio.unan.edu.ni/2986/1/5624.pdf%0Ahttp://fiskal.kemenkeu.go.id/ejournal%0Ahttps://doi.org/10.1016/j.cirp.2016.06.001%0A10.1016/j.powtec.2016.12.055%0A10.1016/j.ijfatigue.2019.02.006%0A10.1

43. Casini, M.: Extended reality for smart building operation and maintenance: a review. Energies **15**(10), 3785 (2022). https://doi.org/10.3390/en15103785
44. Huang, S., Huang, X.: How Green Bankers Promote Behavioral Integration of Green Investment and Financing Teams—Evidence from Chinese Commercial Banks. Sustain. **15**(9) (2023). https://doi.org/10.3390/su15097350
45. Ozougwu, P.E., Madu, C.N., Chukwuorji, J.C., Ozougwu, A.O., Ozougwu, S.U.: Environmental identities and attitude towards crude oil pipeline vandalism in Niger delta oil-producing communities. (2023)
46. Priya, M.M.M., Kothai, P.S., Kohilambal, M.E.: Study on safety practices and their performance in the construction industries. (2016)
47. Bennett, N.J., et al.: Environmental (in)justice in the Anthropocene ocean. Mar. Policy **147**(October), 2023 (2022). https://doi.org/10.1016/j.marpol.2022.105383
48. Costin, A., Wehle, A., Adibfar, A.: Leading indicators—a conceptual IoT-based framework to produce active leading indicators for construction safety. Safety **5**(4), 1–26 (2019). https://doi.org/10.3390/safety5040086
49. Bartulović, D.: Predictive safety management system development. Trans. Marit. Sci. **10**(1), 1–12 (2021). https://doi.org/10.7225/toms.v10.n01.010
50. Muraina, Ajímátanraẹjẹ, A.J.: Gender relations in indigenous Yoruba culture: Questioning feminism action and advocacy. AfricArXiv Prepr., pp. 1–19 (2022). Available: https://osf.io/preprints/africarxiv/x6vma/
51. Mostert, M., Quinn, L.: Using ICTs in teaching and learning : Reflections on professional development of academic staff. J. Educ., **5**(5), pp. 72–84 (2009)
52. Bratianu, C., Stanescu, D.F., Mocanu, R.: Exploring the knowledge management impact on business education. Sustain. **13**(4), 1–16 (2021). https://doi.org/10.3390/su13042313

Chapter 3
Pedagogical Theories in the Digital Age

3.1 Traditional Pedagogy Revisited

Traditional teaching must be revisited before we can explore how the digital revolution affects the humanities. The field of humanities was moulded by educational practices that prioritised memorisation, literary engagement, dialogic teaching, and moral formation long before computational analysis, online learning environments, or digital archives came into being. Although these approaches are sometimes criticised for their shortcomings in modern times, they were crucial in cultivating the critical abilities characterising humanistic study. The mediaeval trivium (grammar, rhetoric, logic) and quadrivium (arithmetic, geometry, music, and astronomy) developed an organised, hierarchical approach to knowledge, which served as the foundation for humanities education. The trivium's emphasis on language mastery (grammar), persuasive expression (rhetoric), and systematic thinking (logic) positioned learning as both an intellectual and moral endeavour, building virtue alongside proficiency. This classical approach persisted until the Renaissance, when humanists such as Erasmus and Petrarch redefined education as civic and ethical training, claiming that reading great texts from Cicero to Augustine gave pupils eternal insight for public service. This perspective focused on the teacher's function as a moral adviser, influencing not just brains but also characters. These historical underpinnings indicate a long-standing conflict in the humanities: between disciplinary rigour and formative purpose, between communicating knowledge and cultivating ethical agency, a contradiction that continues to shape discussions about the humanities' function today [1].

Legacy and Challenges:

1. Enduring Ideals: The conviction that humanistic education promotes critical thinking and civic virtue endures, even if institutional support dwindles.
2. Modern Tensions: Contemporary criticisms of the Western canon and hierarchical teaching call these classical structures into question.

F. A. Alaba and I. G. Ajibola, *Digital Pedagogy: Bridging the Gap Between Software and the Humanities*, Information Systems Engineering and Management 85,
https://doi.org/10.1007/978-3-032-20335-9_3

3. Relevance: Can the humanities' moral objectives coexist with the need for practical, career-oriented education?

Examining these roots reveals how the humanities' historical identity as both a discipline and a vision for human flourishing continues to impact their uncertain future.

3.1.1 Characteristics of Traditional Humanities Pedagogy

The conventional paradigm of humanities education was founded on two interconnected principles: text-centred authority and lecture-based transmission. The adoration of canonical texts, primarily Western works such as Shakespeare's plays, Homer's epics, and Kant's philosophical treatises, was central to its methodology, which included extensive careful reading, line-by-line annotation, and academic interpretation. This paradigm argued that reading "great books" improved both intellectual discipline and cultural literacy, despite its Eurocentric orientation, which often disregarded alternate knowledge traditions. The formal lecture served as the educational antithesis to this textual concentration, with professors acting as authoritative interpreters, delivering selected information to passive student audiences. Together, these techniques created a self-reinforcing system that favoured: content mastery over critical invention, cultural preservation over inquiry, and hierarchical knowledge transmission over collaborative learning. While this approach fostered profound engagement with historical traditions, current educators are grappling with its shortcomings, notably the marginalisation of marginalised voices and the detachment from modern, multimodal modes of knowledge creation and transmission. The struggle between old approaches and changing educational principles continues to drive discussions about the humanities' identity and purpose in the twenty-first century [2] (Fig. 3.1; Table 3.1).

Traditional teaching was based on hierarchical relationships, with the teacher as the authoritative figure. The disciplines were strictly demarcated, with minimal cross-fertilisation. Pedagogical assumptions:

1. Master-apprentice model.
2. The curriculum is standardised.
3. Rigid disciplinary boundaries (e.g., literature vs history).

3.1.2 Strengths of Traditional Pedagogy

The old humanities classroom emphasised depth over breadth, instilling a scholarly attitude that valued prolonged intellectual engagement with complicated texts and concepts. This method encouraged critical investigation of subtle arguments while accepting ambiguity, whether dissecting Kant's categorical imperative or tracking

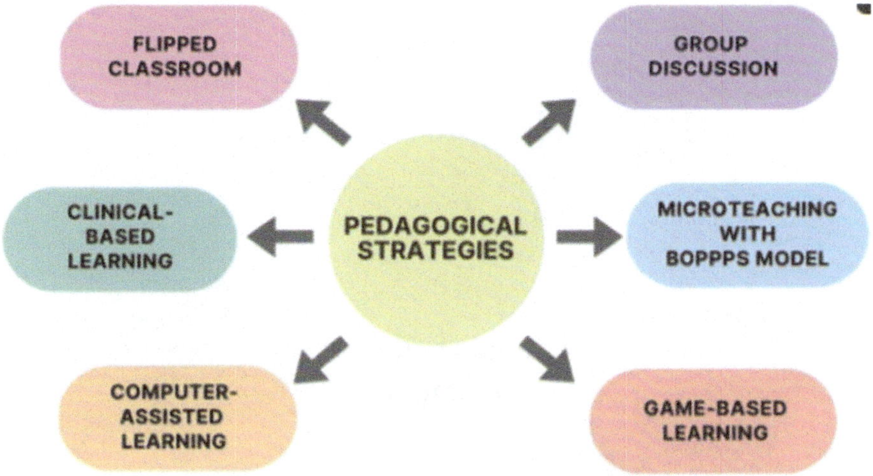

Fig. 3.1 Different innovative pedagogical strategies in dental materials science education [3]

Table 3.1 Comparison of traditional and modern pedagogical modes

Feature	Traditional pedagogy	Contemporary pedagogy
Role of Teacher	Knowledge transmitter	Facilitator of learning
Role of Student	Passive recipient	Active participant
Learning Medium	Texts, lectures	Multimedia, interactive tools
Assessment Methods	Essays, written exams	Portfolios, projects, presentations

moral growth in Shakespearean tragedy. Students gained not just subject knowledge but also critical dispositions such as patience with difficult material, mastery of academic discourse via proper citation procedures, and respect for the historical lineage of ideas. This concept was founded on the assumption that humanistic study affected ethical reasoning philosophy courses studied moral frameworks, literature explored human behaviour via storytelling, and history contextualised ethical quandaries throughout time. These educational ideas claimed that intellectual tradition and moral contemplation were mutually reinforcing, resulting in persons capable of meaningful civic engagement. While contemporary education increasingly emphasises skills-based outcomes, this traditional vision with its emphasis on contemplative depth, ethical formation, and scholarly craftsmanship continues to provide a counterpoint to instrumentalised learning models, despite being criticised for its Eurocentrism and passive knowledge transmission [4] (Fig. 3.2).

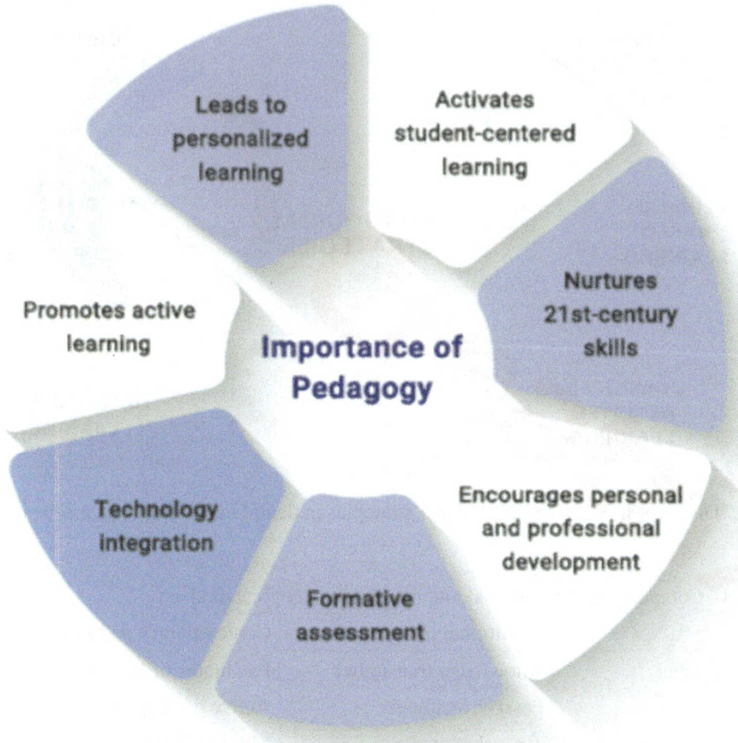

Fig. 3.2 Importance of pedagogy teaching [5]

3.1.3 Limitations and Critiques

Traditional humanities education emphasised depth, critical thinking, and ethical reasoning, but it has been heavily criticised for its Eurocentrism, passive learning methods, and aversion to innovation. The canon's overarching emphasis on Western, white, male authors from Shakespeare to Kant has marginalised foreign viewpoints, supporting colonial ideas and racial inequalities. This discriminatory framework often failed to address non-Western ideologies, Indigenous knowledge systems, or the contributions of women and people of colour, restricting the humanities' ability to conduct genuinely global research. Pedagogically, old techniques emphasised passive absorption over active discovery. The lecture approach and stringent textual exegesis emphasised memorisation over creativity, leaving little possibility for student-led inquiry, collaboration, or project-based learning. This method risks limiting education to mindless repetition rather than promoting autonomous thinking. Furthermore, humanities education was sluggish to adapt to technological change, long opposing digital technologies that may broaden access, permit new kinds of analysis, or link texts to larger media landscapes. This hesitation left pupils unprepared for a quickly changing digital environment in which digital literacy, multimedia storytelling, and data-driven interpretation are becoming more important [6].

3.1.4 Traditional Pedagogy and the Lecture Model: Reassessment

While conventional lectures have been criticised for encouraging passive learning, they continue to be a culturally and intellectually relevant teaching form when organised carefully. A well-crafted lecture is more than just information transmission; it is a live performance of academic thought, demonstrating how professionals create arguments, analyse evidence, and deal with challenging concepts. This ritualised knowledge transmission exposes students to disciplinary frameworks and interpretative traditions that may serve as foundations for further in-depth investigation. However, modern education is rapidly hybridising the lecture style in order to balance exposition with active participation [7]. Strategies include:

1. Pausing for debate: dividing lectures into sections with discussion prompts
2. Embedding multimedia: Using digital archives, video clips, or interactive visualisations
3. Flipped elements: Combining pre-recorded mini-lectures with practical in-class analysis.

This progression maintains the lecture's strengths curatorial clarity and rhetorical power while addressing its flaws by allowing for student engagement, critical questioning, and practical learning. The goal is not to discard the lecture, but to reimagine it as one adaptable instrument among many, acting as a doorway to debate rather than an ending. Considerations for Effective Lecturing:

1. Purpose: Reserve lectures for teaching complicated frameworks, not simply content.
2. Interactivity: Use polls, backchannel talks, or reflection breaks.
3. Accessibility: Provide transcripts/slides for various learners.

The lecture may evolve from a monologue to a provocation, an invitation to think with (and beyond) the provided concepts.

3.1.5 Beyond the Canon: Expanding the Horizons

The humanities are experiencing a profound transformation, from a single, Eurocentric canon to a pluralised curriculum that include postcolonial literatures, feminist theory, Indigenous epistemologies, and works from Africa, Asia, and Latin America. This extension calls upon conventional pedagogies to preserve analytical rigour while accepting decentered knowledge systems, requiring educators to negotiate numerous interpretative frameworks without resorting to hierarchical values. Currently, the profession is reconsidering authorship and authority. Traditional techniques centred on exegesis of "great thinkers," but newer approaches elevate:

1. Reader-response theory, situating meaning as co-creation by audiences
2. Dialogic interpretation in which texts serve as locations of cultural discussion
3. Oral and community-based storytelling: resisting textual domination.

These shifts reflect a larger epistemological shift from transmissive (teacher/text-as-authority) to participatory (knowledge as socially created) paradigms. The problem is to strike a balance between this inclusion and the humanities' traditional focus on critical depth, so that diversity enhances rather than dilutes intellectual engagement.

3.1.6 The Continued Value of Tradition

Traditional pedagogy still has key benefits, even if it has genuine criticisms. In doing so, they learn to take their time, focus on the details, and grapple with difficult moral dilemmas. Rather than rejecting tradition, the difficulty is in finding new meaning in it (Table 3.2).

3.2 Constructivism in Digital Contexts

Constructivism the principle that learners actively generate knowledge through experience and reflection has gained importance as education evolves in the digital age. Digital tools, platforms, and methods provide flexible, interactive environments that follow constructivism. Digital media enhances active participation, learner agency, and meaning co-construction in the humanities, where interpretation, critical inquiry, and cultural context are key. Constructivism holds that learners form new understandings based on their existing knowledge, experiences, and social interactions. Learning is more than just passively absorbing data; it is also an active, contextual process of constructing meaning. Constructivist education emphasises:

1. Learner-centred methods.

Table 3.2 Adapting traditional pedagogy for the digital age

Traditional element	Value retained	Digital reinterpretation
Close reading	Depth of interpretation	Digital annotation tools (e.g., Hypothes.is)
Canonical texts	Historical literacy	Juxtaposed with global and contemporary texts
Lectures	Intellectual modeling	Supplemented by interactive discussions
Writing essays	Argumentation skills	Extended into blogs, digital storytelling

2. Problem-solving and inquiry.
3. Collaborative Learning
4. Reflection and Metacognition

Key theories include:

1. Jean Piaget: Studied individual cognitive development via stages of learning.
2. Lev Vygotsky: emphasised social contact, language, and the zone of proximal development (ZPD) in education.
3. Jerome Bruner: advocated for exploration learning and scaffolding.

3.2.1 Traditional Constructivist Pedagogy in the Humanities

Before digital integration, constructivism in the humanities included seminars, Socratic questioning, and project-based learning. These exercises enabled students to read texts, debate viewpoints, and grapple with historical or philosophical quandaries in collaborative learning settings. Example: In a literary seminar on Toni Morrison's Beloved, students may investigate issues of memory and trauma using historical context, peer debate, and personal interpretation. The teacher must facilitate learning rather than transfer information.

3.2.2 Constructivism and Digital Technologies

Digital surroundings are inherently compatible with constructivist concepts. LMS, online discussion forums, wikis, digital archives, annotation platforms, and virtual reality all allow students to actively participate, interact meaningfully, and generate knowledge in a networked environment [8, 9] (Table 3.3).

Table 3.3 Traditional versus digital constructivist pedagogy

Constructivist feature	Traditional application	Digital context equivalent
Learner agency	Student-led discussions	Interactive simulations, personalised learning paths
Collaboration	Group projects, seminars	Wikis, Google Docs, online peer review platforms
Reflection	Journals, essays	Digital portfolios, blogs, audio reflections
Contextualization	Historical or cultural case studies	Virtual museums, digital archives, geospatial timelines
Scaffolding	Teacher-guided assignments	Automated feedback systems, modular content sequencing

3.2.3 Digital Tools Supporting Constructivist Learning

The ability to participate in real-time, interactive learning regardless of physical location is revolutionising humanities learning via collaborative digital platforms. Collaborative document, map, and idea creation is made easier by tools like Google Workspace, Padlet, and Miro, which promote constructivist learning. At the same time, collaborative tools like wikis and version control systems encourage learning by doing and highlight each user's unique contributions. Digital annotation platforms like Hypothes.is, Perusall, and Littera encourage active participation in texts via the inclusion of hyperlinks, questions, and comments. These developments move the focus away from individual, insular reading and towards group, collaborative analysis of literature [10].

3.2.4 Challenges of Digital Constructivism

Digital constructivism is an educational approach that merges constructivist learning theory with digital technologies. It emphasizes student-centered learning, where learners actively construct knowledge through exploration, collaboration, and digital engagement. While this model aligns well with 21st-century skills and the growing influence of technology in education, it also presents several significant challenges [11–13].

1. One of the primary challenges of digital constructivism is unequal access to technology. Effective participation in digitally mediated learning environments requires reliable internet access, up-to-date devices, and digital literacy. In many parts of the world—particularly in developing countries or low-income areas—students and schools lack the resources needed to fully engage in digital constructivist activities. This digital divide risks reinforcing existing inequalities and limiting opportunities for many learners.
2. Another challenge is the readiness of educators. Digital constructivism demands a shift from traditional teaching roles to that of facilitators or guides. Many teachers are not adequately trained in using digital tools for constructivist learning or in managing more flexible, student-driven classrooms. This lack of preparation can lead to inconsistent implementation, confusion among learners, and a failure to achieve intended learning outcomes. Continuous professional development and institutional support are necessary to help educators effectively transition to this new pedagogy.
3. Curriculum design also poses difficulties. Constructivist approaches require open-ended tasks, interdisciplinary projects, and real-world problems that may not align with rigid curricular standards or standardized testing regimes. Designing digital learning experiences that are both constructivist in nature and aligned with assessment requirements can be a complex task. There is also a

risk of superficial engagement with digital tools—such as using flashy technology without pedagogical depth—resulting in "technocentrism" rather than meaningful learning.
4. Cognitive overload is another concern. Digital constructivist environments often involve multitasking, open-ended exploration, and complex information management. For learners without strong self-regulation skills, this can lead to confusion, distraction, and frustration. Younger or less experienced students may require more structure and scaffolding than constructivist approaches typically provide. Balancing learner autonomy with guided support is essential to prevent students from becoming overwhelmed.
5. Assessment is also a persistent challenge. Traditional exams and quizzes may not adequately capture the depth of understanding or the collaborative and process-oriented nature of constructivist learning. Digital constructivism calls for alternative assessment methods such as portfolios, reflective journals, peer evaluations, and project-based tasks. These methods are often time-consuming and subjective, raising concerns about consistency and scalability, especially in large classes or formal systems.
6. Finally, ethical and privacy issues can arise in digital constructivist environments. Many online tools collect data, track behavior, or encourage the sharing of personal content. Ensuring student data privacy, obtaining informed consent, and promoting safe online behaviours are vital components of responsible digital education.

Hence, while digital constructivism offers powerful opportunities for active, personalized, and collaborative learning, it also presents significant challenges that educators, institutions, and policymakers must address. By investing in infrastructure, training, curriculum reform, and ethical frameworks, the potential of digital constructivism can be more equitably and effectively realized.

3.2.5 The Role of the Educator in Digital Constructivism

Digital constructivism is a pedagogical approach that blends constructivist learning theory with digital technologies. It emphasizes student-centered, experiential learning where learners actively construct knowledge through exploration, collaboration, and interaction with digital tools. Within this framework, the role of the educator undergoes a profound transformation—from being a transmitter of knowledge to becoming a facilitator, guide, and co-learner. Traditionally, educators have been viewed as the primary source of information. In contrast, digital constructivism places learners at the center of the educational process, encouraging them to take ownership of their learning. This shift does not diminish the importance of the educator but rather redefines it. Educators become learning architects, responsible for designing meaningful learning environments that leverage digital tools to support inquiry, problem-solving, and collaboration. One of the educator's key roles is that of a facilitator. This involves guiding students through the learning process rather than delivering content directly. Educators pose challenging questions, provide appropriate resources, and encourage reflection. They help learners connect digital experiences with existing knowledge and real-world contexts. For example, when using digital storytelling tools or simulation software, educators help students interpret and analyze their outputs, thereby deepening understanding [14].

Educators also serve as mentors and motivators, particularly in environments where students must navigate complex digital spaces. They support learners in developing digital literacy, critical thinking, and ethical awareness. With vast amounts of information available online, educators play a crucial role in teaching students how to evaluate sources, manage information, and collaborate effectively in virtual settings. Another essential role is that of a curriculum designer. In digital constructivist classrooms, educators must design tasks that are open-ended, interdisciplinary, and reflective of real-world problems. They must also integrate digital tools in ways that support learning objectives, rather than simply adding technology for its own sake. The educator selects appropriate platforms—such as collaborative wikis, discussion forums, or data visualization software—that foster exploration and creativity while ensuring accessibility and inclusivity.

In addition, educators act as assessors of learning in ways that reflect constructivist values. They implement formative assessment strategies such as reflective journals, peer reviews, digital portfolios, and project presentations. These methods allow for a more holistic view of student progress, emphasizing process over rote memorization. The educator provides continuous, constructive feedback to support learner growth and autonomy. Furthermore, educators must also engage in continuous professional development. The digital landscape evolves rapidly, and staying current with emerging tools and pedagogical practices is essential. Educators must be willing to learn alongside their students, modelling the very curiosity and adaptability that constructivist learning encourages. Thus, the educator in digital constructivism is a dynamic, multifaceted professional who empowers learners to construct their knowledge using digital means. Far from being replaced by technology, the educator becomes even more vital—as a designer, facilitator, mentor, and co-learner. Their guidance ensures that digital tools serve meaningful pedagogical purposes, fostering deeper engagement, critical thinking, and lifelong learning skills [15].

3.2.6 Constructivism in Open Educational Resources (OER)

Constructivist ideas are rapidly being incorporated into OER and Massive Open Online Courses (MOOCs), which empower students to study information at their own speed, connect to global networks, and remix resources. Constructivism, a learning theory that emphasizes active, learner-centered knowledge construction through experience and interaction, aligns naturally with the principles of Open Educational Resources (OER). OER are freely accessible, openly licensed materials—such as textbooks, videos, assessments, and interactive platforms—that support teaching and learning. When designed and used effectively, OER can serve as powerful tools to implement constructivist pedagogy. At the core of constructivism is the idea that learners build their own understanding based on prior knowledge and new experiences. OER support this approach by enabling flexibility, customization, and collaboration, which are crucial for creating personalized and meaningful learning experiences. Unlike traditional, static resources, OER can be adapted to meet the needs of individual learners, different cultural contexts, and varied learning styles. This adaptability empowers both educators and learners to take ownership of content and learning processes [16].

One of the most significant ways constructivism manifests within OER is through open pedagogy, a teaching approach that encourages learners to become co-creators of knowledge. Instead of simply consuming content, students are invited to revise, remix, and create new materials. This process reflects key constructivist principles, such as active participation, reflection, and knowledge construction through authentic engagement. For instance, students might contribute to a collaboratively written online textbook or develop multimedia learning modules that can be shared with global audiences. OER also support collaborative learning, a central tenet of constructivism. Online platforms and open tools encourage learners to work together on shared projects, engage in peer review, and contribute to collective knowledge bases such as wikis or open courseware communities. Through these interactions, learners construct meaning socially, building on each other's insights and experiences. This not only deepens understanding but also fosters a sense of community and shared responsibility for learning. Furthermore, the use of OER in a constructivist framework encourages learner autonomy and self-regulation. Students can access a wide variety of resources at their own pace, explore topics that interest them, and engage in inquiry-based activities. This shift from passive reception to active exploration aligns with the goal of developing independent, critical thinkers—an essential outcome of constructivist education. However, integrating constructivism into OER usage also presents challenges. Many existing OER are still designed for traditional instruction, emphasizing information delivery over exploration or interaction. Educators must therefore critically evaluate and modify OER to ensure alignment with constructivist goals. Additionally, learners may require support in navigating open platforms and contributing meaningfully to collaborative projects, especially in contexts with limited digital literacy or access. In conclusion, OER and constructivism share a common vision of learner empowerment, active engagement, and knowledge creation. When combined, they offer a transformative model of education that is inclusive, participatory, and adaptable. By fostering environments where learners and teachers can co-create and freely share knowledge, constructivist-informed use of OER promotes not only academic growth but also a democratic and open approach to learning in the digital age [17].

3.3 Connectivism and Networked Learning

In the setting of rapid technological advancements and the widespread availability of digital tools, conventional learning theories such as behaviourism, cognitivism, and constructivism have shown inadequacy in comprehensively elucidating the mechanisms of learning inside networked, data-rich contexts. Connectivism, introduced by [18], arises as a reaction to the intricacies of the digital age, serving as a "learning theory for the digital era." Connectivism underscores the significance of networks, digital instruments, decentralised knowledge, and the learner's ability to traverse and link many information sources. In the realm of digital humanities, characterised by cooperation, interdisciplinary, and data-driven research, connectivism provides a theoretical framework and a practical guide. Connectivism holds that learning occurs in networks of people, systems, and information. It envisions knowledge as spread throughout a digital ecosystem, with learning defined as the act of connecting information nodes.

According to [18], the key principles of connectivism emphasise the need for many perspectives for learning and understanding.

1. Learning is the process of linking certain nodes or knowledge sources.
2. Non-human equipment may include learning capabilities.
3. The ability to know more is more important than what is now known.
4. Nurturing and sustaining relationships is required to enable continuous learning.
5. The ability to recognise connections across domains, ideas, and concepts is an essential talent.
6. All connectivism learning aims to provide currency (correct, up-to-date information).

In connectivism approaches, learning does not end at the classroom door. It extends into digital repositories, social media platforms, online journals, interactive databases, collaborative tools, and worldwide communities.

3.3.1 Connectivism Versus Constructivism: A Comparison

Although both theories emphasise active interaction, they diverge in their perspectives on the locus of knowledge and the mechanisms of its acquisition (Table 3.4).

3.3.2 Connectivism in the Digital Humanities

The DH demonstrate connectivist learning concepts, in which information is spread across dynamic, multidisciplinary networks rather than being limited to static texts or

Table 3.4 Connectivism versus constructivism

Feature	Constructivism	Connectivism
Knowledge location	Internal (constructed in the mind)	External (distributed across networks)
Learning emphasis	Meaning-making through experience	Navigating, connecting, and filtering data
Role of technology	Tool for discovery and expression	Environment and co-agent of learning
Interaction model	Learner with peers and teacher	Learner with nodes (human + digital)
Learning goal	Deep understanding and interpretation	Capacity to access, update, and remix

institutional silos. DH researchers collaborate in an ecology of open-access repositories, collaborative platforms, and real-time scholarly exchanges, converting research into a participatory, networked activity.

Key Characteristics of DH as Connectivist Practice:

1. Interdisciplinary Fluidity: Researchers combine methodologies from literature, history, computer science, and design, using platforms such as GitHub for code sharing, arXiv for preprints, and Zotero for collaborative bibliographies.
2. Decentralised Knowledge Nodes: Learning takes place on social media (Twitter/Mastodon discussions, DH Slack groups) as well as official publications, with peer feedback loops speeding innovation.
3. Just-in-Time Resource Sharing: Global networks provide for rapid access to datasets, troubleshooting guidance, and new ideas, eliminating conventional delays in academic communication.

Implications for pedagogy:

1. Teaching students how to use and contribute to these networks becomes a key DH competence.
2. Curation, cooperation, and critical digital literacy are more important for success than memorisation.
3. Assessment focusses on portfolio-based examples of networked involvement (e.g., code changes, forum participation, open peer reviews).

This paradigm sees the humanities as a dynamic network of links rather than a set canon, with competence defined as the ability to tap into and enhance communal wisdom.

3.3.3 Platforms and Tools Enabling Networked Learning

Scholars and students are able to work together to construct, exchange, and improve information due to the proliferation of digital technologies, which are reshaping knowledge ecosystems. MOOCs on platforms such as edX, Coursera, and Future Learn provide learning possibilities that are easily accessible. These MOOCs develop interactive networks by means of forums, peer evaluations, and discussion boards, which encourage nonlinear and rhizomatic learning. Hypotheses is, Transcribe Bentham and Zooniverse are examples of social annotation and crowdsourcing systems that promote communal interaction with texts and archive resources. These tools emphasize shared cognition and scalable involvement. It is also important to note that knowledge management systems like as Notion, Obsidian, and Roam Research enable users to create linked digital gardens by mapping thematic relationships between texts, authors, and historical eras. This makes these systems very useful for digital humanities studies [19].

3.3.4 Educator Roles in Connectivism Contexts

Connectivism, proposed by [20], is a learning theory for the digital age that emphasizes the role of networks, technology, and social connections in the learning process. Unlike traditional learning theories that focus on internal cognitive processes, connectivism sees knowledge as distributed across a network of connections and learning as the ability to navigate and grow those networks. In this context, the role of the educator transforms significantly. In connectivism environments, educators are no longer sole authorities or primary sources of knowledge. Instead, they function as network facilitators who guide learners in accessing, evaluating, and connecting with relevant information sources, communities, and digital tools. The educator helps students build and manage their personal learning networks (PLNs), encouraging them to engage with diverse perspectives through blogs, forums, social media, online courses, and open access resources. A core responsibility of educators in connectivism is curating and organizing resources. Given the vast amount of information available online, learners can easily become overwhelmed. Educators play a critical role in helping students filter high-quality information, identify trustworthy sources, and use tools such as RSS feeds, academic databases, and content aggregators to manage knowledge streams effectively.

Educators must also act as critical thinking coaches. In a networked world, information is constantly evolving and often contradictory. Therefore, teaching students how to question assumptions, cross-verify claims, and reflect on multiple viewpoints is essential. This involves modelling digital literacy, including understanding bias, algorithms, and misinformation. The goal is not just to help learners access knowledge, but to engage with it critically and contextually. Another key function is creating and maintaining learning environments that support collaboration, discussion, and co-construction of knowledge. Educators establish digital spaces—such as LMS, discussion boards, and collaborative wikis—where students can share insights, challenge each other's ideas, and develop collective intelligence. By encouraging peer-to-peer learning and community participation, educators help learners tap into the wisdom of the crowd. Educators in connectivism contexts also need to be technologically agile. They must be familiar with a variety of digital tools and platforms that support networked learning, from communication apps like Slack or Discord to content creation tools like Padlet or Canva. More importantly, they must continuously adapt to emerging technologies and model lifelong learning themselves.

Furthermore, educators must serve as ethical guides, helping students navigate privacy issues, online conduct, and digital citizenship. As learners engage in open, networked environments, the risks of data misuse, trolling, and intellectual property violations increase. Educators must instil values of respect, responsibility, and transparency in digital interactions. Thus, educators in connectivism learning environments play multifaceted roles—as facilitators, curators, guides, and co-learners. Their focus shifts from delivering content to helping learners navigate networks, think critically, and engage meaningfully in a complex, connected world. By fostering autonomy, digital literacy, and collaborative learning, educators help prepare students for continuous, self-directed learning in the digital era.

3.3.5 Challenges and Critiques of Connectivism

Connectivism, despite the fact that it offers a dynamic framework for learning in this digital age, is accompanied with a number of obstacles. Inequality in the digital realm continues to be a major obstacle, as differences in access to bandwidth, devices, and digital fluency may exclude those who do not have the advantage of technical advancement. Furthermore, the overwhelming amount of information that is available online may result in fragmented learning, which encourages surface-level skimming rather than deep involvement. This is a worry that is expressed in the statement made by Audrey Watters, who said that "abundance is not always empowerment." In addition, networked learning environments make it more difficult to use standard evaluation techniques. This is because it is more difficult to monitor individual contributions in collaborative settings, evaluate the depth of learning that goes beyond simple connections, and reduce the likelihood of echo chambers or disinformation.

3.3.6 Implications for Humanities Education

In today's quickly changing digital world, education must prioritise more than simply subject mastery it must encourage critical digital literacy, the capacity to assess and synthesise sources, as well as the ability to map and visualise knowledge networks.

As information develops beyond conventional classroom bounds, lifelong and self-directed learning become critical. Learners must be able to recognise reputable sources, participate in professional learning networks, and maintain a digital scholarly identity that promotes ongoing intellectual progress. By combining these techniques, education shifts from static memorisation to dynamic knowledge-building, equipping students to navigate and contribute effectively in the digital world.

Furthermore, the advent of digital media is changing how authority and authorship are seen in academia. Rather than passively consuming information, learners may actively contribute to knowledge development via editing Wikipedia pages, open-access archives, and peer review procedures. This movement challenges the established canon and fosters multiple viewpoints, pushing students to question material rather than accept it at face value. By adopting these networked learning approaches, school may create a more engaged and intellectually empowered generation capable of making significant contributions in a linked society.

3.4 Critical Pedagogy and Digital Justice

Algorithms, data economies, and digital infrastructures are shaping our society, making critical pedagogy and digital justice vital and transformational. Digital technologies offer access, efficiency, and personalisation, yet they also perpetuate inequities, prejudices, and rhetoric in hidden ways. The emancipatory work of Paulo Freire inspired critical pedagogy, which requires that education liberate. It encourages educators and students to question authority, fight injustice, and co-create knowledge for fairness and justice. This mindset must apply to the virtual classroom, data rules, platform design, and edtech's political economy in the digital age. Paulo Freire's Pedagogy of the Oppressed (1970) questions the conventional "banking model" of education, which views students as passive consumers of information. He argues for dialogic education, co-learning, and conscientização, which involves cultivating critical awareness. Critical pedagogy is based on the concept that education is essentially political, since it either promotes or challenges existing power systems. Learning is seen as a collaborative process rather than a one-way transmission, and contemplation must be linked to action for real societal change. Furthermore, contextual learning emphasises that education should be derived from learners' living experiences rather than abstract, detached teaching.

In the digital age, these concepts take on new meaning as educators and researchers investigate how power functions via digital tools and platforms. Critical pedagogy today requires an examination of whose voices are amplified, who controls the design of educational tools, and how algorithms influence learning experiences. The contemporary Freirean classroom must manage the hurdles given by LMSs, data dashboards, AI tutors, and the geopolitical ramifications of platform ownership, all while ensuring that digital education stays equal and empowering rather than reinforcing existing power structures [21].

3.4.1 Understanding Digital Justice

Digital justice goes beyond access to technology to address the ethical, social, and structural aspects of digital systems. It includes numerous essential ideas, such as access justice, which assures equal access to devices, connection, and digital literacy, and representation justice, which encourages inclusive design and cultural relevance in digital material. Furthermore, algorithmic justice emphasises the need to recognise and mitigate bias in AI and data systems, while sovereignty justice emphasises the ownership and autonomy of data by marginalised people. Finally, participatory justice promotes meaningful engagement in the development and regulation of digital instruments. These principles are especially important in humanities education, as they raise critical questions about whose histories and narratives are digitally preserved, who is excluded from archives, classrooms, and projects, and how digital platforms encode bias in categorisation, metadata, or search rankings. Despite technology's promise to democratise, it also perpetuates existing inequities. For example, the COVID-19 pandemic revealed significant discrepancies in remote learning, particularly in the Global South, where students typically experienced inadequate internet access, a lack of equipment, excessive data prices, and unsupportive home learning environments. Humanities instructors managing digital education in these places faced severe pedagogical upheavals, emphasising the critical need to decolonise digital access and ensuring that technical solutions serve various global populations fairly.

3.4.2 Beyond Devices: The Digital Use Divide

Despite the fact that access is available, there are still discrepancies in terms of digital proficiency, language inclusion, and information that is culturally appropriate. It is common practice to instruct students to utilise technology in the role of consumers rather than as critical producers or co-designers. In discussions about digital equity, much attention is often placed on the "digital divide"—the gap between those who have access to technology and those who do not. While providing access to devices and internet connectivity is a necessary step toward digital inclusion, it is not sufficient. A deeper and more nuanced challenge is the digital use divide, which refers to the disparities in how people use technology, the purposes they pursue, and the outcomes they achieve. This divide reveals that even when access is equal, the benefits of technology can remain unequally distributed [22].

The digital use divide highlights differences in the quality and complexity of digital engagement. For example, a student may own a smartphone and use it to browse social media or watch videos, while another student may use a laptop to code, conduct research, or collaborate on educational platforms. Both have access to technology, but only one is gaining skills and experiences that contribute to academic and professional growth. This divide is often influenced by socioeconomic background, education level, digital literacy, and institutional support. One key aspect of the digital use divide is the distinction between passive and active technology use. Passive use involves consuming content, such as watching videos or scrolling through feeds, while active use involves creating, problem-solving, or engaging in meaningful communication. Research shows that learners from disadvantaged backgrounds are more likely to use technology for passive entertainment, whereas their more privileged peers often engage in productive and skill-building digital tasks. This difference can reinforce existing inequalities, even in classrooms where every student has a device [10].

Digital literacy and guidance from educators play a crucial role in addressing the digital use divide. Without appropriate support, students may not know how to use digital tools for academic or creative purposes. Teachers need professional development to incorporate technology effectively and to teach students how to evaluate information, use productivity software, collaborate online, and engage in critical digital practices. In many schools, however, technology is introduced without sufficient training or pedagogical planning, which limits its impact. Another factor is institutional culture and policy. Schools and organizations must do more than provide hardware; they must foster environments that encourage meaningful digital engagement. This includes integrating technology into the curriculum in ways that promote inquiry, creativity, and problem-solving. It also means ensuring that all students, regardless of background, have opportunities to develop higher-order digital skills—not just technical know-how, but also communication, collaboration, and ethical use [23].

Finally, addressing the digital use divide requires attention to cultural and contextual relevance. Technology use must be aligned with students' realities and interests. When digital tools reflect learners' identities and empower them to express themselves, they are more likely to use technology meaningfully and critically. Therefore, bridging the digital divide is about more than just distributing devices. It requires a shift in how we think about access—from having technology to using it effectively. To close the digital use divide, educators, policymakers, and communities must work together to build equitable, empowering, and pedagogically sound digital environments.

3.4.3 Surveillance, Datafication, and Platform Capitalism

There is a close relationship between the phenomena of surveillance, datafication, and platform capitalism, all of which are influencing the digital economy. The concept of surveillance capitalism relates to the practice of monetising personal data, in which businesses gather, analyse, and sell information about users in order to forecast and affect their behaviour. Large-scale datafication, which is the process of converting human behaviours into quantifiable data, is the foundation upon which this approach, which was pioneered by IT giants, is built. Platforms are able to gather insights from

Table 3.5 Key practices

Principle	Application example
Transparency	Informing students how their data is collected, stored, and used
Consent and agency	Allowing students to opt out of surveillance tools or suggest alternative platforms
Decolonial design	Using locally developed or open-source platforms; centering non-Western knowledge
Collaborative governance	Including students and community members in edtech decisions
Critical media literacy	Teaching students to interrogate algorithms, platforms, and digital narratives

user interactions via the use of algorithms and machine learning, often without using the users' express permission.

These tendencies are further strengthened by platform capitalism, which concentrated economic power in a small number of digital platforms that were dominating at the time. These corporations have complete control over enormous ecosystems, and they dictate the manner in which data is gathered, processed, and monetised. Despite the fact that these technologies provide ease and connectedness, they also pose problems about privacy, autonomy, and the ethical implications of the commercialisation of data. It is becoming more important to have conversations about regulation, transparency, and user rights as the use of algorithms that are not transparent becomes more prevalent in digital places [24].

3.4.4 Towards a Digitally just Pedagogy

One of the goals of critical digital pedagogy is to humanise technology and to reclaim digital spaces as places where resistance and liberation may take place (Table 3.5).

3.4.5 Case Studies in Critical Digital Pedagogy

Students' engagement with technology and the creation of knowledge is being shaped by the fact that digital humanities courses are increasingly tackling themes of data justice and decolonisation from a variety of perspectives. In the course titled "Data Justice in the Classroom," which is offered at New York University, students investigate the algorithmic bias that exists in literary search engines, the colonial heritage that is ingrained in digital archives, and the building of alternative metadata schemas that are founded on decolonial priorities. Students engage in critical reevaluation of conventional taxonomies by means of engaging in hands-on tasks, which include the creation of zines and podcasts that combat digital injustice. Students are prompted to investigate language exclusion in online encyclopaedias, power dynamics in open-access repositories, and the digital erasure of Indigenous oral traditions in the course "Decolonising the Digital" that is being taught at the University of Cape Town. This course integrates African epistemologies into digital pedagogy. The course concludes in the production of digital exhibitions in isiXhosa and Afrikaans via collaborative efforts. These exhibits ensure that local history are maintained and portrayed in a manner that goes beyond the frames of dominant cultures.

3.4.6 Challenges to Implementation

Critical digital pedagogy faces a number of obstacles on the institutional, logistical, and epistemological levels, despite the fact that it has the capacity to bring about significant transformation. The significant focus that the education system places on standardisation, testing, and quantification, which inhibits deeper critical investigation, is often the root cause of institutional opposition by educational institutions. Additionally, educators are constrained in their ability to shape choices about educational technology owing to limited academic freedom. Additionally, economic restrictions usually favour proprietary platforms over open-source alternatives, which restricts accessibility and creativity. The ability of educators is another matter of concern, since a significant number of faculty members do not possess formal training in critical pedagogy and digital ethics. It is common for professional development programs to place a greater emphasis on tool competency than on encouraging critical engagement. As a result, educators are not adequately prepared to include digital justice components into their lessons. Furthermore, student tiredness is a huge obstacle; students may have feelings of being overwhelmed by the demands of activism or struggle with finding confidence in their digital abilities. Therefore, in order to address these issues, care-centered teaching techniques are required. These approaches must strike a balance between criticism and emotional and intellectual support, ensuring that students participate meaningfully without experiencing feelings of heavy weight [25].

3.5 Learning Theories Applied to E-Learning

The growth of e-learning has transformed how knowledge is obtained, created, and shared. However, educational psychology, especially learning theories, continue to have an impact on the design and efficacy of digital learning experiences. Understanding and using these ideas in e-learning situations guarantees that digital tools and platforms promote real learning rather than just displaying data. Behaviorism, a learning theory based on the work of B.F. Skinner and John Watson, defines learning as a change in observable behaviour triggered by external stimuli. It emphasises reinforcement, repetition, and conditioning above internal cognitive processes. This strategy continues to have an impact on e-learning, influencing drill-and-practice programs, multiple-choice quizzes, and gamification assessments. Immediate feedback, repeating workouts, and reward systems (such as points and badges) all correspond with the behavioural notions of reinforcement and conditioning. For example, language learning programs such as Duolingo employ behaviourist concepts to encourage user involvement via XP incentives, progress streaks, and motivating messaging. Despite its success in reinforcing skills and automating learning, behaviorism has been criticised for possibly ignoring higher-order thinking and learner autonomy. In humanities topics that value critical inquiry and reflection, simply behaviourist techniques may fail to foster profound analysis and independent thought. While behaviorism works well in organised learning contexts, a more holistic educational approach is often required to promote creativity and complicated problem solving [26].

3.5.1 Cognitivism and E-Learning

Cognitivism, which emerged in contrast to behaviourism, emphasises the mind's active participation in information processing, concentrating on how students absorb, organise, store, and retrieve knowledge. Scholars like as Jean Piaget and Robert Gagné helped develop this paradigm, which has had a considerable effect on e-learning. Cognitivist ideas are manifested in digital settings via organised information distribution, cognitive load management, and scaffolded learning routes. Learners may efficiently absorb complicated topics using strategies such as modular content (chunking information), infographics and mind maps (visual representation), and interactive tutorials (active processing). Online scientific courses, for example, often include cognitivist tactics, such as progressive information sequencing, the use of animated visuals, and conceptual quizzes to encourage comprehension.

Although cognitivism improves clarity, organisation, and methodical learning, it does have limits. Critics say that it may continue to portray learners as passive receivers rather than active knowledge producers. Humanities educators, in particular, recognise the need for interactive discussion, critical thinking, and interpretative involvement beyond prescribed cognitive frameworks. To address this, instructional design must provide chances for learners to create meaning dynamically rather than merely consuming knowledge in predetermined ways [13].

3.5.2 Constructivism and E-Learning

Constructivism, as created by scientists such as Lev Vygotsky and Jerome Bruner, emphasises that learners actively build knowledge via experiences, reflection, and social interaction, making learning naturally contextual and student-centred. This

Table 3.6 Learning theory in digital spaces

Digital Tool/Method	Social learning feature
Peer-reviewed assignments	Observational learning
Wiki contributions	Collaborative knowledge-building
Webinars with live Q&A	Social modeling
Video tutorials by peers	Vicarious reinforcement

approach is reflected in e-learning environments via problem-based learning, discussion forums, role-playing simulations, and e-portfolios, all of which urge students to interact meaningfully with ideas rather than passively receiving information. For example, onlinc humanities classes may require students to analyse historical events using digital archives and present their results in multimedia forms, promoting critical thinking and interdisciplinary study. However, constructivist e-learning poses difficulties, necessitating active participation, digital literacy, and properly built platforms that provide meaningful connection. When badly executed, it may cause confusion or cognitive overload, especially if learners are left without clear direction. Successful constructivist design needs purposeful scaffolding, ensuring that students can traverse complicated concepts while benefitting from collaborative learning and reflective practice [27].

3.5.3 Social Learning Theory in Digital Spaces

According to Albert Bandura's Social Learning Theory, observation, imitation, and peer interaction influence learning. Learning new abilities and behaviours need context and incentive. Peer review, collaborative editing, and shared media projects in e-learning settings encourage involvement and knowledge-building. Digital platforms use social learning tactics to enable students to actively participate in and improve their knowledge in community-driven contexts [28] (Table 3.6).

3.5.4 Connectivism and Networked Learning

Qian et al. [29] introduced connectivism, which redefines digital learning as a networked process in which knowledge is spread among people, devices, and information sources. Unlike conventional theories that concentrate on individual cognition, connectivism contends that learning occurs via connections between nodes of information, making current knowledge more useful than static understanding. This framework emphasises the need for decentralised knowledge sharing, continuous information synthesis, and the ability to navigate complex digital landscapes. Connectivism ideas are compatible with MOOCs, social media-based learning, and OERs. Hyperlinked materials encourage non-linear learning pathways, online communities facilitate decentralised cooperation, and curated RSS feeds enable self-directed, real-time learning. For example, students using a Twitter hashtag during a digital humanities conference may actively gather ideas from experts all around the globe, demonstrating how connectivism turns education into a participatory, ever-changing network of knowledge [30].

3.5.5 Experiential Learning and Immersive Technology

David Kolb's Experiential Learning Theory emphasises the value of learning via direct experience, reflection, conceptualisation, and experimentation. This cyclical process fosters active engagement and flexibility, which makes it very useful in digital learning contexts. In e-learning, experiential learning is brought to life via immersive technologies such as VR and AR, as well as simulations, role-playing games, case-based learning, and digital storytelling. To expand their knowledge, an art history student may take a virtual museum tour, examining exhibitions, communicating with curators, and keeping a reflective diary. Modern digital technologies boost experiential learning by allowing educators to create immersive and engaging encounters. Platforms such as Unity 3D, Mozilla Hubs, and 360-degree videos enable students to interact with surroundings that promote greater knowledge and empathy. These tools not only enhance the subject content but also motivate students to actively participate in their education, changing passive consumption into meaningful discovery and personal interaction.

3.5.6 Comparative Overview of Learning Theories in E-Learning

Behaviourism focusses on stimulus–response conditioning, making it perfect for drills and memorisation, which are often done via quiz engines and gamified applications. Cognitivism, which focusses on information processing, helps students grasp concepts and develop cognitive abilities via tutorials and infographics. Constructivism, which emphasises learner-driven knowledge production, is best suited for critical thinking and problem-solving activities such as simulations, forums, and problem-based learning (PBL) exercises. Social learning theory emphasises interaction-based learning, making it ideal for collaboration and modelling using wikis, peer review, and webinars. Connectivism, which sees learning as a networked activity, complements informal and lifelong learning via MOOCs, social media, and open educational resources. Finally, experiential learning emphasises hands-on experience, reflection, and practical application by using immersive techniques like VR, AR, and digital fieldwork. This framework is a helpful resource for creating successful digital learning environments [28] (Table 3.7).

Table 3.7 Learning theories in E-learning

Theory	Key Principle	Best for	E-learning Tools
Behaviorism	Stimulus–response conditioning	Drills, memorisation	Quiz engines, gamified apps
Cognitivism	Information processing	Concept mastery, cognitive skills	Tutorials, infographics
Constructivism	Learner-driven knowledge construction	Critical thinking, problem-solving	Simulations, forums, PBL tasks
Social Learning	Learning through interaction	Collaboration, modeling	Wikis, peer review, webinars
Connectivism	Learning through networks	Informal, lifelong learning	MOOCs, social media, OERs
Experiential	Learning through doing	Reflection, practical application	VR, AR, digital fieldwork

3.6 Emotional and Affective Dimensions of Learning

Learning is more than just a cognitive or intellectual process; it is inextricably linked to emotions, motivation, self-perception, and individual identity. The emotional and affective elements of learning, which are sometimes disregarded in conventional instructional design, are being more recognised as important for learner engagement, perseverance, and success. Understanding and planning for emotional experiences is especially important in digital education, when interpersonal indications are minimised. Neuroscientific and pedagogical research has radically changed our view of learning by revealing how emotion and cognition are inextricably linked. Antonio Damasio's research on somatic markers demonstrates that emotions are intrinsic to decision-making rather than peripheral to cognition, but [31]'s studies suggest that emotive involvement improves memory retention and conceptual comprehension. This contradicts conventional pedagogies that prioritise detached analysis, advocating for learning settings that use emotion as a cognitive stimulus.

Bloom's Taxonomy, which was originally centred on cognitive abilities, was broadened to include the affective domain, acknowledging that learning includes attitudes, values, and emotional reactions. This approach progresses from passive receipt (e.g., readiness to interact with a text) to deep internalisation (e.g., using ethical reasoning in real-world situations). For example, in a literature lesson, students may advance from just perceiving a poem's imagery to appreciating its societal criticism and eventually characterising its concepts in their own advocacy.

Motivation theories emphasise the relevance of emotion in learning. The Self-Determination Theory [28] identifies autonomy, competence, and relatedness as key drivers of intrinsic motivation, implying that emotionally supportive classrooms promote deeper engagement. Similarly, Expectancy-Value Theory [32] states that learners engage effort when they anticipate success and value the content, both of which are dependent on emotional and cultural significance.

Implications in Humanities Education:

1. Emotionally Resonant Pedagogy: Using narratives, discussions, and reflective writing to link information to students' real-life experiences.
2. Affective Assessment: Creating assessments that assess not just intellectual abilities, but also ethical thought and sympathetic interaction.
3. Inclusive Motivation: Designing courses to reinforce student autonomy (e.g., assignment selection) and belonging (e.g., community-based learning).

Integrating these insights allows educators to cultivate cognitively rigorous and emotionally meaningful learning, which is critical in an era of widespread disengagement and mental health challenges.

3.6.1 Emotional Engagement in Digital Learning

Online learning offers the difficulty of disembodiment since it often lacks the nonverbal signs, facial expressions, and spontaneous emotional feedback that characterise in-person education. This absence might cause learners to feel alone, disconnected, and unmotivated. To mitigate these consequences, the Community of Inquiry (CoI) Framework prioritises emotional presence alongside cognitive and social presence. Emotional presence refers to how students express and share their emotional states, resulting in a more human and empathic learning environment.

Empathy, belonging, and emotional safety all play important roles in determining the affective climate of digital learning environments. Personal storytelling, reflective writing, and dialogue-based learning may help students understand one another as multidimensional persons rather than simply avatars or text profiles. A strong feeling of belonging improves academic engagement, retention, and accomplishment, and it may be cultivated via tactics such as personalised feedback, inclusive language, icebreaker activities, and student-led discussion forums. Furthermore, emotional safety enables students to take intellectual risks, ask questions, and exhibit vulnerability. This may be facilitated in online environments via polite moderating, clear communication rules, and culturally relevant education. By combining these aspects, educators may create a stimulating and emotionally supportive online learning environment [13].

3.6.2 Emotional Engagement in Digital Learning

When it comes to fostering emotional engagement, which is essential for motivation, retention, and deep learning, digital learning environments present a unique set of challenges and opportunities. Even though technology has the potential to improve emotional connections using interactive storytelling, virtual communities, and personalised feedback, it also runs the danger of producing experiences that are impersonal or alienating if it is not built with careful consideration. Emotional engagement plays a vital role in the success of any learning experience, including digital learning. While much attention in online education is placed on content delivery and technological tools, the emotional dimension—how students feel, connect, and stay motivated—can be a determining factor in their persistence, satisfaction, and overall achievement. As digital learning environments become increasingly prevalent, ensuring that learners feel emotionally connected and supported is essential. Emotional engagement refers to the affective responses students have toward their learning, including interest, enthusiasm, sense of belonging, and connection to content, peers, and instructors. In physical classrooms, such emotions are often fostered through face-to-face interactions, body language, and classroom dynamics. However, in digital learning environments—especially asynchronous ones—such connections are harder to establish and maintain.

One challenge is student isolation, a common barrier in digital learning. Without in-person contact, learners may feel disconnected, which can lead to disengagement or dropout. This makes it crucial for educators to build emotionally supportive environments. Regular feedback, personalised messages, video introductions, and the use of discussion forums can create a sense of presence and care. Instructors who show empathy and approachability in their communication help students feel seen and valued, even across digital distances. Another factor influencing emotional engagement is motivation. Digital learners must often manage their schedules and learning pace, which requires high levels of self-regulation. When students feel overwhelmed, bored, or uncertain, they are less likely to stay engaged. Incorporating interactive elements, such as quizzes, gamified challenges, polls, and multimedia, can increase enjoyment and attention. More importantly, digital learning must connect with learners' goals, interests, and values. Relevance enhances emotional investment and helps students find personal meaning in their studies.

Peer interaction is also central to emotional engagement. Social presence—the feeling of being part of a learning community—is crucial for emotional well-being in online courses. Tools like breakout rooms, collaborative platforms (e.g., Padlet, Miro, Google Docs), and peer feedback forums encourage social connection. When learners collaborate, share experiences, and support one another, they are more likely to feel emotionally engaged. Moreover, emotional intelligence and empathy are important competencies for both learners and educators in digital spaces. Educators need to recognize signs of emotional distress or disengagement and offer timely support. Similarly, students benefit from opportunities to express themselves, reflect on their emotions, and develop self-awareness. Journaling, discussion prompts about personal reactions, and well-being check-ins can help integrate emotional reflection into online learning. Technology itself can aid emotional engagement when used thoughtfully. For example, learning analytics can help identify students who may be disengaged, allowing instructors to intervene. AI-driven Chatbots or virtual assistants can offer encouragement and reminders, though they cannot replace genuine human connection. Thus, emotional engagement is not a peripheral concern in digital learning—it is central to student success. Digital learning environments must be intentionally designed to foster emotional connection, motivation, and a sense of belonging. By prioritizing empathy, interaction, and relevance, educators can create digital spaces where learners not only succeed academically but also feel emotionally supported and inspired [33].

3.6.3 Affective Computing and Emotion-Aware Systems

Rosalind Picard is the one who came up with the phrase "affective computing," which describes the process of developing systems and technologies that can recognise, understand, and react to human emotions. By bridging the gap between technology and human emotions, this cutting-edge profession promises to improve interactions in a variety of fields, such as education, healthcare, and customer service, among others. Through the use of affective computing, digital platforms can provide user experiences that are more deeply personalised and emotionally sensitive. Affective computing is driven by several tools and technologies, which enable computers to have the ability to recognise and respond to different emotional states. While face recognition works by identifying affective states via the use of camera inputs, sentiment analysis can identify emotions in written or spoken language. Chatbots that use emotion AI may personalise their replies depending on the emotional signals they receive, resulting in conversations that are more empathic and engaging. To improve the quality of the learning experience, adaptive learning systems modify the delivery of knowledge depending on the emotions or levels of stress experienced by users. Conversational artificial intelligence is used by platforms such as Replika, for example, to engage users on an emotional level by altering tone and assistance according to the user's perceived mood.

The use of affective computing involves substantial ethical difficulties, including questions of privacy, consent, monitoring, and emotional manipulation, even though it has the potential to provide positive results. The capacity to recognise and exert control over emotions brings with it several hazards that need to be properly addressed. When it comes to effective technologies, educators and developers have a responsibility to respect ethical norms, providing transparency and user autonomy to avoid abuse while appropriately leveraging the advantages of these technologies [34].

3.6.4 Emotional Design in E-Learning

Visual and interface design are very important factors in determining the emotional reactions that students have while they are in digital settings. Colours, fonts, graphics, and animations are all examples of elements that have the potential to greatly influence levels of comfort and engagement. Colours that are warm elicit emotions of vitality and alertness, whilst chilly colours encourage feelings of relaxation and focus. The use of emotive images enhances emotional resonance, which in turn fosters deeper connections with the knowledge, while clear navigation minimises anxiety, which in turn makes the learning process more natural. When compared to a poorly designed interface, which may result in irritation and disengagement, an intelligently designed interface has the potential to reduce cognitive burden and boost user happiness. Narrative and storytelling techniques, in addition to the aesthetics of the interface, contribute to the enhancement of digital learning by fostering emotional involvement and increasing retention. It is possible for educators to create a more immersive experience for their students by introducing journey-based events, characters, and conflicts into their lessons. As an example, a digital humanities course may make use of an interactive narrative module that follows the journey of a displaced refugee. This would assist students in developing historical empathy and global ethical viewpoints. Students are encouraged to engage with complicated societal problems on a personal level via the use of this narrative-driven method, which also helps to increase their understanding [35].

3.6.5 Instructor Presence and Emotional Support

In the context of digital learning, demonstrating concern for the instructor and establishing emotional trust. Within the context of digital and hybrid learning settings, the emotional presence of teachers is an essential component in the development of an environment that is both supportive and engaging. A teaching method known as "expressing care" is one that facilitates the strengthening of emotional connections, the encouragement of student engagement, and the promotion of well-being. This may be accomplished by a number of practical actions:

1. It is possible for teachers to launch weekly or biweekly messages, surveys, or open conversations in order to enquire about the intellectual and emotional well-being of their students for the purpose of doing regular check-ins.
2. Students are given a secure environment in which they may voice their concerns and these check-ins serve as a signal of attention.
3. Providing students with positive feedback, which includes acknowledging their efforts, accomplishments, and successes, helps to encourage their motivation and demonstrates gratitude for their involvement.

The power of words that are encouraging, helpful, and personal may go a long way. When educators share their own learning adventures, struggles, or emotional experiences, they create an atmosphere that is open and respectful to one another. This is accomplished via the sharing of personal reflections. Prompt reactions to discomfort or confusion: A demonstration of real concern and responsiveness is shown when students are receiving timely help when they are experiencing difficulties, whether it be with the material of the course or with emotional obstacles. In the field of education, the notion of vulnerability and authenticity is closely connected to the concept of protection. Rather of seeming as a detached authority figure, a teacher who freely exhibits emotions, humour, or doubt portrays themselves as a human being who is approachable and sympathetic towards their students. Because of this honesty, students may experience a huge boost in both their emotional engagement and their relationship trust. Famous educational author John C. Maxwell once said, "Students don't care how much you know until they know how much you care." This is a quote that has become well-known. This quotation highlights the transforming impact of emotional connection in learning contexts, particularly in situations when physical signs and presence are restricted. Through the cultivation of empathy, honesty, and emotional openness, educators can construct digital classrooms that are both emotionally intelligent and more inclusive [36].

3.6.6 Student Emotions and Learning Outcomes

When it comes to the formation of the learning experience, emotions play a crucial role. To make learning more fun and meaningful, it is important to cultivate positive emotions such as curiosity, pleasure, and pride. These feelings boost engagement, stimulate creativity, and build intrinsic motivation. When students have a sense of excitement or accomplishment, they are more inclined to investigate new ideas, persevere through difficulties, and successfully remember knowledge. On the other side, negative emotions like as worry, irritation, or boredom might be impediments, but they can also serve as vital markers for making modifications to the teaching approach. There is a correlation between anxiety and a decrease in working memory, which makes it more difficult to comprehend information. However, educators may help alleviate this effect by lowering cognitive load and providing confidence. In order to combat boredom, which is caused by a lack of engagement, it is possible to include components that are both innovative and interactive in order to revive interest. Even though it is first upsetting, confusion may really lead to deeper understanding if it is scaffolded appropriately and explained with very specific directions. It is possible for educators to create a learning environment that is more adaptable and helpful if they make an effort to recognise and react to emotional indicators [37].

3.6.7 Designing for Emotional Engagement: A Practical Framework

When it comes to designing for emotional involvement in digital learning, a deliberate strategy that incorporates empathy, belonging, affective computing, and successful interface design is required (Table 3.8).

Table 3.8 Application in E-Learning

Principle	Application in E-Learning
Human connection	Video introductions, synchronous meetings
Emotional scaffolding	Encouragement, coping tools, reflection prompts
Visual affective cues	Thoughtful use of color, icons, avatars
Feedback with empathy	Personal tone, affirming feedback
Authentic assessment	Real-world problems, self-expression tasks

3.7 Digital Scaffolding Techniques

Structured assistance helps learners bridge the gap between autonomous and guided learning in conventional and digital contexts. Scaffolding comes from Lev Vygotsky's socio-cultural learning theory, specifically his Zone of Proximal Development. Adaptive technology, multimedia material, interactive platforms, and collaborative tools have transformed scaffolding in digital education. Vygotsky's Zone of Proximal Development (ZPD) emphasises the gap between what a learner can do on their own and what they can achieve with directed support. Scaffolding gives temporary assistance to learners in this zone, allowing them to acquire skills, gain autonomy, and confidence. By carefully organising learning experiences, instructors may assist students in transitioning from reliance on assistance to autonomous understanding of subjects. Three fundamental characteristics define effective scaffolding: contingency, fading, and transfer of responsibility. Contingency refers to tailoring help to the learner's present performance, ensuring that it is neither overpowering nor inadequate. Fading entails progressively decreasing the degree of help as learners achieve competency, enabling them to depend more on their talents. Finally, the transfer of responsibility enables students to take charge of their education, which promotes independence and self-regulation. While these ideas are still relevant in digital learning contexts, they must be modified to accommodate digitally mediated and asynchronous interactions. The thoughtful deployment of scaffolding tactics in online contexts guarantees that learners get the required advice while gradually growing confidence in their skills [38].

3.7.1 Types of Digital Scaffolding

Cognitive scaffolding is essential for helping learners absorb information, solve problems, and develop conceptual knowledge. In digital contexts, this help may take many forms, such as guided reading questions, evaluation tips or prompts, interactive idea maps, and visual clues or summaries. For example, an online philosophy course may include annotated texts in which significant concepts are highlighted and definitions show on hover, allowing students to more easily understand difficult arguments. Metacognitive scaffolding, on the other hand, enables students to critically reflect on their mental processes, organise their learning, and successfully manage techniques. Reflective journaling applications, digital learning planners, and pop-up self-assessment reminders all help students remain interested and aware of their progress. A history MOOC, for example, may encourage students to stop after a video lesson and think on what they've learnt, fostering stronger links with past knowledge. Procedural scaffolding helps students navigate digital activities and platforms, especially those new to online learning settings. Tutorials, walkthroughs, FAQs, and step-by-step task instructions are vital for guaranteeing learners' confidence in utilising instructional technologies. This form of scaffolding is particularly useful for first-generation learners and individuals who are inexperienced with digital interfaces.

Strategic scaffolding helps learners pick effective problem-solving and decision-making techniques. Decision trees, heuristic prompts, and strategy suggestion tools enable students to approach problems using organised techniques. Using these scaffolding strategies, educators may develop more effective and responsive digital learning experiences that respond to a wide range of student requirements [39].

3.7.2 Tools that Enable Digital Scaffolding

LMSs like Moodle, Canvas, and Blackboard offer necessary scaffolding elements to improve the learning experience. These systems enable teachers to use conditional content release, which ensures that learners acquire knowledge in stages as they understand required ideas. Embedded quizzes and feedback allow students to monitor their comprehension in real time, while sequenced learning routes take them through predefined curriculum phases. Additionally, forums with teacher presence allow for discussion and explanation, promoting a friendly learning environment. Interactive learning tools enhance digital scaffolding by providing specialised help targeted to specific educational requirements. Hypermedia systems such as Perusall and Hypothes allow for layered comments and collaborative reading, making complicated texts more accessible. Simulations, such as those found in Labster and PhET, enable learners to participate in hands-on virtual experiences by guiding them through sophisticated systems in steps. Adaptive learning systems, such as Smart Sparrow and Knewton, tailor training to improve understanding. E-portfolios, such as Mahara and Google Sites, promote reflective learning by allowing students to track their educational experiences.

AI-powered scaffolds improve learning by providing dynamic and personalised help. Artificial intelligence may detect misunderstandings and provide learners with tailored corrections and explanations. AI-powered tools offer personalised hints and real-time adaptation of content difficulty to match students' needs. Additionally, conversational AI chatbots support learners by answering questions, guiding them through complex topics, and maintaining engagement. For example, a writing assistant might help with academic essay creation by proposing phrase openers, improving grammar, and successfully constructing arguments. Through these advancements, digital scaffolding continues to evolve, creating more responsive and adaptive learning environments [40].

3.7.3 Strategies for Designing Effective Digital Scaffolds

Effective digital scaffolding starts with connecting instructional assistance with learning goals. If the aim is critical analysis, scaffolding such as inquiry frameworks or example criticisms might assist students in developing their thinking. Interactive practice sessions provide hands-on interaction, strengthening expertise via real-time application.

Segmenting knowledge into small parts improves learning by enabling pupils to advance incrementally. Each module may include instructional films, linked reflection points, and progress checks to guarantee understanding. This planned strategy avoids overloading students while encouraging ongoing involvement.

Scaffolds are strategically embedded at points of necessity to alleviate irritation and cognitive strain. Rather than frontloading all instructional support, solutions such as just-in-time help or hover-over explanations give aid when learners are having difficulties. This strategy promotes independence while providing prompt assistance.

Finally, adopting numerous modalities ensures that varied learners have access to information and are engaged. The use of text, video, images, and interactive features appeals to a variety of learning styles, improving understanding and retention. Educators may build dynamic, supportive learning environments that allow students to achieve by carefully structuring scaffolding tactics.

3.7.4 Fading and Encouraging Autonomy

Fading, or the slow removal of scaffolding, is an important part of successful learning assistance. This shift from guided assistance to independent work helps learners gain confidence and mastery, ensuring that they acquire the abilities required to manage problems independently. Automated fading uses performance data from digital systems to dynamically change support. Hints, for example, may be eliminated when students frequently respond properly, developing critical thinking and self-reliance. Advanced information may be unlocked if mastery is shown, enabling learners to advance at their own speed. Peer teaching may also be promoted as students achieve competency, which promotes collaborative learning and reinforces comprehension. Reflective fading adopts a metacognitive approach, encouraging learners to assess which aids they still need and which they can handle alone. Self-assessment helps students acquire better self-awareness and optimise their autonomous learning practices. This strategy not only increases autonomy, but also allows students to take control of their educational path. Using intentional fading, instructors may create a

Table 3.9 Balancing scaffolding in digital design

Issue	Problem	Solution
Cognitive overload	Too many instructions or formats	Simplify design; use chunking
Technological exclusion	Tools not accessible to all learners	Offer low-bandwidth or offline alternatives
Loss of autonomy	Students become dependent	Build in gradual fading and reflective tasks

balanced learning environment that promotes progress while eventually developing independence (Table 3.9).

3.7.5 Scaffolding Collaborative Learning

Collaborative learning, where students work together to solve problems, complete tasks, or create products, is a powerful pedagogical approach that supports deeper understanding, critical thinking, and interpersonal skills. However, successful collaboration doesn't occur automatically. It requires careful scaffolding—the intentional structuring and support of learning activities to guide students as they develop the skills and confidence needed to collaborate effectively. Scaffolding in collaborative learning involves a variety of instructional strategies aimed at helping learners move from guided participation to independent and self-regulated collaboration. Instructors play a central role in designing tasks, clarifying expectations, and providing support structures that gradually fade as learners become more capable.

One essential aspect of scaffolding is clarity of purpose. Collaborative tasks must be meaningful, relevant, and appropriately challenging to engage students. Teachers must define clear goals and outcomes while allowing room for autonomy and creativity. When students understand why they are working together and what they are expected to achieve, they are more likely to invest effort and contribute meaningfully. Role assignment is another key scaffolding strategy. By assigning roles such as facilitator, note-taker, timekeeper, or researcher, educators can help structure group dynamics and ensure equitable participation. Roles can rotate over time, allowing students to develop a broad set of collaborative skills. This structure reduces the risk of dominance by certain group members or disengagement by others. Educators should also provide guidelines and rubrics that outline expectations for effective collaboration. These tools can include criteria for communication, problem-solving, respect for diverse perspectives, and accountability. When students know what effective collaboration looks like, they are better prepared to reflect on and improve their group interactions.

In digital or hybrid environments, technology can be used to scaffold collaboration effectively. Tools such as shared documents, project management boards (e.g., Trello, Asana), and discussion forums offer platforms for communication, coordination, and co-construction of knowledge. Features such as version tracking, commenting, and chat support provide transparency and feedback. Another important scaffolding technique is modelling. Teachers can demonstrate collaborative behaviours, such as active listening, constructive feedback, and conflict resolution. They can also share examples of successful group projects and reflect on what made them effective. Through modelling, students gain a clearer understanding of the social and cognitive skills involved in collaboration. Reflection and feedback are integral to scaffolding. After collaborative activities, students should be encouraged to reflect on their group dynamics, challenges faced, and lessons learned. Educators can facilitate debriefing sessions or use peer-assessment tools to promote accountability and metacognition. These practices help students internalize collaborative norms and strategies, improving their future interactions [41].

In conclusion, scaffolding collaborative learning involves thoughtful planning, continuous support, and gradual release of responsibility. It helps students develop not only academic competencies but also the interpersonal and organizational skills essential for lifelong learning and workplace success. By providing structure, guidance, and feedback, educators can create a collaborative learning environment where all students can thrive.

3.8 Gamification and Motivation

Gamification, the application of game mechanics and game-thinking in non-gaming environments, has gained prominence in the educational domain, particularly in digital learning. This technique uses elements like points, badges, levels, leaderboards, and rewards to enhance user engagement and encourage desired behaviours. The adoption of gamification in education is based on the principle that game-based structures can increase motivation and participation among learners by tapping into their natural desires for competition, achievement, and recognition. As educational paradigms shift toward more learner-centred and technology-driven models, gamification provides a bridge between enjoyment and effective learning [18].

3.8.1 Theoretical Foundations of Gamification

Gamification in education is underpinned by several motivational theories. Chief among these is Self-Determination Theory (SDT), which posits that human motivation is driven by the needs for autonomy, competence, and relatedness. In a gamified learning environment, autonomy is offered through choice-based tasks, competence is built through progressive challenges and feedback, and relatedness is fostered through social features such as multiplayer modes and collaborative quests. Another relevant theory is Behaviorism, which emphasises reinforcement and reward systems. Points and badges act as positive reinforcement that encourages repetition of desired behaviours. Additionally, Flow Theory by Csikszentmihalyi explains how gamification can help learners reach optimal experiences where skill and challenge are perfectly balanced, leading to deep immersion in learning tasks [42].

3.8.2 Key Elements of Gamification in Education

Gamification typically includes a combination of the following elements:

1. Points and Scoring: Reward learners for task completion and participation.
2. Badges and Achievements: Visual symbols of success and mastery.
3. Leaderboards: Rank learners to promote healthy competition and social comparison.
4. Levels and Progression: Indicate learner advancement and unlock new content.

3.8.3 Motivational Benefits of Gamification

Gamification fosters both intrinsic and extrinsic motivation. Intrinsically, it appeals to curiosity, mastery, and the enjoyment of overcoming challenges. Extrinsically, it offers visible rewards, social recognition, and status indicators. Learners are more likely to persist through difficult content when motivated by tangible outcomes, such as earning a new badge or rising in leaderboard ranks. Moreover, gamified environments promote a growth mindset by framing failure as part of the learning process, encouraging retrying and iterative learning. In online education, where learner engagement can be low due to isolation and lack of structure, gamification introduces a sense of fun and urgency. It can also personalize learning paths by adapting difficulty levels and content based on user performance, leading to better learner outcomes.

3.8.4 Case Studies and Applications

Case studies and real-world applications serve as powerful tools in understanding the impact and implementation of digital learning strategies. They offer practical insights into how theoretical concepts are translated into educational practice across diverse contexts. Through case studies, educators, researchers, and policymakers can assess the effectiveness, scalability, and challenges of various digital innovations. One prominent example is the "Hole in the Wall" experiment by Sugata Mitra in India, where a computer was installed in a slum wall, allowing children with no prior exposure to technology to explore and learn on their own. This case highlighted the potential of minimally invasive education and self-directed learning. It demonstrated how access to technology, even without formal instruction, can stimulate curiosity, collaboration, and problem-solving among learners, particularly in under-resourced environments. Another case study worth examining is the Flipped Classroom model implemented at Clintondale High School in Michigan, USA. In this model, students watched video lectures at home and used class time for interactive activities, discussion, and problem-solving. The school saw a significant drop in failure rates and disciplinary actions. This approach emphasized active learning and teacher facilitation, showcasing how shifting instructional delivery can enhance student engagement and understanding.

A more recent example comes from Khan Academy, a global nonprofit that uses video tutorials, interactive exercises, and data tracking to provide personalized learning. The platform has been integrated into classrooms and remote learning environments around the world. Case studies from Brazil, India, and the United States show that when used effectively—with teacher support and curriculum alignment—Khan Academy can improve mathematics outcomes and increase learner confidence. In higher education, the University of Edinburgh's use of MOOCs through platforms like Coursera and FutureLearn provides another valuable case. By offering free, high-quality courses to global learners, the university expanded its educational reach and reinforced its reputation. Evaluation of these MOOCs revealed increased digital literacy among participants and fostered international collaboration and peer learning, although issues of completion rates and learner motivation persist.

Case studies also extend to applications of AI in education. For instance, Carnegie Learning's use of AI-driven tutoring systems in teaching mathematics has been tested in several U.S. school districts. Results showed improvements in students' problem-solving skills, especially when AI tools were integrated alongside teacher-led instruction. This illustrates the potential for AI to support differentiated instruction and formative assessment. In low-resource settings, initiatives like Kolibri by Learning Equality offer compelling case studies. Kolibri provides offline access to open educational resources, allowing students in remote areas to benefit from digital content without requiring internet connectivity. Successful implementations in Rwanda, India, and refugee camps demonstrate how context-sensitive technology can promote equity in education. In conclusion, case studies provide concrete evidence of how digital learning tools and pedagogies function in real-world settings. They reveal the importance of contextual factors—such as infrastructure, teacher training, and learner needs—in determining success. By studying these applications, educators and decision-makers can better understand how to implement digital innovations effectively and inclusively across diverse educational environments.

3.9 Inclusive and Universal Design for Learning (UDL)

Inclusive and Universal Design for Learning (UDL) is a forward-thinking educational approach that embraces the diversity of learners. Recognizing that one-size-fits-all teaching methods fall short in addressing varied learning preferences, UDL

Table 3.10 UDL principles and examples

Principle	Description	Example
Multiple means of engagement	Stimulate interest and motivation for learning	Gamification, real-world problem-solving tasks
Multiple means of representation	Present content in diverse ways to support comprehension	Use of multimedia, infographics, and bilingual resources
Multiple means of action and expression	Enable learners to demonstrate their knowledge in various formats	Presentations, portfolios, written reports, and video projects

aims to design flexible learning environments that can accommodate individual learning differences. Grounded in the neuroscience of learning, UDL seeks to minimize barriers while maximizing learning for all students, regardless of their abilities, backgrounds, or learning styles. As the educational landscape becomes increasingly digitized, UDL plays a pivotal role in fostering accessibility and inclusion across platforms and curricula [43]. UDL is deeply rooted in cognitive neuroscience and educational psychology. It draws from research on how the brain learns, identifying three primary networks involved in learning: the recognition network (the 'what' of learning), the strategic network (the 'how' of learning), and the affective network (the 'why' of learning). These networks inform the three core principles of UDL: representation, action/expression, and engagement. UDL aligns with Vygotsky's Zone of Proximal Development and social constructivist theories, emphasizing the importance of scaffolding and differentiated instruction to meet diverse learner needs. UDL is a forward-thinking educational approach that embraces the diversity of learners. Recognizing that one-size-fits-all teaching methods fall short in addressing varied learning preferences, UDL aims to design flexible learning environments that can accommodate individual learning differences [44].

The framework of UDL is guided by three primary principles (Table 3.10):

3.9.1 UDL and Digital Learning Environments

In digital learning environments, UDL becomes even more essential as technology can both support and hinder accessibility. Digital platforms offer unprecedented opportunities to implement UDL principles at scale. With features such as adjustable text size, screen readers, alternative input methods, and closed captions, digital tools can accommodate diverse learners effectively. However, if not intentionally designed, technology can perpetuate exclusion. Hence, inclusive digital learning design must be intentional, ensuring tools and content align with UDL principles.

Benefits of Inclusive and UDL Practices

Implementing UDL in educational settings yields numerous advantages for both students and educators. These benefits include:

1. Improved accessibility for students with disabilities and learning difficulties.
2. Increased student motivation and engagement through personalised learning pathways.
3. Enhanced academic performance and learner autonomy.
4. Promotion of social inclusion and diversity in classrooms.
5. Support for multilingual learners and culturally diverse populations.

Application Strategies for UDL

UDL application involves designing curriculum and learning experiences that provide multiple means of access and engagement. Effective strategies include:

1. Offering multiple assessment formats (e.g., essays, podcasts, infographics).
2. Using flexible seating, schedules, and pacing in classrooms.
3. Providing timely and personalised feedback using assistive technologies.
4. Embedding choices within assignments and learning activities.
5. Integrating collaborative platforms that facilitate peer learning and social interaction.

Challenges and Barriers

Despite its many benefits, the implementation of UDL is not without obstacles. Common challenges include:

1. Lack of training and awareness among educators.
2. Insufficient institutional support and resources.
3. Resistance to change from traditional teaching models.
4. Technological limitations or unequal access to devices.
5. Difficulties in scaling personalised learning for large classes.

Future Prospects and Conclusion

As educational institutions move toward more inclusive and digital models, UDL will continue to shape curriculum development, assessment, and pedagogy. The integration of AI, adaptive learning systems, and learner analytics with UDL frameworks holds promise for creating more responsive and equitable learning experiences. Institutions must invest in professional development, infrastructure, and policy reform to fully realise the benefits of UDL. Ultimately, the goal is to design education systems that are not only inclusive but also empowering, celebrating the unique pot.

3.10 The Learner as Creator

The digital age has redefined the learner's role in the educational process. No longer are students passive recipients of information; they are now empowered creators, capable of shaping their learning environments and producing meaningful content. This shift aligns with broader cultural movements toward participatory culture and user-generated content, enabled by ubiquitous digital technologies. In this chapter, we explore how digital tools and pedagogies support learners as active creators, the theoretical frameworks underpinning this transformation, and the implications for curriculum design, assessment, and equity. The term 'prosumer', a blend of 'producer' and 'consumer', aptly captures the evolving identity of learners in the digital realm. Students today are not just absorbing content; they are remixing, curating, and generating new knowledge. Platforms such as blogs, wikis, podcasts, and video editing software empower learners to author their narratives and share them with global audiences.

Constructivism and Constructionism

This shift towards learner-created content is grounded in constructivist and constructionist theories. Jean Piaget's constructivism posits that learners build knowledge through experience, while Seymour Papert's constructionism extends this idea by emphasising the importance of learners constructing tangible artifacts. In digital education, this translates to students building apps, creating videos, designing infographics, and engaging in collaborative storytelling.

Tools for Creative Expression

Creative expression in digital learning can be enhanced through various tools that empower learners to communicate ideas, explore emotions, and engage with content in dynamic ways. Digital storytelling platforms like Storybird, Adobe Express, and Twine enable users to craft narratives with visuals and interactive elements, fostering engagement through personal storytelling or historical narratives. Multimedia creation tools such as Canva, Powtoon, and Animoto allow learners to design graphics, animations, and presentations, making complex ideas visually accessible and encouraging design thinking. Music and sound production apps like Soundtrap, GarageBand, and BandLab provide avenues for creative audio storytelling and composition, integrating soundscapes into educational projects. Writing and poetry generators, including Poet Assistant, Wordcraft, and Scrivener, help learners refine literary works and experiment with different forms, strengthening their creative confidence. Art and design software like Procreate, Krita, and Adobe Fresco support digital painting and illustration, offering learners opportunities to explore color, texture, and visual storytelling. Meanwhile, virtual and augmented reality creation tools such as CoSpaces Edu, Tilt Brush, and Adobe Aero immerse users in interactive storytelling, simulations, and artistic landscapes, creating highly engaging learning experiences. Project-based learning (PBL) and inquiry-based learning (IBL) provide fertile ground for learners to act as creators. These pedagogies center around real-world problems

Table 3.11 Examples of digital tools supporting learner creativity

Tool type	Examples	Educational use
Multimedia tools	Adobe creative suite, canva	Design graphics, videos, presentations
Coding platforms	Scratch, replit	Create games, applications, interactive content
Publishing platforms	WordPress, medium	Blogging, digital storytelling, opinion pieces
Collaborative tools	Google workspace, padlet	Co-authoring documents, idea boards
Audio/Video yools	Anchor, Audacity, WeVideo	Podcasts, video essays, interviews

and encourage students to explore, research, and present solutions. The end products are often authentic artifacts such as digital portfolios, documentaries, or prototypes [32, 45].

The model of the learner as creator aligns with a broader vision of participatory pedagogy that values voice, agency, and collaboration. As education continues to evolve in response to technological and societal shifts, embracing this model not only enhances engagement but also prepares students for the demands of a dynamic, interconnected world. Empowering learners to create, not just consume, knowledge is both a challenge and an imperative for 21st-century education (Table 3.11).

References

1. Isa, A., Nte, N.D., Baba-Ahmadu, A.: Critical thinking for intelligence analysis for Nigerian students: evaluating the pedagogical challenges in real life scenarios. Indones. Educ. Adm. Leadersh. J. **5**(2), 46–60 (2023). https://doi.org/10.22437/ideal.v5i2.28639
2. Zhao, S., Pan, Q., Gao, D., Cheng, V.: Integrating internet of things and mixed reality to teach performance-based architectural design: a case study of shading devices. Educ. Inf. Technol., (2022). https://doi.org/10.1007/s10639-022-10998-6
3. Ghavifekr, S., Kunjappan, T., Ramasamy, L.: Teaching and learning with ICT tools: issues and challenges from teachers' perceptions. Malaysian Online J. Educ. Technol. **4**(2), 38–57 (2016)
4. Webb, M., Cox, M.: A review of pedagogy related to information and communications technology. Technol. Pedagog. Educ. **13**(3), 235–286 (2004). https://doi.org/10.1080/14759390400200183
5. Schiavenato, M., Chu, F.: PICO: What it is and what it is not. Nurse Educ. Pract., **56**, p. 103194 (2021). https://doi.org/10.1016/j.nepr.2021.103194
6. Zhang, L., Wu, Z.: Enhancing postgraduate academic writing skills through course reform : an action research study. **5**(2), pp. 217–233 (2024)
7. Olowa, T., Witt, E., Morganti, C., Teittinen, T., Lill, I.: Defining a BIM-Enabled learning environment—an adaptive structuration theory perspective. Buildings, **12**(3) 2022. https://doi.org/10.3390/buildings12030292
8. Munawar, H.S., Ullah, F., Qayyum, S.: Big data in construction : current applications and future opportunities (2022)
9. Zhengdao, C.: An Internet of Things-enabled BIM platform for on-site assembly services in prefabricated construction. Autom. Constr. **89**(February), 146–161 (2018). https://doi.org/10.1016/j.autcon.2018.01.001

10. Miorandi, D., Sicari, S., De Pellegrini, F., Chlamtac, I.: Internet of things: Vision, applications and research challenges. Ad Hoc Netw. **10**(7), 1497–1516 (2012). https://doi.org/10.1016/j.adhoc.2012.02.016
11. Pacios Álvarez, A., Ordieres-Meré, J., Loreiro, Á.P., de Marcos, L.: Opportunities in airport pavement management: Integration of BIM, the IoT and DLT. J. Air Transp. Manag., **90**(2019), pp. 1–10 (2021). https://doi.org/10.1016/j.jairtraman.2020.101941
12. Mohammed, B. H., Sallehudin, H., Mohamed, S.A., Satar N.S.M., Bin Hussain, A.H.: Internet of things-building information modeling integration: attacks, challenges, and countermeasures. IEEE Access, **10**, pp. 74508–74522, (2022). https://doi.org/10.1109/ACCESS.2022.3190357
13. Obi, K.C., Iwuanyanwu, O.: Innovation in the assessment of primary school mathematics: an approach to revamping nigeria education. J. Teach. Perspect., **18**(1), pp. 51–66, (2023). Available: http://repositorio.unan.edu.ni/2986/1/5624.pdf%0Ahttp://fiskal.kemenkeu.go.id/ejournal%0Ahttps://doi.org/10.1016/j.cirp.2016.06.001%0A10.1016/j.powtec.2016.12.055%0A10.1016/j.ijfatigue.2019.02.006%0A10.1
14. Korpi, H., Sjögren, T., Mikkonen, K., Piirainen, A.: A systematic review and metasynthesis of qualitative studies on the competencies of health and rehabilitation science educators. NERP, **64**(2), pp. 270–282, (2020). Available: https://www.academia.edu/download/88209414/cr_201tec_202019_20paper_2021.pdf
15. Huttunen, R., Kakkori, L.: Heidegger's critique of the technology and the educational ecological imperative. Educ. Philos. Theory **54**(5), 630–642 (2022). https://doi.org/10.1080/00131857.2021.1903436
16. Irving, K.E.: The impact of educational technology on student achievement: assessment 'of' and 'for' Learning. Sci. Educ., **15**, pp. 13–20, (2006). Available: http://search.ebscohost.com/login.aspx?direct=true&db=eric&AN=EJ773197&site=ehost-live%5Cn, http://www.nsela.org/index.php?option=com_content&view=category&id=51&Itemid=85
17. Purcell, C., et al.: A peer-led, school-based social network intervention for young people in the UK, promoting sexual health via social media and conversations with friends: intervention development and optimisation of STASH. BMC Public Health **23**(1), 1–12 (2023). https://doi.org/10.1186/s12889-023-15541-x
18. Agrawal, S.: A survey on recent applications of cloud computing in education: COVID-19 perspective. J. Phys. Conf. Ser. **1828**(1), 3–9 (2021). https://doi.org/10.1088/1742-6596/1828/1/012076
19. Alraimi, K.M., Zo, H., Ciganek, A.P.: Understanding the MOOCs continuance: The role of openness and reputation. Comput. Educ. **80**, 28–38 (2015). https://doi.org/10.1016/j.compedu.2014.08.006
20. Olika, C.D.: Risk management practices among selected state governments in southwestern Nigeria. Eur. J. Sci. Innov. Technol., **1**(2), pp. 24–31, (2024). Available: https://www.researchgate.net/profile/Hendra-Manurung/publication/356633046_From_Connectivity_to_Digital_Improving_Employee_Readiness_toward_Organizational_Change_in_Digital_Transformation/links/61a5b00671a23a0084c9b061/From-Connectivity-to-Digital-Improvi
21. Saputra, R., Zaid, M.Z.M., Emovwodo, S.O.: The court online content moderation: a constitutional framework. J. Hum. Rights, Cult. Leg. Syst., **2**(3), pp. 139–148, 2022, https://doi.org/10.53955/jhcls.v2i3.54
22. Benzidia, S., Makaoui, N., Bentahar, O.: The impact of big data analytics and artificial intelligence on green supply chain process integration and hospital environmental performance. Technol. Forecast. Soc. Change, **165**, (2021), https://doi.org/10.1016/j.techfore.2020.120557
23. Aljowder, T., Ali, M., Kurnia, S.: Development of a maturity model for assessing smart cities: a focus area maturity model. Smart Cities **6**(4), 2150–2175 (2023). https://doi.org/10.3390/smartcities6040099
24. Mohammed, M.: A Comparative Analysis of Environmental Justice Between Urban Neighbourhoods. J. Incl. Cities Built Environ. **1**(2), 27–40 (2021). https://doi.org/10.54030/2788-564X/2021/v1i2a4
25. Agboola, O., Aremu, M.O., Eze, B.U., Wahab, N.O.: Effect of training and career development on bank employees' performance: evidence from selected banks in Nigeria. EMAJ Emerg. Mark. J. **10**(2), 56–62 (2021). https://doi.org/10.5195/emaj.2020.204

26. Zhang, L.: Smartphone app security: vulnerabilities and implementations. Biomass. Chem. Eng, **3**(2), pp. 1–123 (2019).. Available: http://journal.stainkudus.ac.id/index.php/equilibrium/article/view/1268/1127
27. Ramayasa, I.P.: (2015) Evaluation model of success and acceptance of E-Learning. J. Theor. Appl. Inf. Technol., **3182**(3), pp. 462–469 (2015)
28. Batista, J.P.B., Torre, C., Sousa Lobo, J.M., Sepodes, B.: A review of the continuous professional development system for pharmacists. Hum. Resour. Health, **20**(1), pp. 1–12, 2022, https://doi.org/10.1186/s12960-021-00700-1
29. Qian, P., Liu, Z.G., He, Q.M., Huang, B.T., Tian, D.Z., Wang, X.: Smart contract vulnerability detection technique: a survey. Ruan Jian Xue Bao/J. Softw. **33**(8), 3059–3085 (2022). https://doi.org/10.13328/j.cnki.jos.006375
30. Wu, Y., Jiang, M., Lei, J., Xu, H.: Named entity recognition in chinese clinical text using deep neural network. Stud. Health Technol. Inform. **216**(2), 624–628 (2019). https://doi.org/10.3233/978-1-61499-564-7-624
31. Cinar, A.C., Kara, T.B.: The current state and future of mobile security in the light of the recent mobile security threat reports. Multimed. Tools Appl. **82**(13), 20269–20281 (2023). https://doi.org/10.1007/s11042-023-14400-6
32. Martiskainen, M., Kivimaa, P.: Role of knowledge and policies as drivers for low-energy housing: Case studies from the United Kingdom. J. Clean. Prod. **215**, 1402–1414 (2019). https://doi.org/10.1016/j.jclepro.2019.01.104
33. Waga, D., Makori, E., Rabah, K.: Utilization of cloud computing in education and research to the attainment of millennium development goals and vision 2030 in Kenya. Univers. J. Educ. Res. **2**(2), 193–199 (2018). https://doi.org/10.13189/ujer.2014.020212
34. Dada, E.G., Oyewola, D.O., Yakubu, J.H., Fadele A.A.: Predicting protein secondary structure based on ensemble Neural Network. ITEGAM- J. Eng. Technol. Ind. Appl., Ml, pp 49–56 (2021). https://doi.org/10.5935/jetia.v7i27.732
35. Mohammed, B.H., Husairi, A., Sallehudin, H., Alaba, F.A., Safie, N.: A conceptual framework for securing IoT-BIM. Proc.—AiIC 2022 2022 Appl. Informatics Int. Conf. Digit. Innov. Appl. Informatics Dur. Pandemic, pp. 68–71 (2022). https://doi.org/10.1109/AiIC54368.2022.9914592
36. Samreen, N.F., Alalfi, M.H.: VOLCANO: detecting vulnerabilities of ethereum smart contracts using code clone analysis. **1**(1), pp. 1–18, (2022). Available: http://arxiv.org/abs/2203.00769
37. Fadele Ayotunde Alaba, Hakeem Adewale Sulaimon, Madu Ifeyinwa Marisa, Owamoyo Najeem.: Smart contracts security application and challenges: a review. Cloud Comput. Data Sci., **5**(1), pp. 15–41 (2023). https://doi.org/10.37256/ccds.5120233271
38. Bhatija, V.P., Thomas, N., Dawood, N.: A Preliminary Approach towards Integrating Knowledge Management with Building Information Modeling (K BIM) for the Construction Industry. Int. J. Innov. Manag. Technol. **8**(1), 64–70 (2017)
39. Kaur, S., Mahajan, R.: Hybrid meta-heuristic optimization based energy efficient protocol for wireless sensor networks. Egypt. Informatics J. **19**(3), 145–150 (2018). https://doi.org/10.1016/j.eij.2018.01.002
40. Xie, T., et al.: The untold secrets of WiFi-Calling services: vulnerabilities, attacks, and countermeasures. IEEE Trans. Mob. Comput. **20**(11), 3131–3147 (2021). https://doi.org/10.1109/TMC.2020.2995509
41. Moradi, J., Shahinzadeh, H., Nafisi, H., Marzband, M., Gharehpetian, G.B.: Attributes of big data analytics for data-driven decision making in cyber-physical power systems, In: 2020 14th Int. Conf. Prot. Autom. Power Syst. IPAPS 2020, **2**(12), pp. 83–92 (2019). https://doi.org/10.1109/IPAPS49326.2019.9069391
42. Friedrich, J., Becker, M., Kramer, F., Wirth, M., Schneider, M.: Incentive design and gamification for knowledge management. J. Bus. Res., **106**(2017), pp. 341–352 (2020). https://doi.org/10.1016/j.jbusres.2019.02.009
43. Ignatov, I.I., Gade, P.N.: Data formating and visualization of BIM and sensor data in building management systems. 19th Int. Conf. Constr. Appl. Virtual Real. Enabling Digit. Technol. to Sustain Constr. growth Effic., **5**(11), pp. 141–151 (2019). Available: https://www.researchgate.net/publication/337274994

44. Ignatov, I.I., Gade, P.N.: Data formating and visualization of BIM and sensor data in building management systems. In: 19th Int. Conf. Constr. Appl. Virtual Real. Enabling Digit. Technol. to Sustain Constr. growth Effic., **6**(11), pp. 141–151 (2019). Available: https://www.researchgate.net/publication/337274994
45. Bech-Petersen, S.: Design thinking as a framework for innovation in libraries. Bibl. Forsch. und Prax. **44**(1), 7–11 (2020). https://doi.org/10.1515/bfp-2020-0009

Chapter 4
Tools and Platforms for Digital Humanities Teaching

4.1 Overview of Core Digital Tools

In the evolving landscape of digital pedagogy, Learning Management Systems (LMS) serve as the digital spine of instructional management. These platforms—such as Moodle, Blackboard, Google Classroom, and Canvas—facilitate structured delivery of educational content, streamline communication, and provide robust mechanisms for assessment and feedback. By centralising key aspects of course administration, LMSs allow educators to design, deliver, and manage instructional activities in a flexible, scalable, and user-friendly manner. LMS platforms empower instructors to curate multimedia learning materials, design interactive modules, and create adaptive learning paths tailored to diverse learner needs. For students, these systems offer a single access point to view course materials, submit assignments, participate in discussions, and receive feedback in real time. Instructors can monitor student engagement and performance through integrated analytics dashboards, allowing for timely interventions and support [1, 2].

Moreover, modern LMSs are designed with interoperability in mind, allowing seamless integration with third-party tools such as video conferencing applications (e.g., Zoom, Microsoft Teams), plagiarism detection software (e.g., Turnitin), and collaborative tools (e.g., Google Workspace). These integrations extend the pedagogical capabilities of LMSs beyond static content delivery to include live interaction, peer collaboration, and formative assessment [3].

Key Features of Learning Management Systems

1. **Course and Syllabus Management**: Streamlined uploading, scheduling, and updating of learning modules.
2. **Student Enrollment and Progress Tracking**: Real-time dashboards for monitoring academic engagement and performance.
3. **Communication Tools**: Built-in forums, announcements, and messaging to foster collaborative learning and instructor-student interaction.

F. A. Alaba and I. G. Ajibola, *Digital Pedagogy: Bridging the Gap Between Software and the Humanities*, Information Systems Engineering and Management 85,
https://doi.org/10.1007/978-3-032-20335-9_4

4. **Assessment Integration**: Tools for creating quizzes, grading assignments, and providing individualized feedback.
5. **Content Versatility**: Support for diverse content types including documents, videos, links, and embedded tools.
6. **Mobile Accessibility**: Access through mobile apps, supporting learning on the go.

As higher education and training programs increasingly adopt hybrid and fully online models, LMSs have become indispensable. Their ability to support asynchronous learning, personalized instruction, and digital credentialing positions them at the heart of the modern educational ecosystem (Fig. 4.1).

In the digital age, collaboration is no longer confined to physical spaces. Collaborative platforms have emerged as vital tools in digital pedagogy, enabling students and educators to engage in interactive, real-time, and asynchronous learning experiences. These tools enhance communication, promote teamwork, and support the co-creation of knowledge, making them particularly effective for group-based tasks, project-based learning, and peer assessment. Popular platforms such as Google Workspace (Docs, Sheets, Slides), Microsoft 365 (Teams, OneDrive, Word Online), and Slack facilitate seamless communication and shared productivity. These tools enable users to create, edit, and comment on documents in real-time, promoting transparency, accountability, and dynamic interaction. Educators can use them to assign tasks, monitor group contributions, and provide immediate feedback, while students can brainstorm, draft, and revise content collaboratively.

Fig. 4.1 Features of LMS [4]

The integration of cloud storage further enhances accessibility and version control, ensuring that all users have the most current version of any shared file, regardless of device or location. This is particularly beneficial in hybrid or remote learning environments, where collaboration must continue beyond classroom hours and physical boundaries.

These platforms also support a wide range of media types—text, video, audio, and images—making them versatile for various disciplines in the humanities. Additionally, the ability to integrate third-party apps (e.g., Trello, Miro, Zoom) expands their functionality, allowing educators to create enriched, interactive learning experiences.

Benefits of Collaborative Platforms

1. **Promotes Teamwork and Engagement**: Enhances peer interaction, collective problem-solving, and active participation.
2. **Facilitates Asynchronous and Synchronous Communication**: Allows learners to contribute at their own pace or in real time.
3. **Supports Cloud Storage and Accessibility**: Ensures data security, universal access, and efficient file management across devices.

Overall, collaborative platforms are transforming the learning experience in humanities education by breaking down barriers to interaction, fostering critical thinking through dialogue, and nurturing digital literacy skills essential for 21st-century learners (Fig. 4.2).

Fig. 4.2 Multimedia tools [5]

4.1.1 Assessment and Feedback Tools

Assessment and feedback are essential components of effective learning, especially in digitally enhanced learning environments. The integration of assessment and feedback tools allows educators to evaluate student performance efficiently while fostering deeper engagement and reflective learning. Platforms such as Kahoot, Quizizz, Google Forms, and Socrative offer gamified and interactive formats for quizzes, surveys, and formative assessments. These tools support multiple question types—MCQs, open-ended responses, and polls—and often include features for automated grading, immediate scoring, and analytics dashboards. This allows educators to identify learning gaps promptly and adjust instructional strategies accordingly. In parallel, tools such as Turnitin, Peergrade, and Edmodo are instrumental in providing qualitative feedback. Turnitin supports plagiarism detection and detailed commenting, promoting academic integrity and self-correction. Peergrade and Edmodo enable structured peer reviews and collaborative feedback loops, encouraging students to critically engage with each other's work.

Advantages of Assessment and Feedback Tools:

1. **Automated Grading and Analytics**: Saves time and enables data-driven instruction.
2. **Customisable Question Formats**: Supports various assessment methods aligned with learning objectives.
3. **Real-time Student Performance Insights**: Facilitates timely interventions and personalized support.

By combining assessment and feedback in a digital format, educators not only streamline evaluation processes but also promote transparency, accountability, and student-centered learning.

Content Creation and Curation Tools

In a digitally driven classroom, educators are not just knowledge facilitators—they are also content designers. Content creation and curation tools equip educators with intuitive platforms to produce customized, visually appealing, and interactive learning materials that enhance engagement and comprehension. Applications such as Canva, Genially, Padlet, and Wakelet are widely used for designing presentations, infographics, timelines, digital boards, and multimedia stories. These tools support drag-and-drop interfaces and template libraries, allowing users to create polished instructional content with ease, even without advanced design skills. Beyond creation, these tools also allow educators to curate diverse learning resources, including articles, videos, podcasts, and web links, into cohesive learning modules. Wakelet and Padlet, in particular, function as collaborative spaces where teachers and students can contribute, comment, and organize content thematically [6, 7].

Capabilities of Content Creation and Curation Tools:

Table 4.1 Summary of core digital tools and their functions

Tool type	Examples	Primary functions
Learning management systems	Moodle, Canvas, Google Classroom	Course management, communication, tracking
Collaborative platforms	Google Workspace, Microsoft Teams	Real-time editing, teamwork, and communication
Multimedia tools	YouTube, Edpuzzle, Flipgrid	Video content creation, interactive media
Assessment tools	Kahoot, Quizizz, Google Forms	Quizzes, polls, performance analytics
Feedback tools	Turnitin, Peergrade, Edmodo	Peer review, plagiarism check, and instructor feedback
Content creation tools	Canva, Genially, Padlet	Infographics, presentations, digital storytelling

1. **Design Visual and Interactive Content**: Supports multimodal learning and aesthetic engagement.
2. **Curate Resources and Multimedia**: Enables easy compilation of diverse instructional materials.
3. **Promote Creativity and Digital Storytelling**: Encourages student agency in content production and reflective learning.

These tools enhance both the delivery and personalisation of learning, making them indispensable in humanities education where interpretation, creativity, and narrative are central (Table 4.1).

4.2 Importance in Teaching and Learning

The integration of core digital tools into the teaching and learning process is no longer a supplemental enhancement but a foundational element of modern education. These tools contribute significantly to improving the efficiency, flexibility, and quality of pedagogy, especially in the context of humanities education where discourse, creativity, and critical thinking are central. One of the key benefits of digital tools is their ability to support differentiated instruction, allowing educators to tailor content and assessment methods to accommodate diverse learning preferences—visual, auditory, kinesthetic, or textual. Interactive platforms offer students opportunities for self-paced learning, enabling them to revisit content, access supplementary resources, and manage their learning journey independently. Furthermore, digital tools provide immediate feedback, fostering continuous assessment and timely instructional adjustments. This real-time responsiveness is particularly useful in enhancing learner engagement and achievement. Importantly, these tools help bridge geographical and socio-economic divides. Through mobile accessibility, cloud storage, and asynchronous learning capabilities, students in remote or under-resourced settings gain access to quality education. Tools with inclusive design features—such as text-to-speech, language translation, and captioning—further support equity and accessibility in learning. By strategically integrating these tools, educators can promote student autonomy, collaborative learning, and 21st-century skills like digital literacy, communication, and information synthesis. The result is a more responsive, inclusive, and future-oriented educational experience that prepares learners to thrive in a digital and globalized society [8].

LMS such as Moodle and Canvas represent structured digital environments that facilitate comprehensive course delivery and academic administration. Both platforms support educators in creating, managing, and evaluating course content, student engagement, and instructional resources. Moodle is an open-source LMS widely adopted in academic institutions due to its high level of customization and modular design. It supports a wide array of plugins, enabling functionalities such as quizzes, forums, wikis, and peer assessments. Canvas, developed by Instructure, is known for its intuitive interface, mobile responsiveness, and robust analytics dashboard that empowers data-driven instruction. Both platforms integrate with third-party tools like Google Drive, Microsoft Office, and video conferencing applications, creating an interconnected learning ecosystem. They also include tools for grading, attendance tracking, rubrics, and feedback. Their use in humanities education is especially effective, as they provide structured avenues for discussion, text analysis, collaborative assignments, and multimedia integration—all essential for dynamic and interdisciplinary learning [9].

4.3 Overview of Moodle

Moodle, which stands for Modular Object-Oriented Dynamic Learning Environment, is a powerful and flexible open-source LMS that enables educators to build and deliver online and blended learning environments. Developed by [10, 11], Moodle has since evolved into one of the world's most widely adopted LMS platforms, especially among universities, government institutions, and non-profit organisations. Its open-source nature allows institutions to customise and extend its functionalities through thousands of community-developed plugins. Institutions can host Moodle independently or use cloud-based hosting solutions (e.g., MoodleCloud), providing control over data and customisation. It is built on a modular architecture, allowing different teaching and learning activities, such as quizzes, assignments, forums, and chats, to be added, rearranged, or removed depending on course requirements.

Moodle supports a constructivist approach to learning, emphasising interaction, collaboration, and knowledge construction, making it ideal for humanities education. Its accessibility features, multilingual support, and mobile-friendly design further expand its utility in diverse educational contexts, especially in remote or resource-limited settings.

Key Features of Moodle

Moodle is recognized for its comprehensive suite of features that empower educators to design engaging, interactive, and well-structured courses. The platform supports over 500 plugins and integrates seamlessly with external systems, offering a robust ecosystem for academic delivery.

Notable Features Include:

1. **Course Building Tools**: Drag-and-drop interface for organizing modules, lessons, and topics.
2. **Interactive Activities**: Forums, chats, and wikis encourage student interaction and collaborative learning.
3. **Assessment Modules**: In-built tools for creating quizzes (with multiple question types), timed tests, and assignment uploads with feedback.
4. **Grading and Feedback**: Customizable gradebooks with rubrics, marking guides, and individualised feedback options.
5. **Multimedia Support**: Embedding of audio, video, PDFs, and SCORM packages to diversify content delivery.
6. **Tracking and Reporting**: Logs, completion tracking, and analytics for monitoring student progress and engagement.
7. **Accessibility and Mobile Compatibility**: Responsive themes and a dedicated mobile app for learning on the go.
8. **Integration Standards**: Compliance with SCORM and Learning Tools Interoperability (LTI), facilitating connection with external tools and content providers.

Moodle's versatility makes it especially well-suited for the dynamic and interdisciplinary needs of humanities education, where textual analysis, discussion, and reflective assignments are central. Its community support and continuous updates also ensure that the platform remains responsive to evolving educational needs [12] (Fig. 4.3).

4.3.1 Overview of Canvas

Canvas, developed by Instructure, is a modern, cloud-native LMS known for its user-friendly interface and seamless user experience. Launched in 2011, Canvas quickly gained traction among K-12 and higher education institutions worldwide due to its intuitive design, robust functionality, and minimal IT overhead. Unlike traditional self-hosted LMS platforms, Canvas operates entirely in the cloud, eliminating the need for on-premises infrastructure and allowing institutions to deploy it rapidly with minimal technical complexity. Canvas is built with a focus on accessibility, responsiveness, and integration, making it particularly attractive to institutions seeking a streamlined, turnkey solution that supports both synchronous and asynchronous learning modalities. Its open API architecture, strong adherence to LTI standards, and continual updates have positioned Canvas as a leading LMS in the edtech market.

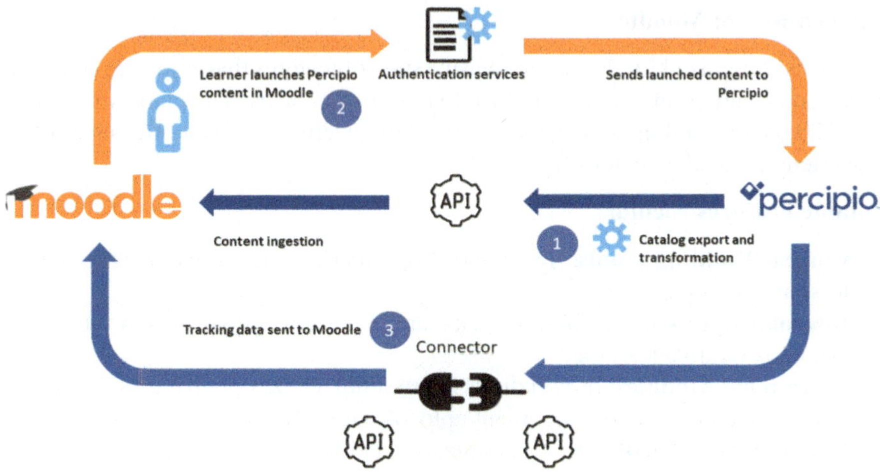

Fig. 4.3 Automated content deployment to the moodle catalogue [13]

The platform is especially valued for its mobile-first design, enabling students and educators to manage their learning activities through dedicated mobile apps, regardless of location or device. This accessibility is especially beneficial in low-resource settings or where learners rely heavily on smartphones and tablets [14, 15].

Key Features of Canvas

Canvas offers a comprehensive suite of tools that cater to the full spectrum of digital teaching and learning activities. Its clean, organized dashboard and flexible architecture make course management intuitive for instructors while providing learners with a clear, structured view of their academic activities.

Notable Features of Canvas Include:

1. **Course Design and Organization**: Instructors can easily build modules, reorder content, and organize lessons using drag-and-drop functionality.
2. **Multimedia Integration**: Embeds audio, video, and links from platforms such as YouTube, Google Drive, and Vimeo.
3. **Synchronous and Asynchronous Tools**: Supports live classes via Zoom and Microsoft Teams, as well as discussion boards, messaging, and announcements.
4. **Assessment and Grading**: Includes rubrics, Speed Grader, and analytics for streamlined evaluation and transparent feedback.
5. **Communication Tools**: In-app messaging, comment threads, and video responses enhance interaction.
6. **Mobile App Accessibility**: Mobile apps for Android and iOS allow students and instructors to access materials, submit assignments, and receive updates in real-time.
7. **Third-Party Integration**: Full compatibility with Google Workspace, Office 365, Turnitin, and countless edtech tools via LTI.

Table 4.2 Comparative analysis of moodle and canvas

Feature	Moodle	Canvas
Licensing	Open-source (free)	Proprietary (subscription-based)
Customizability	Highly customizable via plugins	Limited customization
User interface	Complex but powerful	Modern and user-friendly
Integration	Supports LTI, SCORM, plugins	Extensive API and third-party integrations
Support	Community-driven	Vendor-supported
Mobile access	Mobile app available	Fully functional mobile app
Hosting	Self-hosted or via MoodleCloud	Cloud-hosted

8. **Analytics and Insights**: Provides dashboards for monitoring student progress, engagement, and course completion.

Canvas's sleek design and strong support infrastructure make it ideal for institutions prioritizing ease of use, interoperability, and mobile accessibility, especially in modern humanities education, where interaction, reflection, and digital storytelling are central (Table 4.2).

4.4 Pedagogical Implications

Both Moodle and Canvas provide robust support for a range of pedagogical models including constructivist learning, blended learning, and flipped classroom methodologies. These platforms enable educators to move beyond traditional instruction by offering personalized, student-centered learning pathways. Features such as forums, wikis, and peer review tools facilitate collaborative learning, while integrated analytics help instructors monitor engagement and tailor instruction accordingly. Instructors can use the platforms to provide differentiated instruction, adjusting content format, pacing, and difficulty based on student needs. Real-time feedback, through tools like SpeedGrader (Canvas) or Moodle's assignment module, allows students to reflect on performance and revise work iteratively. This feedback loop is crucial for fostering metacognition and self-directed learning, especially in the humanities, where critical thinking and reflection are key competencies.

These LMSs also support assessment for learning (formative) as well as assessment of learning (summative), enabling educators to track outcomes, identify learning gaps, and refine pedagogical strategies—all of which are vital for data-informed teaching in digital environments.

Implementation Considerations

Choosing between Moodle and Canvas requires institutions to critically assess their operational capacity, educational objectives, and long-term digital learning strategies. Moodle, being open-source, offers extensive customization potential—ideal for institutions with robust IT support and specific instructional needs. It allows administrators to tailor the platform to pedagogical goals but demands ongoing maintenance, hosting, and security management. In contrast, Canvas offers a cloud-hosted, turnkey solution that prioritizes usability and minimal administrative burden. Its intuitive design, mobile optimization, and strong customer support infrastructure make it particularly appealing to institutions with limited IT capacity seeking rapid deployment.

Additional considerations include:

1. **Budget constraints**: Moodle may have lower upfront licensing costs but higher hidden costs in support and customization.
2. **Training requirements**: Both platforms require onboarding, but Canvas's streamlined interface may reduce training time.
3. **Scalability and sustainability**: Institutions planning large-scale or multi-campus implementations must weigh the ability to scale while maintaining platform performance and user experience.

In conclusion, both platforms offer transformative potential. Moodle is best for institutions seeking autonomy and flexibility, while Canvas excels in ease-of-use and efficiency. The ultimate decision should align with institutional vision, technical resources, and educational priorities.

Textual Analysis Tools (Voyant, AntConc)

In the digital humanities, the ability to analyze and interpret large textual corpora is vital. Tools like Voyant Tools and AntConc empower researchers to move beyond close reading by uncovering patterns, trends, and structures in texts that would otherwise go unnoticed. Whether the goal is linguistic research, literary analysis, or socio-political discourse studies, these tools provide scalable and interactive ways to process, visualize, and extract meaning from digital text.

1. **Voyant Tools** is web-based and highly accessible, ideal for quick experimentation and collaborative research. Its interface supports visualization elements such as word clouds, frequency charts, keyword-in-context (KWIC) displays, and trend graphs. It is especially suited for classroom use and for scholars new to digital text analysis.
2. **AntConc**, on the other hand, is a downloadable desktop tool that offers precision and control over corpus linguistics tasks. It is widely used for Concordancing, collocation analysis, and keyword extraction, giving researchers the ability to deeply interrogate text structures and relationships.

These tools enable educators and students in the humanities to approach texts not only as literary artefacts but as data objects—allowing for hybrid methods that blend qualitative insight with quantitative evidence.

4.4.1 Voyant Tools: An Overview

Voyant Tools is a free, browser-based application designed for the visualization and exploration of textual data. Its accessibility and ease of use make it particularly valuable in educational settings where users may have varying levels of technical expertise. Users can upload plain text files or URLs and immediately begin exploring a wide array of visualizations and analytical outputs.

Among Voyant's most commonly used features are:

1. Cirrus (word cloud generator),
2. Trends (word frequency over segments),
3. Summary views (key terms and document metadata),
4. Contexts (KWIC) viewer, and
5. Collocates and correlation graphs.

These tools allow users to identify prominent themes, examine word patterns, and explore stylistic variation across multiple texts. Voyant's interactivity encourages exploratory learning and hypothesis generation, making it a favorite for undergraduate teaching, digital storytelling, and group-based corpus exploration (Table 4.3).

With its suite of widgetsCirrus, Corpus Reader, Trends, and more Voyant offers an interactive playground. As I maneuverer through modules, I could instantly visualize frequency distributions, co-occurrences, and textual patterns without the need for heavy programming skills. It's ideal for educators and collaborative analysis.

Table 4.3 Core features of Voyant tools

Feature	Description
Web interface	Works online in any browser, no installation required
Word clouds	Visual representation of word frequency
Keyword in context (KWIC)	Shows words in their immediate textual context
Trend graphs	Visualizes frequency shifts over the text corpus
Collocation analysis	Examines which words appear together commonly
Export options	Download charts, tables, and texts in multiple formats
Multilingual support	Works with texts in multiple languages

4.4.2 Deep Dive: The Magic of AntConc

In the ever-expanding landscape of digital humanities, AntConc stands out as one of the most versatile and powerful tools for textual analysis and corpus linguistics. Developed by Laurence Anthony, AntConc is a free, downloadable software that allows researchers to delve into the structure and frequency patterns of textual data with remarkable precision. While other tools offer visual flair, AntConc's strength lies in its analytical depth and methodological rigor, making it indispensable for serious academic research.

What Makes AntConc Magical?

At its core, AntConc is a concordance—a tool that displays the occurrences of words or phrases within their surrounding context (Keyword-in-Context or KWIC). This feature is foundational to linguistic analysis, allowing users to examine how language is used across texts. Whether one is investigating thematic patterns in literature, ideological framing in political speeches, or discourse markers in academic writing, AntConc equips users with the capability to parse and quantify such elements across vast datasets.

Beyond Concordancing, AntConc boasts several other powerful modules:

1. **Word and N-Gram Frequency Lists**: Useful for identifying common and rare terms, lexical density, and shifts in language use.
2. **Collocation Analysis**: Helps uncover meaningful word associations, revealing how terms cluster in specific discourses.
3. **Keyword List Generation**: Allows comparison between target and reference corpora to find statistically significant lexical differences.
4. **Concordance Plot View**: Offers a visual timeline of keyword usage throughout the corpus.
5. **File and Cluster Analysis**: Enables users to detect repetition, phrase structure, and stylistic patterns.

AntConc in the Humanities

For humanities scholars, AntConc offers a quantitative lens through which traditional close reading can be enhanced and contextualized. Literature students, for example, can trace motifs and symbols across an author's oeuvre. Historians can analyze shifts in political rhetoric over decades. Linguists can explore discourse variation across genres or populations. These tasks, previously labour-intensive and limited to short passages, are now scalable to corpora containing millions of words. AntConc's plain-text input requirement also encourages researchers to curate and pre-process their own corpora, promoting an intimate understanding of the data. In an educational setting, this can foster critical digital literacy, as students move from passive readers to active investigators of language [16].

Why It Matters

The "magic" of AntConc lies in its ability to unveil the unseen—to transform textual analysis from a subjective act into a reproducible, data-informed exploration. While it lacks the flashy visual dashboards of web-based tools like Voyant, it compensates with granular control, customizability, and academic credibility. It invites users to look beyond surface interpretations and into the linguistic machinery that shapes meaning. In summary, AntConc is more than a piece of software, it is a gateway to computational reading. In a world increasingly shaped by text and data, such tools are not optional add-ons but essential instruments in the modern scholar's toolkit (Table 4.4).

Comparing Voyant and AntConc: Strengths and Limitations

In the realm of digital textual analysis, Voyant Tools and AntConc stand as two of the most prominent tools, widely used by digital humanities scholars, linguists, and students. While both tools facilitate the exploration and interpretation of large corpora, they differ significantly in purpose, functionality, accessibility, and user experience. Understanding their strengths and limitations is crucial for selecting the most suitable tool for specific research or pedagogical goals (Table 4.5).

Table 4.4 AntConc modules at a glance

Module	typical use
Concordance	Searching for word/phrase usage and context
Concordance plot	Visualising word distribution across corpora
File view	Browsing and examining individual text files
Clusters/N-grams	Identifying recurring word groups
Collocates	Finding frequent word partners
Word list	Generating ranked lists of all words in a corpus
Keyword list	Comparing word frequencies between collections

Table 4.5 Voyant versus AntConc

Feature/requirement	Voyant tools	AntConc
Platform	Web-based, no install required	Desktop software (Windows, Mac, Linux)
Usability	Highly intuitive, visual	More technical, for advanced users
Visualization	Flashy: clouds, graphs, charts	Functional: concordance lines, plots
Data handling	Best for small/medium corpora	Handles very large text collections
Export/sharing	Easy export to web, images, CSV	Exports tables, word lists, KWIC entries
Customization	Good, but less flexible	Highly customizable and scriptable
Offline access	No (must be online)	Yes (runs offline entirely)

4.4.3 Spotlight on Google Docs

In the evolving landscape of educational technology, Google Docs has emerged as one of the most transformative tools in contemporary teaching and learning. As part of the broader Google Workspace suite, Google Docs offers far more than just a word processor—it is a real-time collaborative platform that supports critical pedagogical principles such as active learning, co-construction of knowledge, and student agency. One of the defining features of Google Docs is its capacity for real-time collaboration. Multiple users can write, edit, comment, and suggest changes simultaneously, whether working from different rooms or continents. This interactivity transforms passive writing tasks into dynamic learning experiences. Teachers can use the "Comments" and "Suggesting" modes to provide timely, formative feedback directly on students' drafts. Likewise, students can engage in peer review, building communication and critical thinking skills as they analyze and respond to each other's work.

Google Docs encourages a process-oriented approach to writing, where revisions, discussions, and reflections are part of the workflow rather than afterthoughts. The "Version History" tool allows educators and learners to track changes, observe revision patterns, and assess progress over time. This transparency supports metacognitive skills by helping students understand how their thinking and writing evolve. Accessible from any device with internet access, Google Docs supports equity and inclusivity in education. Features such as voice typing, screen reader compatibility, and real-time language translation broaden its usability across different learning needs. Moreover, its cloud-based nature ensures that students working remotely, or those with limited technical infrastructure, can still engage meaningfully in collaborative tasks.

Google Docs integrates seamlessly with other Google tools like Slides, Forms, and Drive, allowing for interdisciplinary and multimodal learning. Humanities educators, for example, can assign tasks that incorporate textual analysis, media embedding, and collaborative annotations—all within a single document. This flexibility promotes creative expression and cross-platform engagement that aligns with the demands of 21st-century literacy. Far from being a simple writing tool, Google Docs is a pedagogical powerhouse. It supports not only productivity and collaboration but also inclusivity, feedback-rich instruction, and reflective learning. In classrooms where learning extends beyond the page and beyond the classroom, Google Docs stands as a foundational platform for empowering both educators and learners in the digital age (Table 4.6).

4.4.4 Overleaf

In the realm of academic research and formal scholarly writing, particularly in fields such as the humanities, sciences, and education, Overleaf has established itself as

Table 4.6 Key features of google docs for collaboration

Feature	Description
Real-time editing	Multiple users editing simultaneously with live updates
Suggestion mode	Proposed changes tracked; easily accepted or rejected
Comments & tagging	Inline comments, resolving threads, and @mentions
Version history	See and restore previous versions by author and timestamp
Live chat	In-document chat to coordinate quickly
Accessibility	Works on PC, Mac, and mobile; screen reader support
Sharing controls	Flexible link sharing and user permissions

an indispensable digital tool for creating professional, publication-ready documents using LaTeX. Combining the power of LaTeX's typesetting capabilities with the accessibility of a cloud-based editor, Overleaf addresses the limitations of traditional word processors while introducing collaborative features essential to 21st-century scholarship. Overleaf allows users to craft highly structured documents that conform to journal and institutional standards. From citations and footnotes to multilingual scripts, tables, and equations, LaTeX on Overleaf ensures typographic precision and structural consistency—attributes critical in academic publishing. What distinguishes Overleaf from offline LaTeX environments is its real-time collaboration functionality. Multiple authors can work on the same document concurrently, leaving comments, tracking changes, and viewing compiled PDFs instantly. This feature is particularly useful for co-authored articles, research theses, group projects, and supervisor feedback.

Despite LaTeX's reputation for being complex, Overleaf simplifies the experience through a split-screen interface, where code appears on one side and the rendered document updates live on the other. It also offers templates for theses, conference papers, CVs, presentations, and grant proposals, making it easy for new users to get started with discipline-specific formatting. Moreover, Overleaf removes the need for software installation or configuration, operating entirely in the browser with cloud-based autosaving and version control. Users can revisit earlier drafts, compare revisions, and restore previous versions, ensuring transparency and continuity throughout the writing process. Overleaf integrates seamlessly with reference managers like Zotero, Mendeley, and JabRef, automating citation formatting and bibliography management. It supports collaborative peer review and supervisor feedback, which are vital in the context of postgraduate research and academic publishing. Its ability to handle multilingual typesetting, special characters, and academic symbols makes it ideal for humanities scholars working with ancient texts, linguistics, philosophy, or cross-cultural studies.

Overleaf merges the technical power of LaTeX with the modern requirements of cloud-based collaboration. It supports scholarly writing that is precise, version-controlled, and collaborative, enhancing both the process and the product of academic work. For educators and students in the humanities and beyond, Overleaf is more than

Table 4.7 Overleaf features for academic teams

Feature	Description
Live LaTeX preview	See formatting, figures, and equations instantly
Commenting	Side-panel comments with tagging and threads
Git integration	Link with external repos for power users
Project sharing	Invite co-authors or make documents public/private
Version control	Snapshot every edit; revert to prior versions fast
Rich export	Download as PDF, TeX, or zip of project files

a writing tool, it is a platform that cultivates professionalism, rigor, and engagement with academic standards (Table 4.7).

4.4.5 Hypothesis

In academic research, particularly within the empirical sciences and social sciences, a hypothesis plays a central role in guiding investigation, structuring inquiry, and formulating conclusions. A hypothesis is essentially a proposed explanation or prediction about the relationship between two or more variables. It is not a guess, but rather an informed, testable statement grounded in theory, prior knowledge, or preliminary evidence.

Types of Hypotheses

There are several types of hypotheses commonly used in research:

1. **Null Hypothesis (H_0)**: Suggests that there is no relationship or effect between the variables being studied. It serves as the default or baseline statement to be tested against.
2. **Alternative Hypothesis (H_1 or H_a)**: Indicates that there is a statistically significant relationship or effect. It represents the researcher's actual expectation or theory.
3. **Directional Hypothesis**: Specifies the direction of the expected relationship (e.g., "X will increase Y").
4. **Non-directional Hypothesis**: States that a relationship exists but does not specify its direction.

Role and Importance in Research

Hypotheses provide focus and direction. They clarify what the researcher aims to test and help in choosing the appropriate research design, methodology, and statistical tests. A well-formulated hypothesis:

1. Is specific and measurable,

Table 4.8 Core capabilities of hypothesis

Feature	Description
Web/page annotation	Highlight and annotate any web page or PDF
Group workspaces	Private groups for classes, labs, or research teams
Reply/threaded notes	Build discussions or ask questions in-context
Permissions	Control visibility from private to public sharing
Tagging	Categorize notes with hashtags for searchability
Export annotations	Download group data for archiving or assessment

2. Is based on prior knowledge or literature,
3. Can be tested using empirical data.

Furthermore, hypotheses play a crucial role in deductive research, where theories are tested through observation and experimentation. They help eliminate ambiguity, ensure scientific rigor, and allow for reproducibility of results. In essence, a hypothesis transforms a research problem into a clear, testable prediction. It bridges the gap between theory and data, allowing researchers to either validate or refute theoretical propositions. Whether in education, health, psychology, or the humanities, a well-defined hypothesis is a cornerstone of systematic inquiry and evidence-based practice (Table 4.8).

4.5 Overview of Digital Exhibitions and Archives

Digital exhibitions and archives are central to the transformation of public scholarship, pedagogy, and research dissemination in the humanities. Platforms like Omeka and Scalar empower scholars to create curated, interactive, and publicly accessible digital projects that preserve and interpret cultural and historical materials. This section explores their capabilities, educational value, and integration into digital humanities curricula. Digital exhibitions are curated digital spaces that present historical, literary, or cultural artifacts with interpretive narratives, often arranged thematically. Archives, by contrast, are digital repositories that emphasize long-term preservation and structured access to collections. Tools like Omeka and Scalar bridge the gap between exhibitions and archives by allowing both interpretive storytelling and archival structuring.

In the digital age, digital exhibitions and archives have become transformative tools in the preservation, dissemination, and interpretation of cultural, historical, and scholarly content. Once the domain of physical museums and institutional libraries, these resources are now accessible globally through web-based platforms, offering immersive and interactive experiences that support both formal education and public engagement.

Digital exhibitions are curated displays of multimedia content—texts, images, audio, video, 3D models—presented in an online environment to simulate the experience of a physical exhibition. These exhibitions may focus on historical events, artistic collections, thematic research projects, or community narratives. They are often created by universities, museums, archives, and research institutions using tools such as Omeka, Scalar, and Google Arts & Culture.

Digital archives, meanwhile, are structured repositories of digitized materials such as manuscripts, photographs, oral histories, ephemera, and scholarly works. They are designed to preserve and provide long-term access to historically significant content, often enhanced with metadata, search functionalities, and user annotations. Examples include the Digital Public Library of America (DPLA), Europeana, and institutional repositories like JSTOR or HathiTrust.

Educational and Scholarly Value

Digital exhibitions and archives serve as rich primary source environments that are especially valuable in humanities education and research. They promote:

1. Critical thinking and inquiry-based learning, as users engage with curated narratives and primary documents.
2. Multimodal literacy, by blending traditional reading with visual, auditory, and interactive content.
3. Global accessibility, allowing students, educators, and researchers to access rare materials without geographical or physical limitations.

For educators, these platforms provide opportunities to integrate digital curation into coursework, enabling students to create their own exhibitions and conduct original archival research. For scholars, digital archives open new avenues for digital humanities methodologies such as distant reading, text mining, and geospatial analysis. Despite their benefits, digital exhibitions and archives face challenges related to copyright, digital preservation, metadata accuracy, and platform sustainability. There is also a need for equitable access to digitization technologies and digital literacy training for both creators and users. Digital exhibitions and archives represent a powerful convergence of technology, culture, and education. They redefine how knowledge is curated, shared, and experienced in the twenty-first century. As they continue to evolve, these platforms will play an increasingly central role in shaping public scholarship, participatory heritage, and inclusive access to human knowledge (Table 4.9).

4.5.1 Omeka: Archival and Exhibition Platform

Omeka is a widely used, open-source web publishing platform specifically designed for creating and managing digital collections, exhibitions, and archives. Developed by the Roy Rosenzweig Centre for History and New Media at George Mason University, Omeka has gained prominence in the fields of digital humanities, public history,

Table 4.9 Digital exhibitions and archives

Feature	Digital exhibitions	Digital archives
Purpose	Interpretation and public engagement	Preservation and research access
Structure	Thematic, curated	Hierarchical or database-driven
Tools used	Omeka, Scalar	Omeka, CONTENTdm, ArchivesSpace
Audience	General public, educators	Researchers, archivists
Interface	Narrative-driven, visual	Metadata-rich, search-oriented

library sciences, and museum studies for its accessibility, flexibility, and emphasis on scholarly metadata.

Purpose and Design Philosophy

Omeka bridges the gap between content management systems and digital curation tools. Unlike general-purpose website builders, Omeka is tailored for scholars, educators, librarians, and curators seeking to showcase digital artefacts in a context-rich environment. Its core design supports Dublin Core metadata standards, making it suitable for academic-grade cataloguing and long-term digital preservation.

Omeka is available in two main versions:

1. **Omeka Classic**—Ideal for smaller projects with curated exhibits.
2. **Omeka S**—Designed for institutions managing multiple collections and users, with linked data support and integration across repositories.

Key Features

1. **Item-Based Content Management**: Every artifact is stored as an "item" with customizable metadata fields, tags, and media files.
2. **Exhibit Building**: Users can curate interpretive exhibits using digital items and text, much like a museum display, arranged through an intuitive drag-and-drop editor.
3. **Plugin Architecture**: Omeka offers a broad ecosystem of plugins for features such as geolocation (Neatline), user-generated content (Collecting), timelines, and social media integration.
4. **Themes and Customization**: Supports theming for visual consistency and branding, with options for responsive, mobile-friendly layouts.
5. **Access and Rights Management**: Allows for open access or restricted content, supporting both public exhibitions and internal archival projects.

Pedagogical and Scholarly Applications

Omeka is increasingly adopted in higher education for digital humanities coursework and collaborative research projects. Students can use it to build digital exhibits based on historical archives, literary analysis, or community oral histories, gaining hands-on experience with primary source interpretation and digital curation. In research contexts, Omeka facilitates public scholarship by allowing academic

Table 4.10 Omeka classic and omeka classic

Feature	Omeka classic	Omeka S
Target user	Independent scholars, small teams	Institutions, digital humanities centres
Metadata Support	Dublin core	RDF, linked data
Usability	User-friendly	Advanced, modular
Customisation	Plugins and themes	Modules and resource templates

projects to reach wider audiences. Whether digitizing rare manuscripts or showcasing a cultural heritage collection, Omeka supports projects that blend scholarly rigor with accessible storytelling.

Strengths and Limitations

Strengths:

1. Tailored for academic and archival needs.
2. Strong metadata support and preservation orientation.
3. Easy exhibit creation with minimal technical skills.
4. Active open-source community and robust plugin ecosystem.

Limitations:

1. Requires hosting knowledge for self-installed versions.
2. Advanced customization may require web development skills.
3. Not as suited for high-volume digitization projects without additional configuration.

Omeka represents a significant advancement in the democratization of digital curation and exhibition. By providing a platform where scholars and students can create structured, accessible, and engaging digital projects, Omeka aligns technology with the pedagogical goals of interpretation, preservation, and public engagement. It continues to serve as a cornerstone in the digital transformation of the humanities (Table 4.10).

4.5.2 Scalar: Multimedia Digital Storytelling

In the expanding landscape of digital scholarship and pedagogy, Scalar has emerged as a powerful platform for multimedia digital storytelling and non-linear narrative construction. Developed by the Alliance for Networking Visual Culture at the University of Southern California, Scalar is designed for authors—especially in the humanities—who wish to integrate text, images, audio, video, and data visualization into cohesive, scholarly, and interactive works. Unlike traditional publishing platforms, Scalar challenges linear writing conventions by enabling users to build richly layered, annotated, and interconnected content structures.

At its core, Scalar is not just a content management system; it is a tool for rethinking narrative structure. It invites authors to construct scholarly arguments through media-rich storytelling, allowing for multiple pathways, nodes, and annotations that reflect the complexity of contemporary knowledge production. This is especially important in the humanities, where interpretation, context, and intertextuality are central to understanding. Scalar's pedagogical design encourages active learning and critical media literacy. By offering a platform where students and researchers can remix, hyperlink, and juxtapose various forms of content, it fosters a deeper engagement with both source material and form. It is particularly well-suited for digital history, visual culture, media studies, and postcolonial studies.

Key Features

1. **Non-linear Structure**: Unlike conventional blogging or LMS platforms, Scalar supports branching narratives and multi-directional reading experiences.
2. **Media Integration**: Seamlessly incorporates media from YouTube, Vimeo, SoundCloud, Internet Archive, and local servers.
3. **Annotation Tools**: Supports deep, inline annotations of images, audio, video, and text, ideal for critical commentary and layered interpretation.
4. **Customizable Templates and Paths**: Users can design custom "paths" through content, allowing readers to navigate through curated sequences or explore freely.
5. **Metadata and Indexing**: Allows for robust tagging and indexing, which enhances searchability and facilitates scholarly referencing.
6. **Open Source and Scholarly Oriented**: Encourages open access publishing and long-form academic work, with features tailored to citation and academic rigor.

Applications in Teaching and Research

In classrooms, Scalar enables student-centered learning by turning learners into creators of interactive projects. Whether designing a digital museum, mapping a historical timeline, or publishing a multimedia essay, students engage in research, authorship, and design—all within a single platform. Faculty can also use Scalar for collaborative publications, peer-reviewed digital journals, and digital dissertations.

In research, Scalar excels in presenting complex arguments or archives where text and media must be in dialogue. It supports visual and spatial storytelling that traditional formats cannot capture, making it ideal for fields like digital art history, ethnography, or performance studies.

Strengths and Considerations

Strengths:

1. Enables richly layered, non-linear digital narratives.
2. Bridges scholarly rigor with creative freedom.
3. Supports a wide variety of media formats and sources.
4. No programming required for basic use.

Considerations:

Table 4.11 Scalar feature

Feature	Scalar
Core functionality	Multimedia-rich, non-linear narratives
Metadata	Supports Dublin Core and semantic tagging
Strengths	Media integration, networked structure
Use cases	Digital books, classroom storytelling, and academic projects

1. Learning curve may be steep for users unfamiliar with non-linear design.
2. Requires hosting or institutional support for long-term preservation.
3. Better suited for project-based work than rapid blogging or casual posting.

Scalar is not just a tool—it's a digital environment for rethinking authorship, scholarship, and pedagogy. By enabling multimedia-rich, interactive, and structurally complex content, Scalar empowers scholars and students to engage in digital storytelling that is both analytically rigorous and experientially rich. As digital humanities continue to evolve, Scalar is a model of how digital platforms can support critical inquiry and creativity in equal measure (Table 4.11).

4.5.3 Project Planning and Workflow

Effective project planning and workflow management are critical components of any successful digital project, especially in interdisciplinary fields like the digital humanities, educational innovation, or institutional advancement. These processes provide the necessary structure to transform abstract ideas into concrete, actionable outcomes, ensuring that goals are met on time and within resource constraints. In digital initiatives—such as building online archives, developing educational platforms, or conducting collaborative research—project planning establishes a clear roadmap that includes objectives, timelines, stakeholders, resource allocation, and evaluation criteria. Without a well-defined plan, projects risk becoming fragmented, delayed, or misaligned with user needs and institutional goals. Planning also fosters accountability and transparency, as tasks can be assigned to individuals or teams, and progress can be tracked systematically. Moreover, project planning allows for the identification of risks and the development of contingency strategies, thereby enhancing the project's resilience in the face of uncertainty.

A project's workflow refers to the sequence of tasks, tools, communication channels, and approvals that guide its implementation. A good workflow balances structure with flexibility. It should clearly define the stages of work—e.g., ideation, design, testing, implementation, review—while allowing space for iterative feedback, revision, and stakeholder input. Workflow design is often supported by digital project management tools such as Trello, Asana, Airtable, or Notion, which allow teams to visualize tasks, deadlines, and dependencies. For academic projects, platforms like

Table 4.12 Project planning and workflow

Stage	Activities
Planning	Define goals, select platform, assign roles
Content gathering	Digitize or collect texts, images, videos
Metadata creation	Apply dublin core or custom schemas
Platform setup	Install themes, plugins/modules, configure structure
Curation	Organize items, create exhibits or paths
Testing and feedback	Usability testing, peer review
Launch and dissemination	Publish online, share via academic and public channels

GitHub (for version control), Google Workspace (for collaboration), and Slack (for communication) are commonly used.

Successful digital projects often require cross-functional teams composed of researchers, designers, developers, curators, writers, and administrators. Clear role definition is essential to avoid duplication and confusion. Project managers or coordinators play a crucial role in keeping the team aligned, resolving conflicts, and ensuring that deliverables are met. Throughout the project, proper documentation—including meeting notes, development logs, user feedback, and change histories—ensures that the project remains transparent and replicable. At the conclusion, project evaluation helps assess whether objectives were achieved, what challenges emerged, and what improvements can be made in future initiatives. Project planning and workflow are not just administrative functions—they are foundational to the creativity, coherence, and sustainability of digital work. In academic and institutional contexts where resources are limited and expectations are high, a strategic approach to planning and execution ensures that digital projects are impactful, timely, and transformative (Table 4.12).

4.5.4 Challenges and Considerations

Despite the transformative potential of digital tools and platforms in education, research, and cultural preservation, the implementation and sustainability of digital projects come with a range of challenges and considerations. These must be carefully addressed during the planning, development, and post-deployment phases to ensure project success, ethical integrity, and long-term impact.

1. **Technical Constraints**: Digital projects often require access to reliable infrastructure, such as high-speed internet, secure servers, data storage, and device compatibility. In under-resourced contexts, these requirements can create substantial barriers. Additionally, project teams may lack the technical skills needed to manage complex tools, platforms, or programming languages, necessitating ongoing technical support or training.

2. **Sustainability and Maintenance**: Many digital initiatives are launched with initial enthusiasm but falter due to a lack of long-term funding, hosting, or maintenance plans. Without sustained investment in platform updates, content moderation, and user support, digital projects risk becoming outdated or inaccessible. Sustainability planning must address not only technical upkeep but also human resource continuity.
3. **Data Management and Preservation**: Digital content—whether born-digital or digitized—must be properly archived, backed up, and documented. Projects lacking metadata standards, file format compatibility, or clear ownership policies may struggle with data loss, duplication, or irrelevance over time. Preservation strategies should comply with best practices in digital archiving, such as open formats, redundant storage, and descriptive metadata (e.g., Dublin Core).
4. **User Accessibility and Inclusivity**: An often-overlooked challenge is ensuring that digital projects are **accessible to diverse users**, including those with disabilities, limited digital literacy, or language barriers. Compliance with accessibility standards (such as WCAG) and inclusive design practices is essential to creating equitable digital environments. Multilingual interfaces, screen-reader compatibility, and mobile responsiveness should be incorporated from the outset.
5. **Ethical and Legal Considerations**: Digital projects that involve user data, historical materials, or community narratives must navigate complex ethical landscapes. This includes obtaining proper consent, respecting cultural sensitivities, managing copyright and intellectual property rights, and protecting user privacy. In educational contexts, especially, student data and participation must be managed according to institutional and legal guidelines (e.g., GDPR, FERPA).
6. **Pedagogical Alignment**: When deployed in teaching contexts, digital tools and platforms must align with **learning outcomes** and curricular goals. Introducing technology for its own sake can lead to distraction or surface-level engagement. Educators must be trained to integrate digital tools in pedagogically sound ways that enhance—not replace—critical thinking, creativity, and collaboration.
7. **Evaluation and Feedback Loops**: Ongoing assessment of the project's usability, effectiveness, and impact is essential. This includes collecting user feedback, analyzing platform analytics, and conducting periodic reviews. However, developing robust evaluation frameworks that balance qualitative and quantitative insights can be challenging and time-consuming.

Thus, while digital projects offer immense value in expanding access, fostering innovation, and preserving knowledge, they are not without challenges. Addressing these concerns proactively—from technical readiness and sustainability to ethical responsibility and pedagogical coherence—can significantly improve project outcomes. Thoughtful planning, inclusive design, and stakeholder collaboration remain central to ensuring that digital efforts are effective, ethical, and enduring.

4.6 Coding for Humanities

In recent years, the boundaries between the humanities and computing have increasingly blurred, giving rise to new modes of inquiry, expression, and analysis. At the heart of this transformation lies coding for the humanities—not as an esoteric technical pursuit, but as a way of expanding the methods and media through which human culture, thought, and experience can be studied and shared. Tools like Twine and Jupyter Notebooks exemplify this intersection, empowering scholars, educators, and students to produce interactive narratives, computational essays, and reproducible research in accessible, engaging formats.

The integration of coding into the humanities represents a paradigm shift in scholarly practices, moving from passive engagement with digital tools to active participation in creating, analysing, and sharing digital knowledge. Coding empowers scholars to ask new types of questions, engage in experimental methodologies, and produce interactive, dynamic, and collaborative forms of scholarship. Coding in the humanities enables scholars to automate repetitive tasks, mine large datasets, model historical events, visualise linguistic patterns, and even simulate philosophical dilemmas. As humanities disciplines increasingly embrace digital methods, coding becomes a vital literacy, much like reading or writing, that broadens the scope and impact of humanistic inquiry.

Twine is an open-source tool for creating non-linear, interactive stories—a form of digital storytelling that resonates deeply with disciplines such as literature, philosophy, history, and cultural studies. Built on simple HTML and scripting, Twine enables users to construct narratives that branch and loop, creating multiple pathways through text. Twine's strength lies in its ability to integrate narrative and interactivity, enabling authors to explore complex themes such as identity, agency, and causality in a way that traditional linear writing cannot. It is particularly effective in pedagogical settings, where students can build their games or stories to critically engage with course content while learning basic coding logic (e.g., variables, conditionals, and hyperlinks).

Educators use Twine to:

1. Teach digital storytelling and game design.
2. Explore historical simulations and counterfactual histories.
3. Foster critical thinking through branching logic and reader choice.

Its low barrier to entry, no programming experience is required—makes Twine ideal for humanities learners new to digital tools.

Jupyter Notebooks: Computational Thinking in Humanities Research

While Twine facilitates narrative interactivity, Jupyter Notebooks offer a powerful environment for computational analysis and reproducible research. Originating in data science and STEM fields, Jupyter has found meaningful applications in the digital humanities, particularly where text analysis, visualization, or metadata processing are involved. A Jupyter Notebook allows users to combine code, data, visualizations, and explanatory text in a single interactive document. Written primarily in Python, it supports natural language processing (NLP), corpus analysis, Stylometry, and other methods central to computational humanities.

Jupyter is used by humanists to:

1. Analyze large text corpora for linguistic or thematic patterns.
2. Visualize historical data with timelines, maps, or graphs.
3. Document workflows for transparency and reproducibility in research.

Its interactive format encourages iterative exploration, making it ideal for researchers who wish to show not only their results, but also how they arrived at them, a key principle of scholarly transparency. Furthermore, the use of Twine and Jupyter demonstrates that coding in the humanities is not simply technical, it is analytical, expressive, and exploratory. Whether building narrative-driven games or analyzing political discourse through algorithms, coding allows humanists to ask new questions, model complexity, and present knowledge in novel ways. Moreover, engaging with code encourages critical digital literacy, an awareness of how software shapes meaning, how data is structured, and how algorithms mediate cultural information. As such, coding becomes a site of reflection as much as creation, fitting squarely within the humanities' traditions of critique and interpretation.

Twine and Jupyter Notebooks represent two ends of a vibrant spectrum where narrative creativity meets computational inquiry. Together, they highlight the growing potential of code as both method and medium in the humanities. Far from displacing traditional humanistic values, these tools invite scholars to reimagine them in digital form, forging paths toward a more interactive, data-literate, and expressive intellectual future. Furthermore, the rise of open-source platforms has democratized access to coding tools, allowing students and researchers with minimal technical background to participate in digital humanities projects. Coding not only enhances the precision and efficiency of scholarly work but also fosters interdisciplinary collaboration with computer scientists, data analysts, and designers.

4.6.1 Twine: Interactive Storytelling

Twine is an open-source tool designed for crafting nonlinear, interactive stories, widely used in education, game design, and experimental literature. Its ease of use and accessibility make it an ideal platform for humanities scholars exploring narrative theory, pedagogy, and public engagement. With a visual interface that maps out story nodes, Twine allows users to construct branching narratives where the reader's choices affect the progression of the story. This capability is especially valuable for reimagining historical events, simulating ethical decisions, or illustrating the consequences of philosophical arguments. Twine supports HTML, CSS, and JavaScript, enabling more advanced users to embed multimedia, create dynamic effects, and develop sophisticated mechanics. However, even beginners can produce rich, meaningful content using its native scripting language, Harlowe,

Table 4.13 Twine features table

Feature	Twine
Use case	Interactive storytelling and simulation
Skill level	Beginner-friendly, supports advanced scripting
Output format	HTML, embeddable in websites or LMS
Strengths	Narrative flexibility, visual interface, open-source
Limitations	Limited analytical capabilities, text-based interface

without any prior programming experience. In educational settings, Twine fosters critical thinking, narrative structure, and digital authorship. Students can simulate debates, create choose-your-own-adventure biographies, or model literary themes. In research, Twine has been used to reconstruct personal stories, cultural memories, and alternate histories, bridging narrative and computation (Table 4.13).

4.6.2 Jupyter Notebooks

Jupyter Notebooks offer a powerful and flexible environment for conducting data-driven research in the humanities. A web-based platform that supports multiple programming languages, most notably PythonJupyter allows users to write code, document their process, and visualize outputs in a single, shareable document. Jupyter is ideal for text mining, sentiment analysis, metadata exploration, and geospatial visualization. It integrates seamlessly with libraries like NLTK, Pandas, and Matplotlib, enabling humanists to process and interpret vast textual corpora or complex datasets. Scholars can clean data, perform statistical analysis, and generate graphs and maps all within one interactive notebook. What sets Jupyter apart is its commitment to reproducible research. Because code, output, and narrative are interwoven, other scholars can trace the logic of the analysis, replicate the steps, and adapt methods to new questions. This transparency fosters a culture of sharing and rigour in digital scholarship.

Educationally, Jupyter introduces students to computational thinking and quantitative methods. Through guided assignments, learners can explore literary trends, digitised archives, or linguistic features while building coding proficiency. It aligns well with data literacy goals and the increasing need for digital fluency in humanities education (Table 4.14).

Table 4.14 Jupyter features table

Feature	Jupyter notebooks
Use case	Text analysis, visualisation, data wrangling
Skill level	Intermediate; some programming knowledge required
Output format	.ipynb, HTML, PDF
Strengths	Open-source, versatile, reproducible, multi-language
Limitations	Initial setup required, limited multimedia support

4.7 Educational and Pedagogical Use

Digital tools support pedagogical approaches that emphasize exploration, problem-solving, and critical reflection. For example, using platforms like Voyant Tools or AntConc, students can engage in textual analysis that moves beyond surface interpretation. By examining word frequencies, patterns, or collocations, they uncover hidden structures and linguistic nuances that deepen their understanding of texts. This data-informed inquiry aligns with constructivist learning theories, where learners actively build knowledge through investigation and interpretation.

The integration of digital tools into the teaching and learning of humanities subjects has significantly reshaped pedagogical strategies, learner engagement, and content delivery. From interactive storytelling to computational text analysis, educational technologies empower students and educators to explore human culture, history, language, and expression in dynamic and multimodal ways. These tools are not merely add-ons to traditional methods; they are catalysts for rethinking what it means to teach, learn, and do scholarship in the humanities.

Tools such as Twine, Scalar, and Canva empower students to become creators of knowledge rather than passive recipients. With Twine, learners can write interactive stories that explore themes of agency, choice, and narrative complexity. Scalar supports multimedia essays that blend text, images, audio, and video, enabling students to articulate ideas through rich digital storytelling. These tools enhance engagement, promote media literacy, and allow for multimodal assessment—where students' digital projects can be evaluated alongside traditional essays and presentations. Collaborative platforms like Google Docs, Microsoft Teams, and Padlet enable real-time co-creation and discussion, both inside and outside the classroom. These tools foster peer learning, enable diverse participation, and support inclusive pedagogies. Students can annotate texts together, offer feedback on drafts, or co-develop digital exhibitions. The asynchronous capabilities of these platforms also cater to different learning styles and allow students to reflect at their own pace.

Digital tools, particularly those that are open-source or web-based, help democratize access to education. For students in remote areas or with physical disabilities, tools like Overleaf, Jupyter Notebooks, and LMS platforms such as Moodle and Canvas make learning more accessible. Screen readers, mobile compatibility, and language localisation features further enhance equity in education by removing barriers to participation. Integrating tools like Zotero (for referencing), GitHub (for

Table 4.15 Educational use comparison

Platform	Pedagogical strengths	Sample assignments
Twine	Narrative creativity, interactivity	Alternate histories, literary games
Jupyter	Data analysis, transparency	Text mining, sentiment analysis

version control), or Jupyter (for computational notebooks) into humanities curricula introduces students to digital research workflows. These skills—organising citations, documenting research processes, collaborating across platforms—are essential for academic rigor, transparency, and career readiness in an increasingly digital scholarly ecosystem.

Both Twine and Jupyter serve as transformative tools in digital humanities pedagogy. They provide platforms for students to explore humanistic questions through creative and analytical lenses, fostering deeper engagement and practical skill development. Twine emphasises creativity and narrative design. Assignments may include rewriting classical myths from alternative perspectives, creating branching ethical dilemmas, or visualising character development in literature. Students learn to build structure, manage reader choice, and reflect on how format shapes meaning. Jupyter introduces analytical rigour and data interpretation. Students might conduct stylometric comparisons between authors, analyse social media data on contemporary cultural phenomena, or visualise migration patterns using historical census data. These assignments promote computational reasoning alongside critical inquiry. When integrated thoughtfully, these platforms support diverse learning styles and help bridge the gap between humanities theory and digital practice. They also prepare students for hybrid careers in digital publishing, education technology, cultural analytics, and beyond.

Hence, the educational and pedagogical use of digital tools in the humanities is more than a matter of convenience—it is a profound transformation of how we think, teach, and learn. These tools align with humanistic values of inquiry, interpretation, and creativity while preparing students for a world where digital fluency is indispensable. When thoughtfully integrated, digital tools amplify the power of the humanities to illuminate the human condition in contemporary, accessible, and innovative ways (Table 4.15).

4.8 Introduction to Mobile Learning in the Humanities

Mobile applications have revolutionised the way knowledge is accessed and disseminated, making learning more flexible, interactive, and accessible across disciplines, including the humanities. These apps facilitate learning on-the-go, empower learners through multimedia engagement, and offer real-time access to a variety of resources. The incorporation of mobile technologies in humanities education supports constructivist learning environments, where students actively engage with content through simulations, games, augmented reality (AR), and collaborative tools. These features allow learners to construct meaning, explore cultural and historical content dynamically, and participate in knowledge creation. Mobile learning aligns with the principles of lifelong learning and inclusive education. Learners from diverse backgrounds, including those in remote or under-resourced regions, can access high-quality educational content via smartphones and tablets. Furthermore, mobile platforms support multimodal learning by integrating text, audio, images, and video, catering to different learning styles.

In the 21st-century educational landscape, mobile learning (m-learning) has emerged as a transformative approach to teaching and learning across disciplines, including the humanities. Defined broadly, mobile learning refers to educational experiences that are facilitated through portable digital devices such as smartphones, tablets, and laptops, enabling learners to access content anytime, anywhere. As access to mobile technology increases globally, so too does the potential for inclusive, flexible, and student-centred learning models, particularly in traditionally text-intensive fields like history, literature, philosophy, languages, and cultural studies.

The humanities have long emphasised critical reading, reflective thinking, dialogical learning, and narrative engagement. These pedagogical practices align well with mobile technologies that offer multimodal content delivery, interactive engagement, and collaborative learning outside the confines of conventional classrooms. Through mobile apps, podcasting, e-readers, augmented reality, and social platforms, learners can explore texts, artworks, oral histories, and cultural phenomena in context and on the move. For example, students can use language learning apps (e.g., Duolingo, Anki), access annotated texts on digital libraries, or contribute to crowd-sourced heritage mapping projects. Augmented reality apps can bring historical events to life during museum visits or fieldwork, while mobile annotation tools foster collaborative literary analysis in real time.

Pedagogical Advantages of Mobile Learning

Mobile learning in the humanities supports:

1. Flexibility and accessibility, allowing students to learn at their own pace and in their preferred environment.
2. Personalisation, through curated content, adaptive quizzes, and learning analytics.
3. Peer interaction, via chat functions, shared documents, and discussion forums.
4. Multimodal engagement, with audio, video, text, and image-rich resources that support diverse learning styles.

Despite its potential, mobile learning presents challenges related to digital distraction, screen fatigue, device inequality, and connectivity limitations. Humanities educators must therefore design mobile learning experiences that are pedagogically sound, context-aware, and ethically responsible. Considerations around accessibility, privacy, and equitable participation are critical to ensuring inclusive implementation. Mobile learning is reshaping how the humanities are taught, learned, and experienced. It represents not a departure from humanistic inquiry, but rather an evolution, making it more responsive to the lives, needs, and learning habits of contemporary students. As mobile technology continues to proliferate, so too does the opportunity to reimagine the humanities as an accessible, interactive, and boundaryless field of study.

4.8.1 Categories of Mobile Apps for Humanities Education

Mobile applications have become instrumental in humanities education, offering learners diverse, engaging ways to interact with subject matter. These apps fall into several pedagogically meaningful categories, each aligned with specific teaching and learning goals. Language learning apps, such as Duolingo, Babbel, and Memrise, use gamified techniques like streaks, point systems, and adaptive learning to promote vocabulary acquisition and grammar skills. Their bite-sized lessons and spaced repetition methods support long-term retention, while their intuitive design enhances user motivation and habit formation. Historical simulation apps like Civilisations AR use AR to bring past events and artifacts to life. These immersive experiences allow learners to virtually place 3D historical objects in their environment, offering an interactive way to explore ancient civilizations, cultural symbols, and key historical figures. Such tools support contextual and experiential learning by bridging the gap between text and tangible understanding.

eBook and audiobook platforms like Audible, Libby, and Google Books provide mobile access to a vast range of classical and contemporary texts. These tools foster reading flexibility by allowing users to listen to or read works of literature during non-traditional study times, such as commuting or multitasking. They are particularly helpful for learners with visual impairments or reading difficulties. Collaborative annotation tools, such as Hypothesis and Perusall, enable students to engage in "social reading." These apps allow users to highlight passages, add comments, and respond to peers directly on the text. This fosters critical dialogue and enhances interpretive skills, particularly in literature, philosophy, and cultural studies. Museum and archive apps like Google Arts & Culture democratize access to global collections. Students can explore curated exhibitions, zoom into high-resolution images of famous artworks, and read context-rich descriptions—all from their mobile devices. These tools are particularly valuable for art history and cultural studies courses, where exposure to visual material is essential.

Together, these categories of mobile apps demonstrate how digital tools are reshaping humanities education. They extend the classroom into everyday life, support diverse learning styles, and foster new forms of engagement with texts, artifacts, and languages. As these tools continue to evolve, their thoughtful integration into curricula will remain key to enhancing learning outcomes and student motivation.

4.8.2 Enhancing Interactivity and Engagement

Interactivity is one of the defining features of effective mobile learning. Unlike static digital content, interactive mobile applications offer learners the opportunity to engage dynamically with material through actions such as clicking, dragging, swiping, and speaking. In humanities education, where engagement with texts, images, ideas, and historical contexts is critical, interactivity transforms passive learning into active exploration. Modern mobile apps use a variety of strategies to enhance engagement. Interactive timelines, clickable maps, voice-enabled quizzes, and branching narratives all contribute to richer learning experiences. For example, in language apps like Duolingo, learners are given tasks such as matching, sentence construction, pronunciation practice, and fill-in-the-blank exercises. These tasks are embedded within a gamified structure that awards points, badges, and access to higher levels for consistent participation.

AR has further revolutionized how learners interact with content. Civilisations AR, developed by the BBC, enables users to place 3D artifacts—like Egyptian statues or Roman armor—into their physical surroundings using a smartphone camera. This makes cultural heritage accessible in a personal and immersive way, helping students to visualize and contextualize abstract concepts. Mobile apps also support real-time collaboration. Platforms like Perusall allow multiple users to annotate a single text simultaneously, fostering live discussions, peer review, and shared inquiry. This collaborative functionality turns texts into dialogic spaces, encouraging learners to co-construct meaning.

Gamification remains a central strategy in enhancing user motivation. Leaderboards, streak counters, daily challenges, and achievement badges tap into behavioral psychology to reinforce consistency and engagement. When used thoughtfully, these features can make learning addictive in a positive way. However, care must be taken to ensure that gamification does not distract from critical thinking or deep reflection. In sum, interactive features in mobile apps have the power to transform how learners approach humanities content. By allowing users to manipulate data, respond to prompts, explore narratives non-linearly, or engage with AR simulations, these apps align well with constructivist pedagogies. The challenge lies in designing or selecting apps that balance fun with academic depth, ensuring that interactivity supports—not substitutes—substantive learning.

4.8.3 Pedagogical Benefits and Use Cases

Mobile learning apps provide numerous pedagogical benefits in humanities education, creating opportunities for more inclusive, flexible, and context-aware learning. These tools facilitate learner-centered approaches by supporting differentiated instruction, enabling personalization, and fostering both independent and collaborative learning. They are particularly well-suited to fields such as literature, history, language, and cultural studies. One major benefit is the flexibility mobile apps offer. Students can access learning materials on demand, revisit difficult concepts, or engage with texts during travel or in non-traditional settings. For example, audiobook platforms like Audible and digital readers like Kindle allow literature students to engage with course texts during daily commutes. This seamless integration into everyday life enhances the learner's connection to the material.

In historical studies, AR and VR apps create immersive simulations that allow students to explore historical contexts and artifacts as though they were physically present. Applications like Civilisations AR or TimeLooper recreate historical sites, helping learners develop spatial awareness and emotional engagement with the past. Collaborative annotation tools like Hypothesis and Perusall promote active reading and dialogue. These tools support social learning by encouraging students to share interpretations, pose questions, and respond to peers within the margins of texts. In literature and philosophy, such tools encourage critical engagement and support argument development. Language learning apps further demonstrate the pedagogical versatility of mobile platforms. Tools like Babbel and Duolingo adapt to learners' skill levels and reinforce language acquisition through repetition, real-time feedback, and contextual examples. These apps provide immersive language environments that are particularly beneficial for multilingual learners or students preparing for study abroad programs.

Use cases continue to evolve. In museums, QR-code-based mobile tours support inquiry-based learning. In classrooms, apps like Padlet and Quizlet support project-based activities and flashcard review. Mobile blogging and micro-publishing tools encourage student reflection and peer feedback in real time. In sum, mobile apps not only expand access to humanities content, but also enhance learner autonomy, motivation, and engagement. Their integration into curricula supports a more student-centered and inquiry-driven model of education that is well-suited to the diverse and interpretive nature of the humanities.

4.8.4 Challenges and Considerations

Despite the many advantages of mobile learning in humanities education, educators must navigate several challenges and considerations to ensure effective, equitable, and ethical implementation. These issues range from infrastructure disparities and pedagogical limitations to privacy concerns and content quality control.

1. A key challenge is the digital divide. Not all students have equal access to smartphones, tablets, or high-speed internet, particularly in rural or underserved communities. This inequality can create disparities in participation and learning outcomes. Institutions must consider strategies to mitigate these gaps, such as providing loaner devices or offering offline-compatible content.
2. Over-reliance on gamification can also undermine deep engagement. While point systems, badges, and leader boards enhance motivation, they may encourage

superficial interaction with content if not paired with reflective activities. Humanities education demands critical analysis and interpretive thinking, which can be diluted if learners are driven solely by extrinsic rewards.
3. Data privacy and security are major concerns, especially with commercial mobile apps that collect behavioral data for personalization. Educators and institutions must ensure that any tools used comply with privacy regulations (e.g., GDPR, FERPA) and do not compromise student data. Transparency about data collection and user consent must be prioritized.
4. Screen time and cognitive overload are additional considerations. While mobile learning encourages flexibility, excessive screen exposure can lead to fatigue, distraction, or diminished comprehension. A well-balanced instructional strategy should blend mobile activities with offline reflection, discussion, or experiential learning.
5. Educator readiness and support are critical to successful integration. Many instructors may lack familiarity with educational technologies or confidence in using mobile apps pedagogically. Ongoing professional development, tech support, and peer-sharing communities can help bridge this gap.
6. Finally, mobile app content quality varies widely. Not all platforms meet academic standards for accuracy, rigor, or cultural sensitivity. Educators must curate and evaluate apps carefully to ensure alignment with course outcomes, disciplinary ethics, and student needs.

In conclusion, the use of mobile apps in humanities education is promising, but not without risks. Addressing these challenges through intentional planning, inclusive policies, and continuous evaluation is essential to maximizing their pedagogical potential while safeguarding the well-being and rights of learners.

References

1. Schwab, K.: The fourth industrial revolution: what it means and how to respond. World Economic Forum (2016). https://www.weforum.org/agenda/2016/01/the-fourth-industrial-revolution-what-it-means-and-how-to-respond/
2. Korot, E., et al.: Enablers and barriers to deployment of smartphone-based home vision monitoring in clinical practice settings. JAMA Ophthalmol. **140**(2), 153–160 (2022). https://doi.org/10.1001/jamaophthalmol.2021.5269
3. Ghosh, A., Edwards, D.J., Hosseini, M.R.: Patterns and trends in Internet of Things (IoT) research: future applications in the construction industry. Eng. Constr. Archit. Manag. **28**(2), 457–481 (2021). https://doi.org/10.1108/ECAM-04-2020-0271
4. de-Lima-Santos, M. F., Mesquita, L., de Melo Peixoto, J. G., Camargo, I.: Digital news business models in the age of industry 4.0 digital Brazilian news players find in technology new ways to bring revenue and competitive advantage. Digit. J. **0**(0), 1–25 (2022). https://doi.org/10.1080/21670811.2022.2037444
5. Garcia-Teruel, R.M.: A legal approach to real estate crowdfunding platforms. Comput. Law Secur. Rev. **35**(3), 281–294 (2019). https://doi.org/10.1016/j.clsr.2019.02.003
6. Pilkevych, I., Fedorchuk, D., Naumchak, O., Romanchuk, M.: Fake news detection in the framework of decision-making system through graph neural network. In: 2021 IEEE 4th International

Conference on Advanced Information and Communication Technologies AICT-Proceedings, pp. 153–157 (2021). https://doi.org/10.1109/AICT52120.2021.9628907
7. Afsari, K., Eastman, C.M., Castro-Lacouture, D.: JavaScript Object Notation (JSON) data serialization for IFC schema in web-based BIM data exchange. Autom. Constr. **77**, 24–51 (2017). https://doi.org/10.1016/j.autcon.2017.01.011
8. Zhengdao, C.: An Internet of Things-enabled BIM platform for on-site assembly services in prefabricated construction. Autom. Constr. **89**(February), 146–161 (2018). https://doi.org/10.1016/j.autcon.2018.01.001
9. Ekechi, C.C., Chukwurah, E.G., Oyeniyi, L.D., Okeke, C.D.: A review of small business growth strategies in African economies. Int. J. Adv. Econ. **6**(4), 76–94 (2024). https://doi.org/10.51594/ijae.v6i4.1071
10. Biswas, A., Roy, A.: Blockchain and Ipfs Based Secure Cloud Banking System Using Smart Card, pp. 1–34 (2022)
11. Slesinger, I.: The limits of control: technological agency, urban terrain, strategy and the state in the 2014 Gaza War. Polit. Geogr. **93**, 1–39 (2022). https://doi.org/10.1016/j.polgeo.2021.102530
12. Xiao, L.: Integrating building information modeling and prefabrication housing production. Autom. Constr. **100**(January), 46–60 (2019). https://doi.org/10.1016/j.autcon.2018.12.024
13. Liu, Y., Han, W., Zhang, Y., Li, L., Wang, J., Zheng, L.: An internet-of-things solution for food safety and quality control: a pilot project in China. J. Ind. Inf. Integr. **3**, 1–7 (2016). https://doi.org/10.1016/j.jii.2016.06.001
14. Alam, M.N.: A critical analysis of internal and external environment: case study of apple Inc. J. Int. Bus. Manag. April 2016 (2021). https://doi.org/10.37227/jibm-2021-08-149
15. Igboin, B.O.: Small fires causing large fires: an analysis of boko haram terrorism–insurgency in Nigeria. Religions **13**(6) (2022). https://doi.org/10.3390/rel13060565
16. Nagao, A.: A Genre-based Approach to Teaching Descriptive Report Writing to Japanese EFL University Students, vol. 26, no. 3, pp. 1–28 (2022)

Chapter 5
Coding for Humanists

5.1 Why Humanists Should Code

Coding for humanists is a shift in the humanities that uses programming skills and computational thinking to investigate, interpret, and communicate humanistic knowledge. This approach enhances the reach and depth of humanistic inquiry by enabling new forms of research, visualization, and public engagement. Humanists typically use languages such as Python, R, HTML, CSS, and JavaScript, often in conjunction with tools like Python Notebooks, Twine, RStudio, and GitHub. These tools are not ends in themselves but means to deepen scholarly exploration. The digital transformation of the humanities has encouraged scholars to engage with coding as a technical skill and a powerful intellectual tool. Coding allows humanists to experiment with, manipulate, and analyse data in ways that traditional methods often cannot support. It allows for innovative research, greater transparency, and deeper engagement with the tools that shape contemporary knowledge production. Coding is not about abandoning the qualitative depth of humanistic inquiry. Instead, it provides a complementary methodology that empowers scholars to ask new questions, explore large-scale data, and visualize insights in compelling ways. As digital tools become standard in education, archives, and publishing, the ability to code becomes a form of scholarly literacy for the 21st-century humanist [1, 2].

Teaching coding in humanities contexts fosters critical digital literacy, enabling students to ask new questions and communicate findings in interactive formats. It also promotes interdisciplinary collaboration, transparency, and professional skill development, opening career pathways in digital publishing, archival management, UX design, and more. Coding also supports public humanities by allowing researchers to develop accessible digital exhibits, mobile apps, and educational resources that reach global audiences. However, coding in the humanities comes with challenges, including the need for a supportive learning environment and time to develop skills, as well as ethical issues such as algorithmic bias, data privacy, and platform sustainability. In summary, coding for humanists is about expanding their toolkit to include

F. A. Alaba and I. G. Ajibola, *Digital Pedagogy: Bridging the Gap Between Software and the Humanities*, Information Systems Engineering and Management 85,
https://doi.org/10.1007/978-3-032-20335-9_5

```
import math

def mandelbrot(z , c , n=40):
    if abs(z) > 1000:
        return float("nan")
    elif n > 0:
        return mandelbrot(z ** 2 + c, c, n - 1)
    else:
        return z ** 2 + c

print("\n".join(["".join(["#" if not math.isnan(mandelbrot(0, x + 1j * y).real) else " "
                 for x in [a * 0.02 for a in range(-80, 30)]])
                 for y in [a * 0.05 for a in range(-20, 20)]])
    )
```

Fig. 5.1 Learning coding for humanists

digital methods that complement and enhance traditional approaches, empowering scholars to engage more deeply with texts, histories, and cultures while preparing students for a digital future. As the digital humanities continue to grow, coding will remain a vital means of fostering innovation, interdisciplinary research, and inclusion in humanistic scholarship [3]. Learning to code encourages a systematic and logical approach to problem-solving. Humanists, often trained in critical analysis and interpretive nuance, can benefit from the rigor and clarity coding demands. It allows them to design structured methodologies, build reproducible workflows, and approach problems with computational precision. This integration strengthens critical thinking by bridging abstract concepts with concrete implementations. For instance, text encoding with XML or TEI standards forces scholars to consider the structure and semantics of texts in highly specific ways, enhancing both interpretation and preservation [4] (Fig. 5.1).

5.1.1 Access to Data-Driven Research

The exponential growth of big data within the humanities—spanning digitised archives, social media content, historical corpora, and bibliometric databases—has transformed the scope and scale of academic inquiry. However, harnessing these vast datasets requires more than just theoretical knowledge; it demands practical, technical proficiency, particularly in coding and computational methods. For researchers in the digital humanities, coding skills serve as a gateway to meaningful data exploration, providing tools to scrape, clean, analyse, and visualise complex information. Programming languages such as Python and R are widely adopted in the digital humanities for their robust libraries and flexibility in handling large textual and numerical datasets. For example, scholars can perform sentiment analysis on 19th-century novels to detect emotional trends or shifts in cultural narratives over time. With tools like NLTK or spaCy in Python, researchers can break down textual data into components, classify linguistic features, and identify recurring patterns or anomalies that would be invisible to manual inspection.

Moreover, coding enables the construction and analysis of citation networks, helping to trace intellectual influences and disciplinary intersections across decades. Through network visualisation tools such as Gephi or using Python libraries like NetworkX, humanities scholars can map relationships between scholars, themes, and institutions with precision. Historical document collections can also be systematically compared using topic modelling algorithms like Latent Dirichlet Allocation (LDA), offering fresh insights into discourse shifts across periods. Without the ability to code, researchers often become dependent on third-party tools that may lack transparency or customisability. This creates limitations in the reproducibility of findings and the depth of interpretation. Mastery of code ensures that scholars can validate their methodologies, tailor analyses to specific questions, and innovate beyond the confines of pre-set functionalities. Ultimately, coding is not merely a technical add-on but a methodological asset that amplifies scholarly agency, rigour, and creativity in the data-driven humanities [5, 6].

5.1.2 Empowering Digital Project Development

In the evolving landscape of digital scholarship, the ability to build and manage digital projects independently represents a significant advantage for humanities scholars. Coding equips researchers with the tools and autonomy to design, prototype, and deploy their own interactive platforms, archives, visualisations, and scholarly applications. This shift from passive users to active creators' fosters innovation, encourages interdisciplinary collaboration, and enhances the visibility and accessibility of humanities research. Digital projects in the humanities often involve interactive storytelling, data visualisation, and public engagement tools. Platforms like Twine allow scholars to create branching narratives that blend literary theory with game design, enabling users to experience history or literature through immersive, nonlinear paths. Jupyter Notebooks combine code, data, and narrative in a single document, making it ideal for presenting computational research with clarity and transparency. Tools such as D3.js, a powerful JavaScript library, enable the development of dynamic, browser-based visualisations that respond to user input and reveal hidden trends in cultural or historical data [7].

Coding also reduces dependence on external developers or institutional IT departments, which often have limited availability or different priorities. By gaining control over the design and implementation process, scholars ensure that their projects reflect their academic goals and user needs. This self-sufficiency accelerates experimentation, supports iterative design, and makes the development process more responsive and agile. Furthermore, digital projects developed through coding can reach broader audiences through the web, increasing the societal impact of humanities scholarship. Virtual exhibits, mobile applications, and digital archives created with coding tools serve not just as research outputs but as educational resources, community engagement platforms, and cultural preservation tools. For example, a scholar of medieval history can build an online repository of illuminated manuscripts with zoomable images, annotations, and linked metadata, all coded into an accessible web platform. In summary, coding empowers humanists to bridge the gap between research and digital expression. It allows them to imagine new forms of scholarly communication that are interactive, accessible, and enduring, placing control over both content and medium firmly in their hands [8].

5.2 Basic Programming Concepts for Beginners

Programming is the process of creating a set of instructions that tell a computer how to perform a task. In today's technology-driven world, programming is no longer a niche skill reserved for computer scientists—it's increasingly becoming a foundational literacy. Whether building websites, analyzing data, or automating tasks, programming offers powerful tools to solve problems and express ideas. For beginners, understanding basic programming concepts is the first step in unlocking this potential. These instructions, often called code, are written in programming languages such as Python, Java, or C++. For beginners, understanding the foundational concepts of programming is crucial for learning more advanced topics. This section introduces the core principles that form the building blocks of any programming language, ensuring that learners gain a strong grasp of programming logic and structure [9].

At the heart of programming is the concept of instructions or commands that tell a computer what to do. These instructions are written in programming languages like Python, JavaScript, or Scratch, which serve as a bridge between human thinking and machine logic. Learning to program begins with grasping a few core concepts that apply across most languages [10].

1. One of the most essential concepts is the variable. A variable is like a container that stores information, such as numbers, text, or dates. For example, if you want to remember a person's name, you might create a variable called name and assign it a value like "Alice". Variables help programmers manage and reuse information throughout a program.
2. Another key concept is the data type. Data comes in different forms—numbers (integers or decimals), text (strings), true/false values (booleans), and collections (lists or arrays). Understanding how to work with these data types is crucial, as each type behaves differently and serves specific purposes.
3. Conditionals, or decision-making statements, allow a program to respond to different situations. Using commands like if, else, or elif (in Python), you can create logic such as: "*If the user is logged in, show the homepage. Otherwise, show the login screen.*" This adds flexibility and intelligence to a program's behavior.
4. Loops are another foundational concept. They enable repetition of instructions, saving time and effort. For instance, if you wanted to print a greeting for 100 users, a loop would help you do that with just a few lines of code. Common types

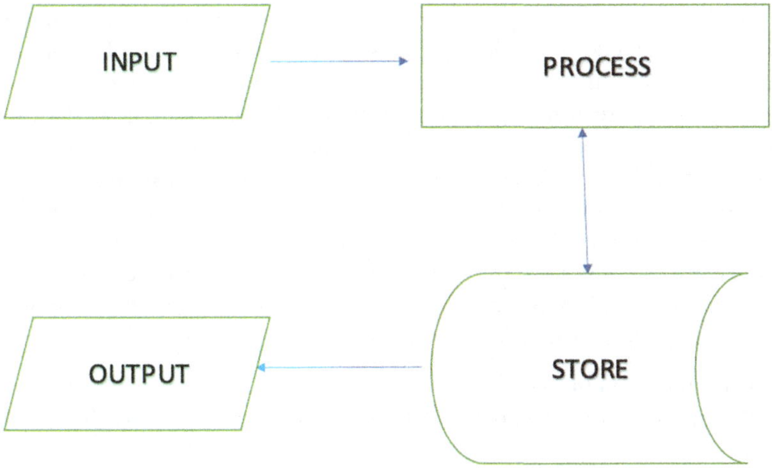

Fig. 5.2 Basic programming concept

include for loops (used when the number of repetitions is known) and while loops (used when a condition must be met to stop repeating).

5. Functions help organize code into reusable blocks. Instead of repeating the same code in different places, you can define a function once and call it whenever needed. This makes programs easier to read, debug, and maintain. For example, a function called calculateTax() could be used in different parts of a shopping app to apply the same tax logic consistently.
6. Finally, debugging is the process of identifying and fixing errors in code. Every programmer, no matter how experienced, writes code that doesn't work the first time. Learning to read error messages, test small sections of code, and think logically are important skills for effective debugging.

Therefore, learning to program starts with understanding a few basic concepts: variables, data types, conditionals, loops, functions, and debugging. With these tools, beginners can begin to build simple yet powerful programs. As confidence and experience grow, these foundations support more complex projects and real-world applications. Programming encourages creativity, problem-solving, and critical thinking skills that are valuable in every field (Fig. 5.2).

5.2.1 Variables and Data Types

In the world of programming, two of the most fundamental building blocks are variables and data types. They are essential concepts that enable programmers to store, manage, and manipulate information in a structured and logical way. Understanding how they work is crucial for anyone beginning their coding journey.

A variable is a named storage location in a computer's memory that holds a value which can be changed during the execution of a program. Think of a variable as a labeled box where you can store something like a number or a piece of text. You can open the box to read the value or replace it with something new. For example, in Python, you can create a variable by writing age = 25. Here, age is the variable, and it holds the value 25. Variables are useful because they make programs flexible and readable. Instead of writing a value multiple times, you can use a variable. For instance, if a program calculates the price of items with tax, storing the tax rate in a variable means you only need to change it in one place if it updates later.

Every variable in programming has a data type, which tells the computer what kind of information the variable holds. Common data types include:

1. Integers (int)—Whole numbers, like 10, −3, or 0.
2. Floating-point numbers (float)—Numbers with decimal points, like 3.14 or −0.75.
3. Strings (str)—Sequences of characters, such as "Hello", "2025", or "Name".
4. Booleans (bool)—True or False values, used for logical decisions.
5. Lists/Arrays—Collections of items stored in a single variable, like [1–3] or ["apple", "banana", "cherry"].

Each data type has its own set of operations. You can add two integers together, but you can't add an integer and a string without converting one of them first. This is why it's important to know the type of data you're working with. In dynamically typed languages like Python, the type is assigned automatically. In statically typed languages like Java or C++, you must declare the data type when defining the variable (e.g., int age = 25;).

Variables and data types also help prevent errors and improve performance. For example, using the correct type ensures that a program runs efficiently and avoids issues like trying to divide a string by a number. Additionally, some data types take up more memory than others, so choosing wisely can affect how well a program performs, especially with large datasets.

Thus, variables and data types are essential for writing clear, functional, and efficient programs. They allow programmers to store and manage data effectively while maintaining logical structure and clarity in their code. Mastering these concepts lays a strong foundation for learning more advanced programming topics, such as conditionals, loops, and functions.

5.2.2 Operators

Operators are unique characters or symbols that carry out operations on operands, which can be variables or values. Because they tell the computer how to work with data, they are cornerstones of mathematics and computer science. Arithmetic operators, used for simple computations, and logical and bitwise operators, used for more difficult tasks in advanced programming, are examples of more sophisticated operators. The number of operands needed and the sort of operation they execute are the two main criteria for classifying operators. Programmers may do a wider range of computational tasks with clarity and efficiency when they have a firm grasp of these areas. Using operators effectively is key to performing calculations and building logic within a program [11] (Table 5.1).

Table 5.1 Categories of operators

Category	Description	Examples	Usage scenario
Arithmetic operators	Perform basic mathematical calculations like addition, subtraction, multiplication, division, and modulus	+, −, *, /, %	Calculating sums, differences, averages, remainders
Relational (comparison) operators	Compare two values or expressions and return a Boolean result (true/false)	==, !=, >, <, >=, <=	Checking conditions, control flow (if statements)
Logical operators	Combine multiple Boolean expressions and return a Boolean result	&& (AND), || (OR), ! (NOT)	Complex condition checking, decision-making
Assignment operators	Assign values to variables, sometimes combining with arithmetic or bitwise operations	=, +=, -=, *=, /=	Updating variable values during computations
Bitwise operators	Perform operations at the bit level on integer types	&, |, ^, ~, <<, >>	Low-level programming, performance optimization
Unary operators	Operate on a single operand, such as negation or increment/decrement	−, ++, −, !	Negating values, incrementing counters
Ternary operator	A shorthand conditional operator that takes three operands	condition ? expr1: expr2	Compact if-else statement
Other operators	Includes special operators like sizeof, typecasting, or pointer operators in certain languages	sizeof(), (type), * (pointer dereference)	Language-specific operations

5.2.3 Control Structures and Functions and Modularity

Control structures are fundamental elements in programming that determine the flow of execution in a program. They allow the program to make decisions, repeat tasks, and execute instructions conditionally. Without control structures, programs would simply run line-by-line, with no flexibility or adaptability to changing inputs or conditions.

There are three main types of control structures: sequential, selection (conditional), and iteration (looping).

1. **Sequential structure** is the default flow of a program, where instructions are executed one after another. This is the simplest type of control flow and forms the backbone of all programs.
2. **Selection structures** allow a program to choose between different paths of execution based on a condition. The most common selection structure is the **if statement**. For example:

```
if temperature > 30:
print("It's a hot day")
else:
print("The weather is moderate")
```

Also, the program checks whether the condition (`temperature > 30`) is true. If it is, it executes the first block; otherwise, it runs the second. More complex decisions can be handled using `if...elseif...else` statements.

3. **Iteration structures**, or loops, are used to repeat a block of code multiple times. The two most common loops are the **for loop** and the **while loop**. A `for` loop is used when the number of repetitions is known in advance, such as:

```
for i in range(5):
print("Hello")
```

A `while` loop runs as long as a condition is true:

```
while count < 5:
print(count)
count += 1
```

Control structures are essential for handling dynamic data and enabling user interaction. For instance, they allow a program to respond to user input, validate data, or perform repetitive calculations without manual duplication of code. In more advanced applications, **nested control structures**—where one structure is placed inside another—are used to handle complex logic. For example, a loop might contain an `if` statement to decide what action to take on each iteration. Understanding and using control structures properly is key to writing efficient, logical, and flexible programs. They empower programmers to create software that reacts, adapts, and automates, forming the foundation of intelligent systems.

Functions and Modularity

Functions are reusable blocks of code designed to perform a specific task. They are a key part of modular programming, which breaks down complex problems into smaller, manageable pieces. This modular approach makes code more organized, easier to understand, and simpler to debug or extend. A function typically has a **name**, a set of **parameters**, and a **body** that contains the instructions. It may also return a value. Here's a basic example in Python:

```
def greet(name):
return "Hello, " + name
```

Calling `greet("Alice")` returns `"Hello, Alice"`.

Functions are essential for **code reusability**. Instead of writing the same code multiple times, a function can be written once and used as many times as needed. This minimizes redundancy and ensures consistency. If the logic changes, it only needs to be updated in one place. Functions also improve **readability**. When code is divided into logical parts, each part is easier to understand. Instead of reading 100 lines of continuous code, a reader can focus on understanding a few well-named functions like `calculate_tax()`, `display_results()`, or `validate_input()`.

Modularity extends this principle beyond individual functions. A modular program is structured as a collection of independent units or modules. In large-scale applications, each module can represent a component, such as a user interface, data processing, or database access. These modules can be developed, tested, and maintained separately, making collaboration easier in team environments. Most programming languages include **standard libraries** with built-in functions to handle common tasks. Programmers can also create their **custom functions** tailored to specific problems. Functions can be further categorized into **pure functions** (no side effects, output depends only on input) and **impure functions** (interact with external states like files or databases). In object-oriented programming, functions are often referred to as **methods**, and they are part of objects. Regardless of the paradigm, the principle remains: breaking tasks into functions and modules leads to cleaner, more efficient, and scalable code.

In summary, functions and modularity are cornerstones of effective programming. They promote clarity, reduce duplication, facilitate maintenance, and support collaboration qualities that are essential in both beginner projects and enterprise-level software development.

5.2.4 Debugging and Error Handling

In programming, mistakes are inevitable. Even experienced developers frequently encounter bugs, errors or flaws that prevent a program from running correctly. Debugging and error handling are essential skills that help programmers identify, understand, and resolve these issues. Mastering them not only improves the reliability of software but also builds confidence and problem-solving ability in new programmers. Debugging is the process of locating and fixing errors in code. Bugs can take many forms, including syntax errors, logical errors, and runtime errors.

1. **Syntax errors** occur when the code violates the rules of the programming language, such as forgetting a colon in Python or a semicolon in Java. These are usually detected by the compiler or interpreter and are often easy to fix.
2. **Logical errors**, however, are subtler. These occur when the program runs without crashing but produces incorrect results. For example, using the wrong formula in a tax calculator may not cause an error message, but it will give wrong outputs. These bugs require careful analysis of the logic behind the code and often involve tracing the flow of execution step-by-step.
3. **Runtime errors** occur while the program is running and can be caused by things like dividing by zero, accessing an out-of-range list index, or trying to open a non-existent file. These errors may not be obvious during development but can cause the program to crash unexpectedly.

To debug effectively, programmers use various techniques and tools. One basic method is printing values at key points in the program using statements like `print()` in Python or `console.log()` in JavaScript. This helps trace variable values and identify where things go wrong. More advanced tools include debuggers integrated into code editors, which allow programmers to set breakpoints, step through code line-by-line, and inspect variable states during execution. In addition to debugging, programmers must also implement error handling to make their programs more robust and user-friendly. Rather than allowing a program to crash, error handling anticipates and manages potential problems gracefully. This is done using constructs like `try...except` in Python, `try...catch` in JavaScript, or `try...catch...finally` in Java.

For example:

```
try:
number = int(input("Enter a number: "))
result = 10 / number
print("Result:", result)
except ZeroDivisionError:
print("Error: Cannot divide by zero.")
except ValueError:
print("Error: Invalid input. Please enter a number.")
```

Also, the program anticipates two common issues: dividing by zero and entering non-numeric input. Instead of crashing, it provides clear messages to help the user correct their input. Good error handling also includes logging errors for future analysis and writing clear, descriptive error messages that help users and developers understand what went wrong. Thus, debugging and error handling are critical aspects of software development. Debugging ensures that code behaves as expected, while error handling prepares programs to deal with unexpected situations gracefully. Together, they improve code quality, user experience, and developer productivity. For beginners, learning to debug systematically and implement basic error handling is a vital step toward becoming a skilled and confident programmer.

5.2.5 Teaching Python and R in Humanities

In recent years, the integration of computational tools in the humanities has become increasingly vital. As the digital transformation of education continues, the integration of programming into the humanities is no longer a futuristic ideal but a current necessity. Languages like Python and R, traditionally associated with data science and computational fields, are increasingly being taught to students and scholars in disciplines such as history, literature, philosophy, and cultural studies. This trend reflects a growing recognition that digital tools can enhance research, broaden access to information, and open new analytical frontiers in the humanities.

Python and R are particularly well-suited for humanities education due to their accessibility, versatility, and supportive communities. Python, with its clean syntax and readable structure, is ideal for beginners. It supports a wide range of applications, including text analysis, web scraping, natural language processing (NLP), and data visualization—tasks that are becoming more central in digital humanities. R, on the other hand, is a powerful tool for statistical analysis and data visualization, making it a great fit for historians, sociologists, and researchers working with quantitative or mixed-methods data.

Teaching these programming languages in humanities courses offers several pedagogical benefits. First, it helps develop critical digital literacy. Students learn to think algorithmically, assess the reliability of digital tools, and understand the mechanics behind the technologies that shape contemporary society. They become empowered not just as users of software but as informed creators who can question and manipulate the tools they use. Second, programming encourages interdisciplinary learning. When students in literature use Python to analyze patterns in Shakespeare's plays, or when art historians use R to examine trends in exhibition data, they blend traditional inquiry with computational methods. This integration supports innovation, allowing scholars to ask new kinds of questions and generate novel insights.

Practical projects can include analyzing sentiment trends in 19th-century novels, visualizing citation networks in philosophical journals, or mapping migration patterns using historical census data. Tools like Jupyter Notebooks (for Python) and RStudio provide user-friendly interfaces where students can write code, annotate findings, and display results all in one document. These platforms support reproducible research and make complex analysis transparent and collaborative. However, teaching Python and R in the humanities does come with challenges. Instructors must balance technical instruction with domain-specific analysis, and students may feel intimidated by the perceived difficulty of coding. To address this, humanities educators are encouraged to use project-based, incremental learning strategies. This includes starting with small, relevant tasks (e.g., counting word frequency in a poem) and building toward more complex analyses.

Ultimately, the goal is not to turn humanities students into full-time programmers, but to equip them with foundational coding skills that enhance their research and adaptability. As data becomes more central to all disciplines, the ability to interpret, visualize, and manipulate information is crucial. Hence, teaching Python and

R in the humanities empowers students to participate fully in the digital age. It cultivates analytical, creative, and technical skills while enriching traditional humanistic inquiry. By embracing these tools, the humanities can evolve not by abandoning their roots, but by expanding their methods and reach.

5.2.6 Rationale for Teaching Programming in the Humanities

The integration of programming into humanities education may seem unconventional at first, given the field's historical focus on critical analysis, interpretation, and qualitative inquiry. However, as the world becomes increasingly digitized, there is a compelling and growing rationale for equipping humanities students and scholars with basic programming skills. Programming is not merely a technical activity; it is a mode of thinking, creating, and understanding that can significantly enrich humanistic inquiry. One of the strongest arguments for teaching programming in the humanities is the rise of digital research methods. Scholars today encounter vast amounts of digital text, images, audio, and data. Programming enables them to engage with these materials systematically. For instance, text mining techniques can uncover patterns, trends, and connections across hundreds of novels or historical documents that would be impossible to detect manually. Languages like Python and R provide tools for analyzing literary themes, mapping historical events, or visualizing cultural trends over time.

Programming also supports the creation of interactive and public-facing digital projects. Humanities scholars increasingly use web-based platforms, digital archives, virtual exhibits, and mapping tools to present their work to wider audiences. By learning to code, they gain the ability to build and customize these platforms themselves. This reduces reliance on external developers and enhances academic autonomy, creativity, and control over the presentation of research. Moreover, programming cultivates computational thinking, a valuable cognitive skill that complements traditional critical thinking. It encourages logical structuring of arguments, clarity of expression, attention to detail, and problem-solving—all of which are central to both programming and humanities scholarship. Writing code requires precision and iterative improvement, much like constructing a well-argued essay or interpreting a complex text.

From a pedagogical perspective, teaching programming in the humanities also helps foster interdisciplinary learning. It bridges the divide between the sciences and the arts, encouraging students to apply technical tools to humanistic problems and vice versa. This not only makes humanities students more adaptable and versatile but also prepares them for a wide range of careers in media, research, education, publishing, cultural heritage, and the tech industry. Additionally, in an era dominated by algorithms, automation, and artificial intelligence, understanding how these systems work is essential for ethical reflection and social critique, roles traditionally

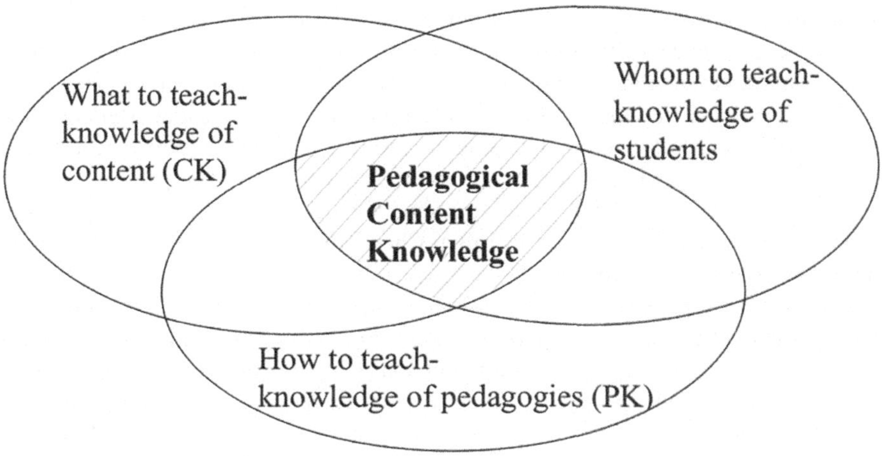

Fig. 5.3 Rationale for teaching programming in the humanities

championed by the humanities. Teaching programming empowers students to critically examine the digital infrastructures that shape our lives, from search engines and recommendation systems to surveillance technologies and data privacy concerns.

Finally, learning programming promotes digital literacy and lifelong learning. It helps humanities students become not just consumers of digital tools, but informed creators and critics. As academic institutions place increasing emphasis on employability and innovation, programming provides students with practical skills that are both intellectually enriching and professionally valuable. Therefore, the rationale for teaching programming in the humanities is both practical and philosophical. It enhances scholarly research, empowers creative expression, bridges disciplines, and equips students to engage critically with the digital world. Rather than replacing the core values of the humanities, programming expands their scope and relevance in the twenty-first century (Fig. 5.3).

5.2.7 Pedagogical Approaches

Teaching Python and R in the humanities requires thoughtful and inclusive pedagogy that balances technical skills with interpretive insight. Unlike in computer science, where programming is often taught through abstract logic and algorithm design, effective instruction for humanities learners must be grounded in meaningful contexts and narrative-based learning. To make coding both approachable and relevant, educators must blend conceptual understanding with hands-on application.

One of the most effective methods is PBL. PBL encourages students to apply coding skills to real-world humanities problems, such as analyzing patterns in political speeches, visualizing historical migration data, or constructing digital literary maps. Projects not only offer tangible outcomes but also reinforce a sense of purpose and ownership. By linking code to familiar themes—identity, culture, history—students see the value of programming beyond its technical scope.

Flipped classrooms are another useful approach. In this model, students first engage with tutorials, videos, and readings outside of class, which prepares them with the foundational knowledge needed to tackle problems during in-person sessions. Class time is then used for collaborative coding, debugging, and creative experimentation. This setup allows for active learning, where students benefit from peer support and real-time instructor feedback.

Scaffolded assignments are also essential. Starting with basic tasks—like variables, string manipulation, and loops—students gradually progress to more complex challenges such as data cleaning, sentiment analysis, or topic modeling. Each stage builds confidence and competence, ensuring that learners are not overwhelmed by complexity too soon. Scaffolded learning also allows instructors to differentiate instruction and provide targeted support.

A critical component across all these methods is to emphasize narrative over syntax. Instead of focusing solely on technical correctness, educators should frame coding within the pursuit of humanistic inquiry. For example, rather than teaching a loop as a mechanical concept, it can be introduced as a way to scan and interpret recurring themes in a corpus of literature. This approach demystifies programming and connects it to the critical and creative practices students already value.

By fostering a supportive environment and making code personally and intellectually meaningful, instructors can reduce student anxiety and increase motivation. This is key to making Python and R accessible tools in the humanist's toolkit.

Comparing Python and R for Humanities Research

When selecting a programming language for humanities research, Python and R stand out as the most popular and accessible choices. Each offers distinct advantages, and the choice between them often depends on the research question, the data type, and the methodological approach.

Python is a versatile, general-purpose language with an intuitive syntax that makes it ideal for beginners. Its popularity in digital humanities stems largely from its powerful NLP capabilities. Libraries such as NLTK, spaCy, and TextBlob allow researchers to tokenise texts, analyze sentiment, extract named entities, and perform part-of-speech tagging across large corpora. Python also excels in web scraping (e.g., using BeautifulSoup), allowing scholars to gather contemporary and historical data from online sources such as archives, newspapers, or social media platforms. Because of its readability and widespread use, Python is often the first language taught in humanities coding courses.

R, on the other hand, is particularly well-suited for statistical analysis and data visualization. Its strength lies in packages like ggplot2, tidyverse, and dplyr, which make complex data manipulation tasks efficient and elegant. Humanities scholars who work with structured datasets, such as census records, survey data, or bibliometric information, will find R's statistical rigor and graphic capabilities extremely useful. For visualizing trends, distributions, or networks, R provides precise control

Table 5.2 Comparing Python and R

Feature	Python	R
Syntax readability	Beginner-friendly	Slightly steeper learning curve
Data visualization	Matplotlib, Seaborn	ggplot2, plotly
NLP libraries	NLTK, spaCy	quanteda, tm
Community resources	Extensive (especially for NLP tasks)	Extensive (especially for statistics)
IDEs	Jupyter, VS code	RStudio

and high-quality output, which is valuable for both academic publication and public-facing projects.

While Python supports a broader range of programming tasks, R remains unmatched in domains that require detailed statistical modelling. That said, many researchers adopt a hybrid approach, using Python for data collection and text processing, and R for statistical analysis and visualization.

Ultimately, both languages are open-source, well-documented, and supported by vibrant communities, which makes them excellent choices for humanities scholars. The decision should be guided by project goals: choose Python for textual analysis and versatility, and R for data-heavy research and visualization. Equipping students with both languages—or at least an understanding of when to use each maximises their analytical capabilities and prepares them for interdisciplinary collaboration in digital humanities research (Table 5.2).

5.2.8 Tools, Platforms and Assessment Strategies

As programming becomes increasingly integrated into humanities education, selecting the right tools, platforms, and assessment strategies is crucial to ensure effective teaching and meaningful learning outcomes. The goal is not only to teach coding syntax but also to foster analytical thinking, creativity, and digital literacy within humanistic contexts.

Tools and Platforms

A wide range of digital tools can support the teaching of programming to humanities students. These tools must be accessible, user-friendly, and aligned with pedagogical goals.

1. For Python, platforms like Jupyter Notebooks offer an ideal environment for combining code, text, and visualizations in a single document. This format is especially well-suited to humanities disciplines, where narrative and analysis are intertwined. Jupyter encourages documentation alongside coding, supporting reflective and transparent scholarship.

2. For R, RStudio provides a powerful interface tailored for statistical computing and visualization. Its integration with packages like ggplot2 and tidyverse makes it especially valuable for researchers working with historical data, surveys, or bibliometric analysis.
3. Other platforms like Google Colab (for Python) allow cloud-based coding without the need for software installation, making them ideal for classrooms with varying access to technology. Tools such as Twine (for interactive storytelling), Voyant Tools (for textual analysis), and Hypothesis (for collaborative annotation) offer creative ways to apply code and digital methods in humanistic inquiry.
4. For collaboration and version control, GitHub introduces students to real-world practices of open-source development and project documentation. Using GitHub in classroom settings teaches the value of reproducibility and teamwork—skills increasingly valued across disciplines.

Assessment Strategies

Assessment in coding for the humanities should reflect the blended nature of technical skill and critical thinking. Traditional testing on syntax may have limited value; instead, formative and summative assessments should emphasize problem-solving, creativity, and the application of programming to humanities-related challenges.

1. One effective strategy is project-based assessment. Students might build a digital timeline, conduct sentiment analysis on historical speeches, or create visualisations of literary themes. These projects demonstrate not only programming ability but also interpretive insight and research design.
2. Portfolio-based assessment is also valuable. A digital portfolio allows students to collect and reflect on their work over time, showcasing their progress and versatility. Portfolios can include code samples, screenshots of outputs, written reflections, and commentary on the research process.
3. Peer review and collaborative assessment promote critical engagement and communication skills. Students can review one another's code and project ideas, offering constructive feedback and learning from different approaches.
4. Scaffolded assignments that grow in complexity—from simple tasks like text cleaning to more advanced applications like topic modeling or geospatial mapping—help instructors track skill development while reducing student anxiety.

Finally, rubrics for assessment should evaluate both technical execution (e.g., code functionality, clarity, reusability) and humanistic interpretation (e.g., relevance of data, critical insight, narrative integration). This dual-focus approach ensures that programming is not treated as an isolated skill but as an integrated method of scholarly inquiry.

5.3 Web Technologies

In the digital age, understanding how websites and web applications function is a valuable skill—not just for developers, but also for educators, researchers, and students across disciplines. At the heart of the modern web are three foundational technologies: HTML (HyperText Markup Language), CSS (Cascading Style Sheets), and JavaScript (JS). Together, they form the backbone of front-end web development, enabling the creation of dynamic, interactive, and visually engaging digital experiences.

HTML—Structure

HTML is the fundamental building block of the web. It provides the structure and content of a webpage using a system of tags. These tags define elements such as headings, paragraphs, images, links, and lists. For example, <h1> defines a main heading, while <p> creates a paragraph.

In the context of the humanities, HTML can be used to create digital exhibits, publish scholarly articles online, or build annotated editions of texts. Because HTML content is semantic, it also improves accessibility and searchability, allowing digital humanists to make their work more discoverable and inclusive.

CSS—Styling

CSS is responsible for the appearance and layout of HTML elements. It controls fonts, colors, spacing, alignment, and responsiveness. By separating content (HTML) from presentation (CSS), developers can maintain clean and maintainable code. For instance, CSS allows a digital archive to switch from a dark theme to a light theme, adjust text size for accessibility, or format footnotes in a scholarly article. Responsive CSS frameworks like Bootstrap or Tailwind CSS make it easier to create mobile-friendly designs, which is essential for reaching broader audiences.

JavaScript—Interactivity

JavaScript is a powerful scripting language that enables interactivity and dynamic functionality. With JavaScript, webpages can respond to user actions in real time—like displaying a popup message, validating form input, or filtering search results without reloading the page. JavaScript is particularly useful for creating interactive timelines, quizzes, text visualizations, and data dashboards—tools that are increasingly common in digital humanities and educational projects. Libraries like D3.js allow for the creation of rich, interactive data visualizations, while frameworks such as Vue.js or React power more complex applications.

Educational Value and Humanities Applications

Learning HTML, CSS, and JavaScript introduces students to computational thinking and digital creativity. These languages are beginner-friendly and provide immediate visual feedback, which makes them excellent entry points into programming. For humanities students, web technologies offer tools to present research in engaging, public-facing formats, such as online archives, digital storytelling projects, or interactive maps. Additionally, knowledge of web technologies enhances digital literacy and opens opportunities for collaboration with technologists in interdisciplinary projects. Whether creating a personal portfolio, contributing to an open-access journal, or designing a digital edition of a historical manuscript, web skills empower scholars to control both content and medium.

Hence, HTML, CSS, and JavaScript are essential technologies for building the modern web. For humanities scholars and educators, they offer the means to create interactive, accessible, and visually compelling content. Mastering these tools not only supports digital scholarship but also fosters a deeper understanding of how digital culture is constructed and experienced.

5.3.1 Pedagogical Rationale and Curriculum Integration

As digital technologies continue to reshape how knowledge is produced, shared, and preserved, integrating programming and digital literacy into the humanities curriculum has become an urgent educational priority. The pedagogical rationale behind this integration rests on the need to equip students with both the critical thinking and technical competencies required to engage meaningfully with the digital world. Far from diminishing the value of traditional humanistic inquiry, such integration enhances it—allowing students to explore texts, cultures, and histories in new, data-rich, and interactive ways.

One of the core principles driving this shift is the idea of authentic, project-based learning. Programming encourages students to become active participants in the creation and analysis of knowledge, rather than passive consumers. Whether building a digital archive, visualizing migration patterns, or creating an interactive timeline, students learn by doing—developing technical skills alongside research and interpretive skills. This constructivist approach aligns with well-established pedagogical models, which emphasize learner autonomy, experimentation, and reflection.

Moreover, incorporating programming into the humanities helps bridge the gap between theory and practice. Concepts such as narrative structure, historical causality, and cultural representation can be explored not only through section and discussions but also through code, data visualizations, and digital tools. For instance, using Python to perform sentiment analysis on Victorian novels, or using R to examine trends in political speeches, allows students to test hypotheses and make arguments based on empirical patterns—deepening their understanding of both the content and the methods.

From a curriculum design perspective, integration can take several forms. One approach is the development of stand-alone digital humanities courses, which introduce programming languages (such as Python, R, HTML, or JavaScript) in the context of literary analysis, history, or cultural studies. Another is the embedding of coding modules into existing courses, such as teaching basic text mining in a literature course or digital mapping in a history seminar. This model ensures that technical instruction is always anchored in disciplinary relevance.

Collaborative, interdisciplinary teaching is also key. Humanities faculty can work alongside computer science or digital media instructors to co-design and co-teach courses. This not only enriches the learning experience but also models the kind of interdisciplinary collaboration that is increasingly valued in academic and professional settings.

Assessment strategies should reflect the dual focus on technical proficiency and critical engagement. Students can be evaluated through digital projects, annotated code notebooks, a reflective section, and group presentations. Rubrics can assess not just whether the code works, but how effectively it supports inquiry, interpretation, and communication.

In summary, the integration of programming into the humanities is pedagogically sound and future-focused. It cultivates critical, creative, and computational thinkers who are prepared to navigate and contribute to a rapidly changing digital world. By embedding digital skills into the heart of humanistic education, educators foster a new kind of literacy—one that combines the analytical depth of the humanities with the innovative tools of the digital age.

5.3.2 Tools and Platforms for Instruction

Effectively teaching web technologies such as HTML, CSS, and JavaScript in humanities or interdisciplinary settings requires not only sound pedagogy but also accessible and engaging tools. A variety of platforms have been developed to lower the barriers to entry, enhance collaboration, and support real-time feedback and publishing, critical components for building student confidence and creativity.

CodePen and JSFiddle are two popular online code editors that allow students to write and preview HTML, CSS, and JavaScript code directly in the browser. These platforms provide a low-pressure, installation-free environment where learners can experiment with ideas and instantly see the results of their changes. For educators, they are excellent tools for sharing live examples, offering feedback, or hosting coding challenges in class.

Visual Studio Code (VS Code) is a more advanced Integrated Development Environment (IDE). It is widely used in professional web development but is also suitable for students due to its user-friendly interface, smart code suggestions, and extensive library of extensions. VS Code fosters best practices such as folder structuring, version control with Git, and modular development. For students interested in taking their projects beyond the browser, VS Code is an essential next step.

To make student work accessible and shareable, GitHub Pages is a valuable tool for hosting live projects online. GitHub's integration with Git also introduces students to version control, collaborative development, and open-source culture—key components of real-world digital practice.

Finally, the Mozilla Developer Network (MDN) is an invaluable resource. It offers comprehensive, well-organised tutorials, documentation, and references for HTML, CSS, JavaScript, and web standards. MDN empowers students to learn independently and supports instructors in crafting accurate and up-to-date lesson materials. Together, these tools create an ecosystem where students can build, share, and iterate on digital projects while developing technical fluency and critical digital literacy.

5.3.3 HTML: Structuring Humanities Content

HTML is the foundational language of the web and plays a crucial role in organizing and presenting content online. In humanities contexts, HTML offers a powerful, flexible way to structure and annotate texts, create digital editions, and present research in accessible, user-friendly formats.

At its core, HTML defines the semantic structure of digital content. Tags such as <header>, <article>, <section>, and <footer> give meaning to different parts of a webpage, allowing both humans and machines (such as search engines or screen readers) to understand how the content is organized. This semantic clarity is especially important in humanities disciplines, where text structure and meaning are deeply intertwined. For example, a digital edition of a historical manuscript can use <blockquote> for cited text, <em> for editorial emphasis, or <abbr> to clarify archaic abbreviations. With HTML, scholars can annotate texts, embed metadata, and link primary sources—all while preserving the visual and intellectual integrity of the original material.

HTML also promotes accessibility and inclusivity. When used correctly (e.g., with alt text for images and proper heading levels), it ensures that digital content is navigable by screen readers and assistive technologies, aligning with universal design principles and the ethical imperatives of humanities scholarship. Furthermore, HTML empowers students and researchers to publish and share their work online without relying on proprietary platforms. A student creating a digital exhibit or annotated literary timeline can use HTML to build a custom interface that reflects their scholarly interpretation and creative vision.

In summary, teaching HTML within the humanities equips learners with the tools to structure and communicate knowledge in digital environments. It bridges traditional scholarly values—rigour, interpretation, and clarity—with the digital fluency needed for 21st-century research and outreach (Table 5.3).

5.3.4 Challenges and Solutions in Teaching Web Tech

As web technologies such as HTML, CSS, and JavaScript become increasingly integrated into education—particularly within digital humanities and interdisciplinary programs—educators encounter a range of challenges. These obstacles stem from

Table 5.3 Structuring humanities content

Element	Description	Example use in humanities
	Represents self-contained content	Digital section, blog entries
	Groups thematic content	Chapters in an online book
/	Embeds images with captions	Annotated historical images
	Quotation from another source	Quoting primary sources

both pedagogical and technical domains, as many students (and instructors) come from non-technical backgrounds. However, with thoughtful planning and appropriate tools, these challenges can be effectively addressed.

Challenge 1: Varying Technical Backgrounds: One of the most common issues is the diverse range of prior experience among students. Some may be encountering code for the first time, while others might already be proficient in web development. This disparity can create frustration on both ends of the spectrum.

Solution: Implement scaffolded learning strategies. Start with basic HTML and simple styling tasks, and gradually introduce more complex concepts like responsive design or DOM manipulation. Differentiated assignments or optional extension tasks can keep advanced learners engaged while supporting beginners. Collaborative work also allows peer learning to flourish.

Challenge 2: Intimidation and Lack of Confidence: Students from humanities backgrounds may feel intimidated by technical tasks, viewing programming as a skill meant only for computer scientists. This psychological barrier can hinder learning before it even begins.

Solution: Emphasize conceptual relevance and creativity over technical mastery in the early stages. Use relatable, low-stakes projects like creating a personal webpage, building an online timeline for a historical event, or designing an interactive poetry gallery. Providing plenty of encouragement, using inclusive language, and highlighting successful examples from non-technical students can boost confidence.

Challenge 3: Software Installation and Setup Issues: Installing software like VS Code, Git, or Node.js can present obstacles—particularly on institutional computers or for students using different operating systems.

Solution: Begin with browser-based platforms like CodePen, JSFiddle, or Glitch, which require no installation. These environments are ideal for teaching HTML, CSS, and JavaScript basics. Once students are comfortable, gradually transition to desktop tools like Visual Studio Code, guiding them through setup with visual aids or video tutorials.

Challenge 4: Assessing Technical and Creative Work: Evaluating web projects can be complex, as they involve both functional code and aesthetic or interpretive design

Table 5.4 Challenges and solutions in teaching web tech

Challenge	Solution
Students overwhelmed by code	Use visual builders and scaffolded tasks
Accessibility compliance	Teach semantic HTML and WCAG standards
Device/browser compatibility	Emphasize testing across platforms
Project scope creep	Encourage use of wireframes and project planning

choices. Traditional assessment methods may not fully capture student effort or creativity.

Solution: Use rubrics that assess both technical execution (e.g., code structure, use of semantic HTML, functionality) and creative/analytical elements (e.g., theme relevance, clarity of message, originality). Incorporate self-reflection components where students explain their design choices and problem-solving processes. Portfolios are especially effective for showcasing iterative growth over time.

Challenge 5: Keeping Up with Rapid Technological Change: Web technologies evolve rapidly, and educators may struggle to keep course materials current.

Solution: Focus on foundational principles—such as semantic structure, cascading styles, and interactivity—rather than fleeting tools or trends. Encourage students to consult up-to-date resources like the Mozilla Developer Network (MDN) and to develop habits of independent learning and problem-solving that will serve them as technologies continue to evolve.

Therefore, while teaching web technologies in non-technical disciplines comes with challenges, these can be overcome with inclusive pedagogy, accessible tools, and learner-centred design. By addressing fears, celebrating creativity, and building strong foundational knowledge, educators can empower students to become confident and competent digital creators (Table 5.4).

5.4 Computational Text Analysis

Computational text analysis refers to the systematic use of computational tools and methods to analyse large volumes of text data. It is an interdisciplinary field that combines linguistics, computer science, digital humanities, and data science to extract meaningful patterns, insights, and knowledge from textual sources. In the context of digital humanities, computational text analysis allows researchers to move beyond traditional close reading by employing distant reading techniques that reveal trends and structures in large corpora. Computational text analysis is the process of using algorithms and digital tools to examine large volumes of text systematically. Unlike traditional close reading, which focuses on detailed interpretation of a few selected passages, computational analysis allows researchers to explore patterns, themes, and relationships across entire corpora. This method is increasingly vital in the digital humanities, where scholars engage with everything from literary archives and historical documents to social media and news databases. At its core, computational text analysis seeks to answer humanistic questions through quantitative methods. For example, how has the portrayal of women changed in literature over the past two centuries? What emotional tone dominates political speeches during times of crisis? How do word choices reflect cultural or ideological shifts in newspapers? These are questions that can be approached by combining literary insight with digital tools such as NLP, machine learning, and data visualization [12].

The process typically begins with text pre-processing—the cleaning and formatting of raw text data. This may involve removing punctuation, converting text to lowercase, eliminating stop words (like "the" or "and"), and tokenizing (breaking text into individual words or phrases). These steps standardize the data and prepare it for meaningful analysis. One common method is frequency analysis, which identifies the most commonly used words or phrases in a text or set of texts. This can reveal dominant themes, keywords, or stylistic tendencies. More advanced techniques include sentiment analysis, which measures the emotional tone of text, and topic modeling, which groups words into clusters that represent hidden thematic structures. Tools like Voyant Tools, AntConc, and programming languages such as Python (using libraries like NLTK or spaCy) and R (using packages like quanteda or tidytext) support these types of analysis. These tools allow scholars to visualize data as word clouds, frequency graphs, correlation networks, and more, turning abstract textual data into concrete insights [13].

Importantly, computational text analysis does not replace traditional interpretive skills—it complements them. Scholars must still decide which texts to include, how to frame their research questions, and how to interpret the patterns revealed. The most meaningful computational analyses arise from a thoughtful integration of quantitative methods with qualitative insight. There are also ethical considerations to bear in mind. When working with user-generated content such as tweets or online reviews, issues of privacy, consent, and representation become critical. Researchers must ensure transparency and sensitivity in how data is collected, processed, and presented. In educational settings, computational text analysis offers students a powerful entry point into digital literacy, data-informed reasoning, and interdisciplinary exploration. Assignments might include analyzing themes in a novel, tracking the evolution of political language, or comparing media coverage across time periods. These projects encourage students to think critically about both content and method, preparing them for scholarly work in a digital world [14, 15].

Therefore, computational text analysis is a transformative tool for the humanities. It allows for large-scale exploration of text, supports new forms of inquiry, and bridges the gap between technology and interpretation. As digital texts multiply and access to data increases, computational methods will remain essential for making sense of our textual past and present (Table 5.5).

Table 5.5 Key techniques and methods

Technique	Description	Tools/Software
Word frequency analysis	Counts and analyses the frequency of words in a text	Voyant Tools, AntConc
Collocation analysis	Identifies words that frequently appear together	AntConc, NLTK
Sentiment analysis	Detects the emotional tone behind a body of text	Python (TextBlob), R
Topic modeling	Discovers abstract topics in a corpus	MALLET, Gensim
Named entity recognition	Identifies proper nouns like names, places, and organizations	SpaCy, Stanford NLP
Stylometry	Examines writing style to attribute authorship or detect patterns	R (stylo package)

5.4.1 Tools for Computational Text Analysis

Computational text analysis has revolutionized the way scholars engage with written language, allowing them to examine patterns, themes, and trends across large bodies of text. To support this evolving field, a wide array of digital tools and platforms have been developed—ranging from user-friendly interfaces for beginners to powerful libraries for advanced research. These tools play a central role in enabling researchers to extract meaning from text using computational methods, whether in literary studies, history, linguistics, media analysis, or cultural research.

1. Voyant Tools: Voyant Tools is one of the most accessible and widely used platforms in the digital humanities. It is a web-based application that allows users to upload text or a corpus of texts and explore word frequency, distribution, collocations, and topic patterns through interactive visualizations. Tools like word clouds, trend graphs, and keyword-in-context (KWIC) views make Voyant ideal for classroom settings and exploratory research. Its intuitive interface makes it a powerful entry point for humanities scholars with little to no programming experience.

2. AntConc: AntConc is a free desktop concordance tool designed for linguistic and textual analysis. Developed by Laurence Anthony, it allows users to perform detailed investigations into word frequency, collocations, concordance lines, and n-grams. Though less visually dynamic than Voyant, AntConc is particularly valuable for researchers working with corpora in historical linguistics, discourse analysis, or Stylometry. It supports a range of encoding formats and is commonly used in both research and teaching.

3. Python (NLTK, spaCy, TextBlob): For more advanced and customizable analyses, **Python** is a leading choice. Libraries such as:

1. **NLTK (Natural Language Toolkit)** – offers tools for tokenization, stemming, tagging, parsing, and semantic reasoning.

2. **spaCy** – known for its speed and efficiency, spaCy excels at tasks like named entity recognition, dependency parsing, and part-of-speech tagging.
3. **TextBlob** – ideal for beginners, providing a simplified interface for sentiment analysis, classification, and translation.

These libraries are open-source and supported by strong online communities, making them excellent resources for both students and researchers.

4. R (tidytext, quanteda): **R**, widely known for statistical computing, also supports rich textual analysis through packages like:

1. **tidytext** – integrates well with the tidyverse suite, enabling seamless text mining and sentiment analysis.
2. **quanteda** – designed specifically for managing and analyzing textual data, offering tools for document-feature matrix creation, topic modelling, and keyword extraction.

R is especially suited for projects that combine qualitative text with quantitative data analysis, such as comparing literary trends with historical census data.

5. Gephi and Cytoscape: For scholars interested in network analysis of textual data, such as citation networks or character interactions, Gephi and Cytoscape provide powerful visualization environments. These tools allow users to map relationships and structures that emerge from texts, adding new dimensions to narrative or bibliometric studies.

In summary, the landscape of computational text analysis is rich and diverse. From entry-level platforms like Voyant Tools to advanced programming environments in Python and R, these tools empower humanities scholars to explore texts in innovative, data-driven ways. The choice of tool depends on the research question, the size and type of corpus, and the technical proficiency of the user, but all serve the shared goal of deepening our understanding of language, culture, and society.

5.4.2 Applications in Humanities Research

Computational text analysis is transforming the landscape of humanities research by offering new methods for exploring large volumes of textual data with speed, scale, and precision. From literature and history to philosophy, linguistics, and cultural studies, this approach allows scholars to identify patterns, track changes over time, and generate fresh insights that would be difficult—if not impossible—to achieve through traditional close reading alone.

One of the most widespread applications of computational text analysis is in literary studies. Researchers use techniques such as word frequency analysis, sentiment analysis, and topic modeling to study entire collections of novels, plays, or poems. For example, scholars have used these methods to trace emotional arcs across 19th-century fiction, examine gendered language in Shakespeare's plays, or detect stylistic fingerprints that help attribute anonymous works to known authors. These approaches provide empirical support to interpretive arguments and open the door to distant reading—analyzing texts from a bird's-eye view rather than page by page. In historical research, computational analysis helps scholars explore large archives of newspapers, legal documents, letters, and speeches. Using tools like named entity recognition and keyword extraction, historians can identify how public discourse about topics such as race, war, or religion has evolved over time. For example, tracking the frequency and context of terms like "freedom" or "equality" in American political rhetoric can reveal how the meanings of these concepts have shifted across different eras and social movements.

Philosophical and religious texts also benefit from computational analysis. Scholars can use network analysis to map the relationships between ideas, figures, and concepts across centuries of writing. This can reveal how specific doctrines developed, how ideas travelled between authors, and how intellectual movements gained influence. Text similarity measures can also help trace citations and paraphrases in texts where formal attribution may be lacking. In linguistics and discourse studies, computational tools are used to analyze grammar, syntax, and usage across different dialects, time periods, and genres. For instance, corpus linguists can study how modal verbs (e.g., "must", "should", "might") are used in legal texts compared to everyday language. This kind of research has implications not only for language teaching and preservation but also for understanding how power and authority are expressed through language.

Another important area of application is digital cultural heritage. By processing and indexing large digital archives, museums and libraries can make historical texts more accessible to both scholars and the public. Computational analysis also enables the creation of searchable databases, interactive exhibits, and educational tools that allow broader audiences to engage with the humanities in dynamic ways. Finally, computational text analysis supports interdisciplinary research, linking the humanities with data science, computer science, sociology, and media studies. This cross-pollination fosters innovative methods and broader perspectives, ensuring that humanistic inquiry remains vital in a rapidly digitizing world.

In summary, computational text analysis expands the methodological toolkit of the humanities, enabling new forms of research, interpretation, and public engagement. It reinforces the relevance of humanistic inquiry in the digital age by blending critical reflection with computational power.

5.4.3 Challenges and Limitations

While computational text analysis has opened new frontiers in humanities research, it also presents a number of challenges and limitations that scholars must navigate carefully. These issues range from technical and methodological constraints to deeper epistemological and ethical concerns. Understanding these limitations is essential to applying digital tools responsibly and meaningfully within humanistic inquiry.

1. One of the most immediate challenges is access to clean, structured data. Many historical and literary texts are stored in scanned image formats or poorly digitized documents, requiring significant pre-processing through optical character recognition (OCR). OCR errors can introduce noise into the data, especially in older texts with unusual fonts or non-standard spelling. Inconsistent formatting, missing metadata, and language variation further complicate analysis. Without careful data cleaning and verification, results may be misleading or inaccurate.
2. Another key limitation is the reduction of meaning through quantification. Humanities research often values nuance, ambiguity, and context—qualities that are difficult to capture through computational methods. For example, sentiment analysis might label a passage as "positive" based on keywords, yet miss irony, sarcasm, or cultural subtext. Similarly, word frequency analysis may identify common terms but overlook their symbolic or rhetorical function. This risk of over-simplification underscores the importance of combining quantitative techniques with close reading and interpretive analysis.
3. Bias in datasets and algorithms poses a further concern. Many NLP tools are trained on contemporary or Western-centric corpora, which may not perform accurately on historical, multilingual, or culturally diverse texts. This can lead to skewed results or reinforce existing biases. In historical research, for example, texts from marginalized voices may be underrepresented or classified inaccurately, affecting the inclusivity and validity of findings. Scholars must therefore critically assess the sources of their tools and data, ensuring transparency and ethical use.
4. The technical barrier is another limitation. Many humanities scholars and students lack formal training in programming or statistical analysis, which can create obstacles to adopting computational methods. While there are user-friendly tools available, meaningful and flexible use of computational text analysis often requires some level of coding skill, as well as an understanding of underlying models and assumptions. This raises the need for interdisciplinary collaboration, targeted training, and institutional support.
5. Interpretive challenges also arise in determining what the data reveals. Patterns in text—such as recurring keywords or clusters—do not speak for themselves. Scholars must decide which patterns are significant, how to contextualize them, and how to avoid cherry-picking results to fit preconceived narratives. This calls for methodological rigor and an openness to complexity, contradiction, and uncertainty.
6. Finally, computational analysis may shift focus away from marginal or underrepresented texts simply because they are less likely to be digitized or included in large corpora. There is a risk that the availability of data, rather than scholarly interest or relevance, will shape research agendas—what some scholars call the "tyranny of the digital."

In summary, while computational text analysis offers powerful tools for humanities research, it must be used critically and ethically. Its limitations do not diminish its value, but they do require careful attention to data quality, methodological transparency, and the interpretive traditions of humanistic scholarship. When balanced thoughtfully, computational methods can enrich, not replace, the critical depth of the humanities.

5.5 Spatial Humanities

Spatial Humanities is an interdisciplinary field that combines traditional humanistic inquiry with digital mapping technologies to study how space, place, and geography influence human culture, history, and meaning. By integrating tools like GIS, historical cartography, and spatial data analysis, scholars in spatial humanities explore the geographic dimensions of texts, events, and cultural practices. This approach enables new forms of research, visualization, and interpretation that transcend the limitations of narrative and linear analysis.

At its core, spatial humanities seek to answer questions such as: *Where did historical events occur and why? How did physical geography shape cultural production? What can the spatial distribution of certain ideas or populations tell us about the past?* It builds on the foundational premise that space is not just a backdrop for human activity but an active component of historical and cultural processes.

One of the most prominent tools in this field is GIS, which allows researchers to layer data onto digital maps and examine spatial patterns and relationships. For example, historians can map the movement of armies during a war, migration routes over centuries, or the spread of a religious tradition across continents. Literary scholars might trace the geographical references in a novel or analyze how urban environments are represented in different literary genres. GIS not only helps visualize such phenomena but also facilitates quantitative spatial analysis, revealing correlations that are difficult to detect through traditional methods. Another key component of spatial humanities is spatial narrative—the use of maps and visual storytelling to represent historical or cultural developments over time and space. Projects in this area often involve interactive maps, story maps, and virtual reconstructions that invite users to explore historical contexts dynamically. For example, a project might reconstruct a medieval city and link buildings to archival records, allowing users to virtually navigate the space while learning about its social and political structures [16, 17].

Spatial humanities also engage with critical cartography, which questions the assumptions and power dynamics embedded in traditional map-making. Scholars analyze how colonial maps shaped worldviews, how borders reflect political ideologies, and how marginalized communities have been represented—or erased—in spatial records. In doing so, the field aligns with broader movements in postcolonial and digital humanities that seek to deconstruct dominant narratives and amplify diverse perspectives. Applications of spatial humanities span numerous disciplines. In archaeology, it supports site analysis and landscape archaeology. In history, it helps track epidemics, trade routes, or diasporas. In literature, it uncovers spatial themes and settings. Even philosophy and religious studies use spatial methods to examine sacred spaces and ritual geographies [18, 19].

5.5.1 Tools and Technologies Used for Spatial Humanities

Spatial humanities is a dynamic field that explores how space, place, and geography shape human history, culture, and identity. To conduct research in this area, scholars use a wide range of digital tools and technologies that allow them to analyze, map, and visualize spatial data. These tools enable researchers to move beyond traditional narrative forms and investigate relationships and patterns that are not immediately visible in texts or archives. The technologies used in spatial humanities vary in complexity and purpose, but all serve the shared goal of integrating geographic information with humanistic inquiry.

1. Geographic Information Systems (GIS)

At the heart of spatial humanities research is **Geographic Information Systems (GIS)**. GIS allows users to create, manage, analyze, and visualize spatial data. With GIS software, scholars can overlay historical maps, plot geographic references from texts, analyze spatial relationships, and model how places have changed over time. Popular GIS platforms include:

- **ArcGIS**: A professional-grade GIS tool widely used for mapping, spatial analysis, and visualization. It supports historical reconstructions, demographic studies, and the integration of archival data with modern maps.
- **QGIS (Quantum GIS)**: A powerful open-source alternative to ArcGIS, QGIS is accessible to scholars and students without institutional licenses. It supports plugins and customization, making it highly adaptable for historical and cultural research.

2. Story Mapping Tools

Story maps combine maps with narrative text, images, video, and interactive elements. These tools are ideal for communicating research to broader audiences. Scholars can create engaging, multimedia-rich presentations that guide users through historical journeys or thematic explorations. Common platforms include:

- **ArcGIS StoryMaps**: A tool within the ArcGIS ecosystem that integrates text and spatial data into immersive, interactive stories.
- **Knight Lab's StoryMapJS**: A user-friendly, open-source tool ideal for humanities projects that involve linear, location-based storytelling—such as tracking a character's journey in a novel or mapping a historical figure's travels.

3. Historical GIS and Mapping Platforms

Several platforms are specifically designed for **historical mapping** and cultural heritage projects. These include:

- **Recogito**: Developed by Pelagios, Recogito allows scholars to annotate and geotag historical texts, manuscripts, and maps. It supports collaborative work and connects spatial annotations with linked open data.
- **Pleiades**: A community-built gazetteer of ancient places, often used in classical studies to map references from ancient texts.

4. Remote Sensing and 3D Modeling

In archaeological and environmental humanities, tools such as **LiDAR**, **satellite imagery**, and **photogrammetry** are used to gather high-resolution data of landscapes and structures. Software like **Agisoft Metashape** or **SketchUp** allows researchers to create 3D models of historical sites, enabling virtual exploration and analysis.

5. Web Mapping and Visualization Tools

For publishing spatial projects online, scholars often use:

- **Leaflet.js** and **Mapbox**: JavaScript libraries for building custom, interactive web maps.
- **Google Earth Engine**: A platform for analyzing satellite data, useful for environmental history or landscape studies.

In summary, the tools and technologies used in spatial humanities empower researchers to combine geographic analysis with humanistic interpretation. Whether through GIS, story mapping, or 3D modeling, these tools enable the creation of spatial narratives that illuminate historical processes, cultural patterns, and place-based meanings in new and compelling ways.

5.5.2 *Applications in Literature, History, and Culture*

Spatial humanities reveal the often-overlooked geographical dimensions of human experience, allowing scholars to explore how space and place shape narrative, culture, and social change. In literature, mapping has been used to analyze the geography of narrative worlds. For example, in James Joyce's *Ulysses*, mapping Leopold Bloom's journey through Dublin helps uncover layers of meaning related to modern urban life, memory, and identity. By visualizing literary settings, researchers can trace character movement, investigate symbolic geographies, and examine the spatial politics embedded in fiction. In history, spatial analysis supports the visualization of key events such as trade networks, military campaigns, and the spread of colonial empires. Historical GIS allows for dynamic mapping of data across time, revealing patterns of migration, urban development, or conflict that are difficult to see in static narratives. It also enhances public history efforts, enabling interactive timelines and digital archives. Archaeology benefits immensely from spatial methods. GIS is used to document excavation sites, analyze the distribution of artifacts, and understand how ancient communities interacted with their environment. Spatial data can inform theories about settlement patterns, resource use, and cultural exchange. In sociology and anthropology, spatial tools help interpret how communities organize space. Ethnographic mapping and spatial ethnography explore neighbourhood boundaries, rituals, and routes of everyday life. Mapping projects can uncover social divisions, patterns of mobility, and cultural memory encoded in place.

5.5.3 Challenges in Spatial Humanities

Despite its innovative potential, spatial humanities face several significant challenges. One major limitation is the accuracy and availability of historical spatial data. Historical maps may be distorted, incomplete, or produced with colonial or political bias. Coordinates and place names can be ambiguous or lost, making precise geolocation difficult. Technical barriers also persist. GIS and mapping tools often require specialized knowledge that many humanities scholars lack. Although open-source tools like QGIS have lowered the entry barrier, learning to work with geospatial data still demands training in data formats, projection systems, and spatial analysis techniques. Another risk is the misinterpretation of spatial data. There is a tendency to assume that maps convey objective truth, but spatial representations are always mediated by cultural, political, and methodological assumptions. Scholars must avoid deterministic readings, where space is seen as the sole driver of historical or cultural outcomes, and instead focus on how space interacts with other factors. Finally, there are ethical considerations, particularly when mapping sensitive cultural sites or Indigenous lands. The digital dissemination of spatial data can unintentionally expose sacred or protected areas. Responsible mapping practices require consultation with affected communities and a commitment to ethical data stewardship [20, 21].

5.5.4 Educational Use and Pedagogy

Spatial humanities are increasingly finding a place in educational settings, where it serves as a dynamic and interdisciplinary tool for teaching. By introducing students to mapping tools like ArcGIS StoryMaps, Google Earth, or Recogito, educators foster digital literacy and spatial thinking—skills essential for 21st-century scholarship. Mapping assignments encourage students to engage with texts or data in new ways. Instead of just reading about a historical event or literary journey, they can visualize it on a map, revealing connections between place, narrative, and context. This multimodal analysis enhances critical thinking and promotes creativity in how students interpret and present knowledge. Spatial projects also promote collaborative and interdisciplinary learning. Students from different disciplines—history, literature, geography, computer science—can work together, combining their expertise to develop meaningful spatial narratives. These experiences mirror real-world research environments and encourage teamwork, communication, and problem-solving. Educators can scaffold these experiences by starting with simple map-reading or annotation exercises and progressing to full digital projects. When paired with reflective writing and critical theory, spatial humanities projects can deepen students' understanding of how space shapes cultural meaning [20].

5.5.5 The Future of Spatial Humanities

The future of spatial humanities is vibrant and technologically rich. As geospatial technologies evolve, the field is poised to adopt cutting-edge innovations like 3D modelling, AR, and real-time geographic storytelling. These tools promise more immersive and interactive experiences, allowing users to virtually explore reconstructed historical environments, walk through digital recreations of ancient cities, or engage with layered spatial narratives in real-time. Technologies such as LiDAR, drone mapping, and photogrammetry are already being used to document archaeological landscapes in detail. In education and public engagement, AR apps and virtual tours bring spatial history and literature alive, making research more accessible and engaging for non-specialist audiences. Moreover, advances in machine learning and spatial data mining will enable the discovery of new patterns in historical and cultural data at unprecedented scales. However, as the field advances, it must maintain a critical and ethical stance, ensuring that technological innovation serves humanistic inquiry rather than overshadowing it. The spatial humanities of the future will not only interpret space but create new kinds of place-based knowledge, bridging digital tools, narrative insight, and cultural understanding in ways that redefine both scholarship and public engagement [22, 23].

5.6 Code as Cultural Expression

While traditionally viewed as a neutral, functional tool used for computation, code is increasingly recognized as a medium of cultural expression. In the context of digital humanities, this shift marks a significant transformation in how programming is understood—not merely as a set of instructions for machines, but as a textual and aesthetic practice that reflects ideologies, social norms, and identity. Much like literature, architecture, or visual art, code can convey values, tell stories, and even resist dominant systems of thought. This perspective is rooted in the development of software studies and critical code studies, which examine the social, cultural, and philosophical dimensions of computer code. Scholars such as Wendy Hui Kyong

Chun, Matthew Fuller, and Lev Manovich argue that code, like any human-produced artefact, is embedded with meaning. Critical code studies treat code not only as functional but as *readable*, *interpretable*, and *contestable*. Code becomes a site of narrative, power, and creativity—open to rhetorical analysis and cultural critique [24].

Philosophically, this aligns with post-structuralism thought, where texts (and by extension, code) are shaped by context, ideology, and power. In this view, code is not fixed in meaning; it is always written and interpreted through the lens of its socio-political environment. For instance, the algorithmic logic behind predictive policing software or recommendation engines is not culturally neutral—it reflects assumptions about behavior, risk, and identity, often reinforcing existing social biases. Historically, the emergence of open-source movements, hacker cultures, and digital art collectives has reinforced this cultural framing of code. Communities like the Free Software Foundation advocate for not just functional openness, but ideological freedom, questioning corporate control and promoting collective participation. The ethos of Creative Commons and copyleft licensing reflects a cultural stance against ownership hierarchies and toward communal expression and access.

Even programming languages, themselves can be seen as cultural artefacts, shaped by and shaping the communities that use them. Each language embodies distinct design philosophies, stylistic norms, and assumptions about the nature of problem-solving:

1. **Python**, known for its readability and simplicity, encourages clarity and inclusivity. Its community promotes accessibility and openness, making it a popular choice in education and collaborative research.
2. **Perl**, with its motto "There's more than one way to do it," reflects a hacker ethos that values creativity, individuality, and a playful relationship with rules. Its flexibility speaks to a culture that celebrates experimentation over convention.
3. **Java**, on the other hand, emphasizes strict object-oriented principles, modularity, and security. It caters to large-scale enterprise development, reflecting values like structure, control, and maintainability that align with corporate software needs.

Each of these languages, in their syntax and design, encodes assumptions about logic, efficiency, creativity, and even power relations. Understanding code as cultural expression allows scholars and students to critically engage with programming not just as a technical skill, but as a language of power, resistance, and identity (Fig. 5.4).

Table 5.6 summarises cultural characteristics of selected programming languages.

5.6.1 Cultural Narratives in Code Projects

In the evolving field of digital humanities and creative coding, code is not merely a tool for automation or analysis—it is a powerful medium for crafting and communicating cultural narratives. Through interactive storytelling, data visualizations, digital art, and open-source activism, code-based projects increasingly serve as platforms

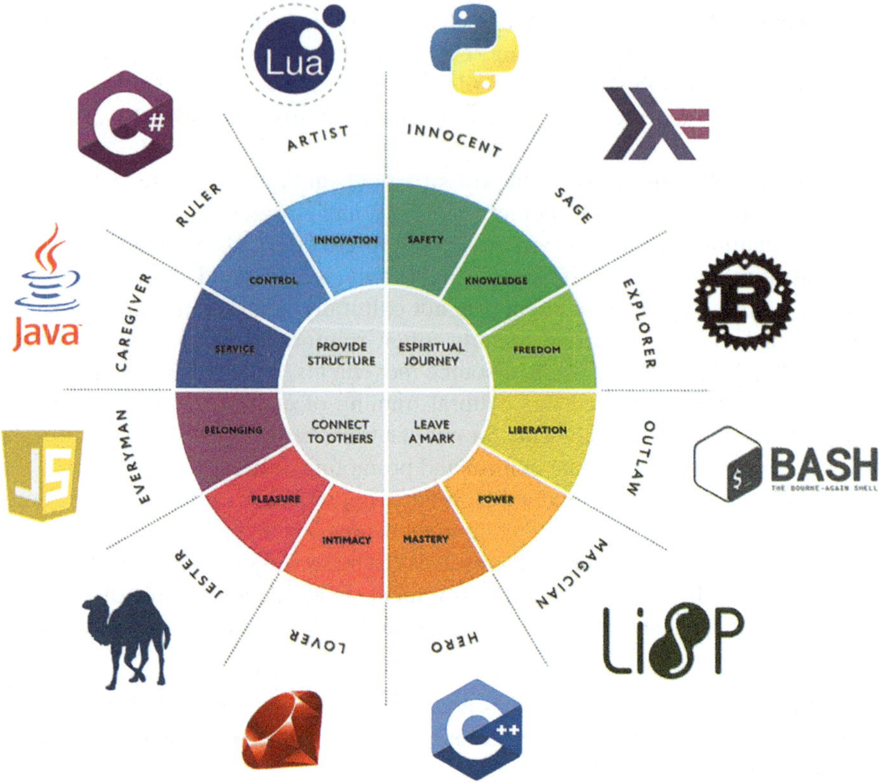

Fig. 5.4 Programming language and humanities [25]

Table 5.6 Cultural characteristics of programming languages

Language	Key traits	Cultural association
Python	Clean syntax, readability	Transparency, education
Perl	Flexibility, multiple ways	Hacktivism, creativity
Java	Strict structure, OOP	Enterprise, standardisation
JavaScript	Asynchronous, browser-native	Web culture, interactivity
Haskell	Functional purity	Academic rigor, mathematical elegance

for expressing social identities, historical memory, political critique, and community values. In this way, coding becomes a narrative practice—one that is both interpretive and generative. Cultural narratives are stories that communicate the beliefs, values, and experiences of a group. In code projects, these narratives may appear explicitly, such as in digital exhibits that tell the history of a marginalized community, or implicitly, such as in the design of user interfaces, data structures, or algorithmic behaviour that reflect particular worldviews. These coded expressions can preserve endangered languages, simulate ancestral journeys, visualize diasporas, or explore themes like gender, race, and belonging.

One example of cultural storytelling through code is the use of interactive maps to trace historical or literary journeys. Projects like *Mapping the Republic of Letters* or *Slave Voyages* visualize the movement of people, ideas, and goods, making complex spatial–temporal narratives accessible and emotionally resonant. Similarly, artists and scholars use web-based platforms like Twine, Processing, or P5.js to create interactive narratives that respond to user choices, drawing on the structure of code to model branching, non-linear storytelling traditions found in oral cultures. Open-source code repositories also carry cultural significance. Projects that promote Indigenous data sovereignty, for instance, are often coded with protocols and permissions that reflect cultural respect and community consent. Initiatives such as the Mukurtu content management system allow Indigenous communities to tell their own stories on their terms, embedding ethical and narrative principles directly into the software architecture [26].

Even the aesthetic choices in code projects—colour schemes, font usage, animation, data representation can reflect cultural identity and politics. For example, a digital artwork visualizing environmental data from the Amazon rainforest might use design elements inspired by Indigenous cosmologies, embedding local knowledge systems into the visual and interactive logic of the code. At a more conceptual level, critical code studies urge us to read the code itself, not just its output, as a site of narrative and meaning. Code comments, naming conventions, modular structures, and choices of logic reflect the values and assumptions of the coder. A codebase written in a collective workshop might show how different contributors' voices, intentions, and cultural positions intersect in a collaborative narrative form. However, not all cultural narratives in code are liberatory. Code can also reflect dominant ideologies and biases, especially when written without critical reflection. Algorithmic discrimination, surveillance systems, and content moderation tools often encode unexamined cultural assumptions that marginalize or exclude [27].

In summary, cultural narratives in code projects demonstrate how programming can be used not just for solving problems, but for telling stories, shaping memory, and expressing identity. Recognizing code as a narrative form allows programmers, artists, and scholars to engage in meaningful digital expression that bridges technology and culture in profound ways (Fig. 5.5 and Table 5.7).

Code, Identity, and Inclusion

Code can reflect and shape identity. The demographics of who codes and who is excluded have profound implications. Cultural barriers in tech include gender bias, racial disparities, and socio-economic exclusion. Inclusive coding practices aim to bridge these gaps through pedagogy, accessibility, and design. Coding bootcamps for underrepresented communities, gender-neutral programming tools, and culturally sensitive UX design are all ways in which code becomes a site for social inclusion. Moreover, representation matters seeing diverse coders changes perceptions about who belongs in tech. Efforts like Black Girls Code, Girls Who Code, and Coding for

Fig. 5.5 Community cultural development

Table 5.7 Examples of code projects as cultural narratives

Project	Description	Cultural message
FemTechNet	Feminist coding network	Empowerment through inclusive tech
Cyborg foundation	Tech for body augmentation	Post-human identity and accessibility
Mozilla firefox	Open-source browser	User rights and freedom
Linux	Open-source OS	Community-driven innovation
Maori keyboard app	Indigenous language tool	Cultural preservation through tech

Veterans exemplify this shift. These initiatives do not merely teach technical skills; they redefine who gets to create digital culture.

Future Directions: Towards a Critical Code Literacy

Recognising code as cultural expression leads to the concept of critical code literacy the ability to read, write, and critique code with awareness of its social implications. This goes beyond syntax and logic to include ethical, political, and aesthetic dimensions. Educational institutions are beginning to integrate these ideas into digital humanities, media studies, and even computer science curricula. Courses in "Code and Culture" or "Critical Code Studies" encourage students to see coding as a humanistic practice.

References

1. Mahapatra, C., Sheng, Z., Kamalinejad, P., Leung, V.C.M., Mirabbasi, S.: Optimal power control in green wireless sensor networks with wireless energy harvesting, wake-up radio and transmission control. IEEE Access, Special Sections on Energy Harvesting and Scavenging: Technologies, Algorithms, and Communication Protocols, vol. 5, pp. 1–1 (2016). https://doi.org/10.1109/ACCESS.2016.2644607.
2. Hodtani, H., Heravi, H.: RNA-based liquid biopsy in head and neck cancer. Cells **12**(14) (2022). https://doi.org/10.3390/cells12141916
3. Fu, Q., Rahman, A.A.A., Jiang, H., Abbas, J., Comite, U.: Sustainable supply chain and business performance: the impact of strategy, network design, information systems, and organizational structure. Sustain. **14**(3) (2022). https://doi.org/10.3390/su14031080
4. Sun, B., Gui, C., Song, Y., Chen, H.: A novel network coding and multi-path routing approach for wireless sensor network. Wirel. Pers. Commun. **77**(1), 87–99 (2014). https://doi.org/10.1007/s11277-013-1496-y
5. Fatmawati, I., Fauzan, N.: Building Customer Trust through Corporate Social Responsibility : The Effects of Corporate Reputation and Word of Mouth, vol. 8, no. 3, pp. 793–805 (2021). https://doi.org/10.13106/jafeb.2021.vol8.no3.0793
6. Alaba, F.A., Jegede, A., Eke, C.I.: Robust data security framework for IoT. IJAMML **1**(December), 5–23 (2020)
7. Nkuda, M.: Strategic agility and competitive advantage: exploration of the ontological, epistemological and theoretical underpinnings. Br. J. Econ. Manag. Trade **16**(1), 1–13 (2017). https://doi.org/10.9734/bjemt/2017/30979
8. Wong, W.P., Soh, K.L., Le Chong, C.: Differentiated service consumption and low cost production: striking a balance for a sustainable competitive advantage in Malaysia. Int. J. Prod. Econ. **181**, 450–459 (2016). https://doi.org/10.1016/j.ijpe.2015.09.029
9. Steffen, S., Bichsel, B., Baumgartner, R., Vechev, M.: ZeeStar: private smart contracts by homomorphic encryption and zero-knowledge proofs. In: Proceedings of IEEE Symposium on Security and Privacyvol, pp. 179–197 (2022). https://doi.org/10.1109/SP46214.2022.9833732
10. Górski, T.: Reconfigurable smart contracts for renewable energy exchange with re-use of verification rules. Appl. Sci. **12**(11) (2022). https://doi.org/10.3390/app12115339
11. Ghelani, D., et al.: Cyber security threats, vulnerabilities, and security solutions models in banking cyber security threats, vulnerabilities, and security solutions models in banking (2022). https://doi.org/10.11648/j.xxx.xxxxxxxx.xx
12. Daud, I., et al.: The effect of digital marketing, digital finance and digital payment on finance performance of indonesian smes. Int. J. Data Netw. Sci. **6**(1), 37–44 (2022). https://doi.org/10.5267/J.IJDNS.2021.10.006
13. Zou, D., Huang, X., Kohnke, L., Chen, X., Cheng, G., Xie, H.: A bibliometric analysis of the trends and research topics of empirical research on TPACK. Educ. Inf. Technol. **27**(8), 10585–10609 (2022). https://doi.org/10.1007/s10639-022-10991-z
14. Alaba, F.A., Amina, M., Ibrahim, U., Sulaimon, H.A., Zainab Aliyu, A.: Exploring the role of the internet of things (IoT) on economic growth in Nigeria: challenges and opportunities. Eng. Technol. Trends **2**(6) (2024)
15. Relations, E., Repository, P.F.: An exploration of the practices of locational flexibility in developing economies: insights from the Nigerian higher education sector abstract, vol. 42, pp. 166–179 (2021)

16. Bakari, A.U.: The effects of internal security on educational challenges and development in Nigeria. Sub-Sahara African Acad. Res. Publ. 249–258 (2020)
17. Yang, X., Hu, M., Tukker, A., Zhang, C., Huo, T., Steubing, B.: A bottom-up dynamic building stock model for residential energy transition: a case study for the Netherlands. Appl. Energy **306**(PA), 118060 (2022). https://doi.org/10.1016/j.apenergy.2021.118060
18. Carneiro, J., Rossetti, R.J.F., Silva, D.C., Oliveira, E.C.: BIM, GIS, IoT, and AR/VR integration for smart maintenance and management of road networks: a review. In 2018 IEEE Int. Smart Cities Conf. ISC2 2018 (2019). https://doi.org/10.1109/ISC2.2018.8656978
19. Wang, H., Pan, Y., Luo, X.: Integration of BIM and GIS in sustainable built environment: a review and bibliometric analysis. Autom. Constr. **103**, 41–52 (2019)
20. Callway, R., Le Gouais, A., Bird, E.L., Chang, M., Kidger, J.: Integrating health into local plans: a comparative review of health requirements for Urban development in seven local planning authorities in England. Int. J. Environ. Res. Public Health **20**(5) (2023). https://doi.org/10.3390/ijerph20054079
21. Cucca, R., Friesenecker, M.: Potential and limitations of innovative housing solutions in planning for degrowth: the case of Vienna. Local Environ. **0**(0), 1–15 (2021). https://doi.org/10.1080/13549839.2021.1872513
22. Ruano, P. et al.: Corporate social responsibility as the organization's commitment against stakeholders. Intech, no. tourism, p 13 (2016). https://www.intechopen.com/books/advanced-biometric-technologies/liveness-detection-in-biometrics
23. Nomura, A., Takeji, Y., Shimojima, M., Takamura, M.: Digitalomics: towards artificial intelligence/machine learning-based precision cardiovascular medicine (2025). https://doi.org/10.1253/circj.CJ-24-0865
24. Wakefield, S., Chandler, D., Grove, K.: The asymmetrical anthropocene: resilience and the limits of posthumanism. Cult. Geogr. **29**(3), 389–404 (2022). https://doi.org/10.1177/14744740211029278
25. Bailey, J., Blignault, I., Renata, P., Naden, P., Nathan, S., Newman, J.: Barriers and enablers to aboriginal and torres strait islander careers in health: a qualitative, multisector study in western New South Wales. Aust. J. Rural Health **29**(6), 896–908 (2021). https://doi.org/10.1111/ajr.12764
26. Li, J., et al.: A remote monitoring and diagnosis method based on four-layer IoT frame perception. IEEE Access **7**(3), 144324–144338 (2019). https://doi.org/10.1109/ACCESS.2019.2945076
27. De Gaetani, C.I., Mert, M., Migliaccio, F.: Interoperability analyses of BIM platforms for construction management. Appl. Sci. **10**(13), 1–15 (2020). https://doi.org/10.3390/app10134437

Chapter 6
Designing Digitally-Enriched Curricula

6.1 Principles of Curriculum Design

The systematic organisation and structuring of educational material, learning experiences, and assessment procedures to promote successful student learning is what is known as curriculum design. The underlying ideas that guide the construction of curricula are investigated in this chapter. It provides insights into historical views, theoretical frameworks, and practical implementations within contemporary educational systems. Learners' intellectual, emotional, and social development is significantly influenced by the curriculum that they follow, which should be well-structured [1] (Fig. 6.1).

6.1.1 Historical Context of Curriculum Design

Over many centuries, philosophical, political, and socio-economic forces have had a significant amount of impact on the development of an educational curriculum. During the early stages of education, the curriculum was carefully limited, with an emphasis placed on the teaching of moral principles and classical knowledge. During the mediaeval period in Europe, for example, education was centred on the trivium, which included grammar, rhetoric, and logic, as well as the quadrivium, which included mathematics, geometry, music, and astronomy. The cornerstone of a liberal arts education was comprised of these subjects, which placed an emphasis on intellectual discipline and the development of reason. During the Renaissance and Enlightenment eras, humanist philosophers campaigned for wider learning, which included the study of literature, science, and the arts. This led to the expansion of the curriculum. Thinkers such as Comenius and Rousseau laid the framework for contemporary educational philosophy by placing an emphasis on natural learning processes and the development of the child as a whole. John Dewey's progressive

F. A. Alaba and I. G. Ajibola, *Digital Pedagogy: Bridging the Gap Between Software and the Humanities*, Information Systems Engineering and Management 85,
https://doi.org/10.1007/978-3-032-20335-9_6

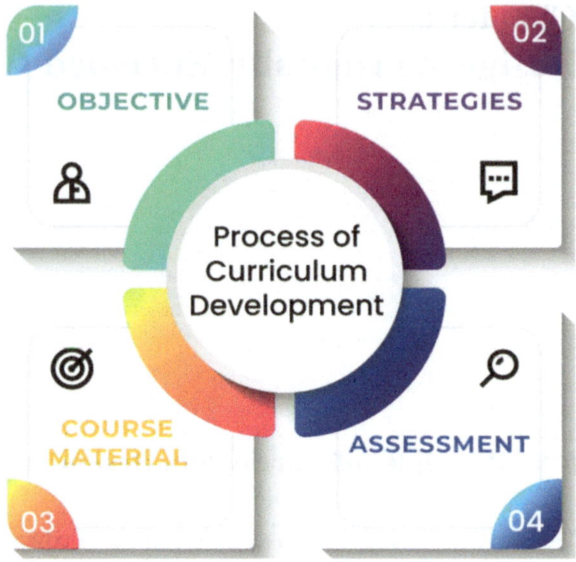

Fig. 6.1 Curriculum development process

education approach brought about a significant change in the way that curricula were designed at the beginning of the twentieth century. Dewey maintained that education had to be based on experiences that are relevant to actual life and should be aimed towards teaching democratic participation. His concepts advocated for a curriculum that was centred on the child and placed an emphasis on problem-solving, inquiry-based learning, and the integration of disciplines according to the requirements and fascinations of the students [2].

The effect of behaviourism and scientific management on the process of curriculum design occurred in the middle to late twentieth century. This resulted in the development of models that were organised and objectively driven, such as Tyler's reasoning and Bloom's tree of knowledge. There was a strong emphasis placed on learning objectives, performance assessment, and standardised material in these methods. Over the last several decades, many factors, including globalisation, technology breakthroughs, diversity, and inclusive education, have contributed to the ongoing evolution of curriculum design. Constructivist and competency-based curricula, which emphasize critical thinking, creativity, and digital literacy, have been more popular in recent years. To reflect modern values and the requirement for skills that are relevant in the twenty-first century, the curriculum models that are used today are becoming more flexible, multidisciplinary, and learner-centred. A continual negotiation between tradition and innovation, authority and agency, standardisation and personalisation is shown throughout the historical path of curriculum design because it requires constant negotiation [3] (Fig. 6.2 and Table 6.1).

Fig. 6.2 Types of curriculum development [4]

Table 6.1 Evolution of Curriculum Approaches

Period	Approach	Key characteristics	Key proponents
Medieval Era	Trivium and quadrivium	Focused on grammar, logic, rhetoric, arithmetic, geometry, music, and astronomy	Boethius, Alcuin
19th Century	Classical curriculum	Emphasis on Latin, Greek, and moral philosophy	Thomas Arnold
Early 20th Century	Progressive education	Student-centered, experiential learning	John Dewey
Late 20th Century	Competency-based education	Focus on outcomes and measurable skills	Benjamin Bloom
21st Century	Integrated and digital curriculum	Use of technology and interdisciplinary learning	Various educational theorists

6.1.2 Key Theoretical Frameworks

Several educational theories influence curriculum design. These include:

1. Constructivism (Jean Piaget, Lev Vygotsky): Emphasizes learning as a process of building on prior knowledge.
2. Behaviorism (B.F. Skinner): Focuses on observable changes in behavior as evidence of learning.
3. Cognitivism (Jerome Bruner): Centers on how learners process and structure information.
4. Humanism (Carl Rogers): Values personal growth and student agency in learning.
5. Social Learning Theory (Albert Bandura): Recognizes the role of observation and imitation in learning.

Designing a curriculum often involves blending these theories to meet the diverse needs of learners.

6.1.3 Fundamental Principles of Curriculum Design

Effective curriculum design is guided by several core principles [5] (Fig. 6.3 and Table 6.2):

Curriculum Models

Different models guide the development and implementation of curricula:

1. Tyler's Model (1949): Emphasizes setting clear objectives, selecting learning experiences, organizing content, and evaluating outcomes.
2. Taba's Model (1962): Advocates a bottom-up approach where teachers play a key role in curriculum development.

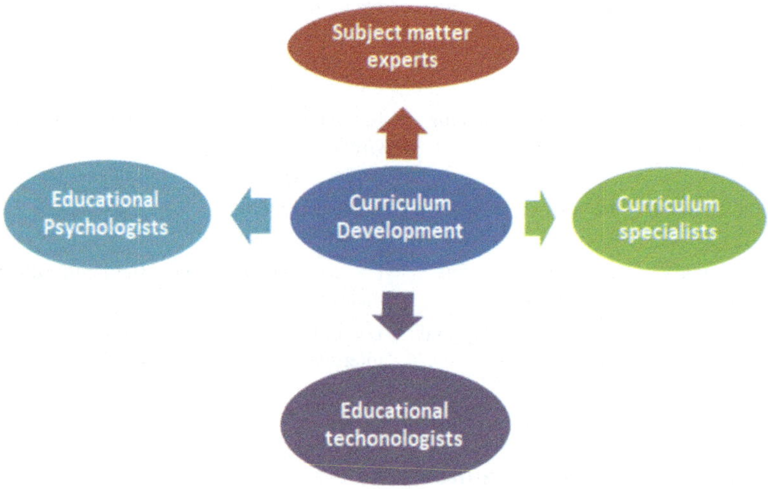

Fig. 6.3 Key principle of curriculum design

Table 6.2 Core principles of curriculum design

Principle	Description	Application example
Relevance	Curriculum should meet students' present and future needs	Incorporating real-world case studies in lessons
Coherence	Logical flow and consistency across topics and levels	Vertical and horizontal alignment of subjects
Balance	Equitable emphasis on knowledge, skills, and attitudes	Mixing academic and co-curricular content
Flexibility	Accommodates diverse learning styles and needs	Offering elective modules and differentiated instruction
Progression	Content should build on prior knowledge and increase in complexity	Spiral curriculum structure

Table 6.3 Types of curriculum alignment

Type	Focus	Example
Horizontal	Same grade/level	Common assessments for all Year 8 science classes
Vertical	Across grades	Building algebra skills from Year 7 to Year 9
Instructional	Within lessons	Using projects that assess the same skills taught

3. The Wheeler Model (1967): An extension of Tyler's model with cyclical planning and continuous feedback.
4. Kerr's Model (1968): Focuses on objectives, content, learning experiences, and evaluation as interrelated elements.
5. The Spiral Curriculum (Bruner): Revisits core ideas at increasing levels of complexity.

Each model has strengths and limitations depending on the educational context.

Curriculum Alignment

Curriculum alignment ensures that educational objectives, teaching methods, and assessments are in sync. There are three primary types of alignment:

1. Horizontal Alignment: Ensures consistency across the same grade level.
2. Vertical Alignment: Ensures a logical progression from one grade level to the next.
3. Instructional Alignment: Aligns objectives with instructional strategies and assessments (Table 6.3).

6.1.4 Cultural Relevance and Inclusivity

Modern curriculum design must address diversity and promote inclusion. Key considerations include [6]:

1. Cultural Relevance: Embedding local and global cultural contexts into learning materials.
2. Linguistic Inclusivity: Supporting learners who speak different languages.
3. Gender Sensitivity: Avoiding stereotypes and promoting equity.
4. Special Educational Needs (SEN): Providing accommodations and differentiated content.

Inclusive curricula recognize the diverse backgrounds and abilities of learners, fostering equity and access.

Technology Integration in Curriculum Design

In the digital age, curriculum design must incorporate technology to enhance learning outcomes. Key strategies include [7, 8]:

1. Blended Learning Models: Combining face-to-face and online instruction.

2. Digital Literacy: Embedding skills such as coding, information evaluation, and digital communication.
3. LMS: Platforms like Moodle or Google Classroom to manage content and assessments.
4. Interactive Media: Use of simulations, gamification, and virtual reality for experiential learning (Fig. 6.4 and Table 6.4).

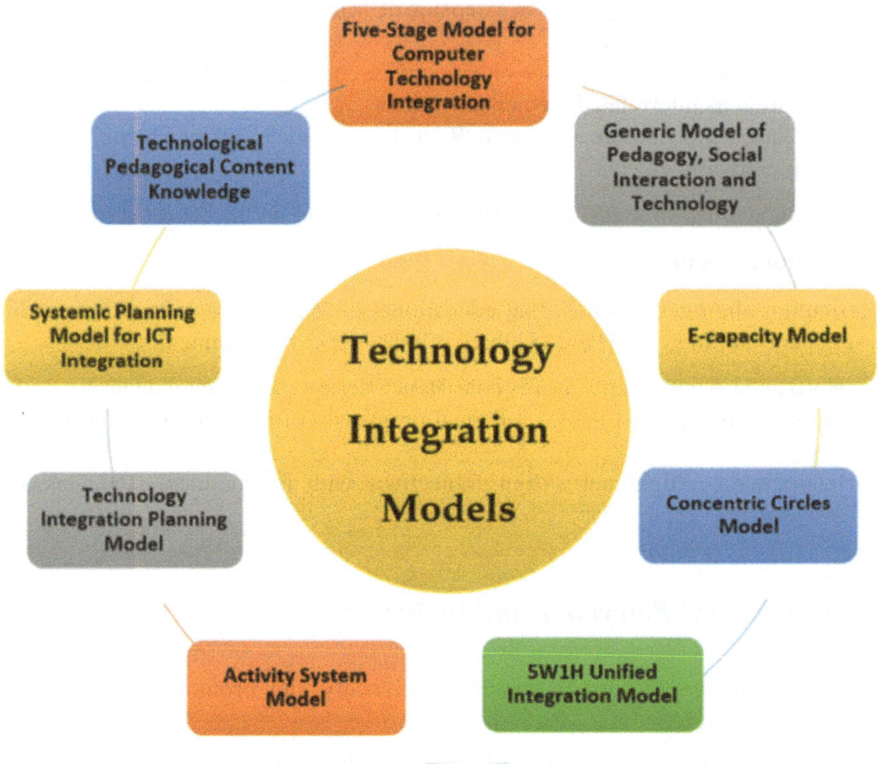

Fig. 6.4 Technology integration models [9]

Table 6.4 Examples of technology integration

Technology	Educational use	Benefits
LMS (e.g., Moodle)	Course delivery, assignments	Centralized content management
Simulation software	Science and engineering labs	Risk-free experiential learning
Virtual reality	Immersive history or geography lessons	Enhanced engagement and understanding
Gamified apps	Language learning tools	Motivation and immediate feedback

6.2 Assessment and Feedback Mechanisms

Assessment is an integral part of curriculum design. It should be aligned with learning objectives and instructional strategies. Types of assessments include [10]:

1. Formative Assessment: Ongoing checks for understanding (e.g., quizzes, reflections).
2. Summative Assessment: Final evaluations of learning (e.g., exams, projects).
3. Diagnostic Assessment: Identifies students' prior knowledge and misconceptions.
4. Peer and Self-Assessment: Encourages reflection and ownership of learning.

Effective feedback is timely, specific, and actionable, supporting both teachers and learners in the learning process.

Curriculum design is a dynamic and multi-faceted process that requires careful consideration of educational goals, learner needs, societal demands, and instructional methods. By grounding design decisions in sound theory and practice, educators can craft curricula that are coherent, inclusive, flexible, and responsive to the changing landscape of education. Future curriculum development must continue to embrace innovation while remaining anchored in core pedagogical principles [11].

6.2.1 Identifying Digital Learning Outcomes

Digital learning outcomes are explicit statements of what learners are expected to know, do, or value at the end of a digital learning experience. In an era dominated by technology, traditional learning objectives must be redefined to reflect the competencies required in a digitally driven world. Identifying appropriate digital learning outcomes ensures that technology integration in education is purposeful, effective, and aligned with curriculum goals. Digital learning outcomes refer to the knowledge, skills, attitudes, and values learners should acquire through digital means such as online courses, virtual labs, simulations, multimedia resources, and collaborative digital tools. These outcomes go beyond content mastery to include digital literacy, communication, collaboration, creativity, and critical thinking [12, 13] (Fig. 6.5).

6.2.2 Theoretical Frameworks Supporting Digital Learning Outcomes

Digital learning, whether in formal education or lifelong personal development, is most effective when grounded in robust theoretical frameworks. These frameworks help educators design instruction, assess outcomes, and support learners in digital environments. By aligning pedagogy with theory, digital learning becomes more

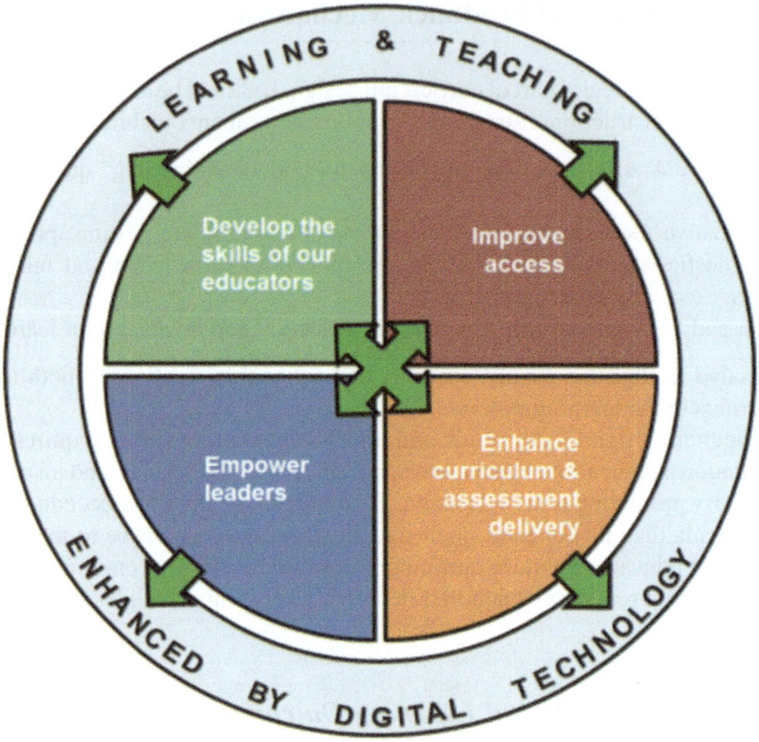

Fig. 6.5 Enhance teaching and learning through digital skills [14]

purposeful, inclusive, and transformative. Several well-established learning theories support and explain how students engage with, process, and benefit from digital learning experiences.

1. Constructivism

Constructivism, rooted in the work of Jean Piaget and Lev Vygotsky, posits that learners construct knowledge through active engagement and personal experience. In digital contexts, this translates to project-based learning, interactive simulations, and exploratory environments where students can experiment and reflect. Tools like virtual labs, game-based learning platforms, and digital storytelling allow learners to build meaning through hands-on experiences. Vygotsky's concept of the Zone of Proximal Development (ZPD) further supports the use of digital scaffolding—like hints, tutorials, and AI-driven feedback—to guide learners just beyond their current level of competence [15].

2. Connectivism

Developed by George Siemens and Stephen Downes, Connectivism is a learning theory specifically tailored for the digital age. It emphasizes the role of networks, digital tools, and community interaction in learning. In this view, knowledge is distributed across devices, platforms, and social connections. Learning happens when individuals make connections between information sources, both human and technological. This theory underpins MOOCs, forums, collaborative wikis, and other web-based learning environments where learners navigate, curate, and create knowledge in real-time networks [16].

3. Universal Design for Learning (UDL)

UDL provides a framework for creating inclusive digital learning environments that accommodate the diversity of learners. Based on cognitive neuroscience, UDL emphasizes multiple means of representation, expression, and engagement. In a digital setting, this may include captioned videos, interactive text, customizable reading formats, and multimodal assessments. UDL ensures that digital learning outcomes are accessible to all students, regardless of ability, learning style, or background, by removing barriers at the design stage [17].

4. Bloom's Digital Taxonomy

An update of Bloom's traditional taxonomy of learning objectives, Bloom's Digital Taxonomy reframes the cognitive domains (Remembering, Understanding, Applying, Analyzing, Evaluating, and Creating) through digital tools and tasks. For example [18]:

1. **Remembering** = bookmarking, highlighting,
2. **Applying** = editing code, running simulations,
3. **Creating** = building websites, producing videos.

This taxonomy helps educators align digital activities with cognitive goals, ensuring that technology use supports deeper learning rather than superficial engagement.

6.3 SAMR Model

The SAMR model (Substitution, Augmentation, Modification, Redefinition) guides the integration of technology into pedagogy. At the substitution level, tech replaces traditional tools (e.g., eBooks for textbooks). At the redefinition level, it enables new tasks previously inconceivable, like virtual collaboration across continents. Aligning digital learning outcomes with the higher levels of SAMR can promote innovation, creativity, and real-world problem-solving. In summary, theoretical frameworks like Constructivism, Connectivism, UDL, Bloom's Digital Taxonomy, and the SAMR Model provide essential guidance for designing and evaluating digital learning experiences. They help ensure that technology is used not as a gimmick, but as a meaningful tool to enhance learning outcomes, foster engagement, and support equity in diverse learning environments [19] (Table 6.5).

Table 6.5 Frameworks supporting digital learning outcomes

Framework	Description	Key elements
Bloom's digital taxonomy	Updates Bloom's cognitive domain to fit digital learning contexts	Remembering, understanding, applying, analysing, evaluating, creating
SAMR model	Explains the levels of technology integration	Substitution, augmentation, modification, redefinition
TPACK	Integrates technology, pedagogy, and content knowledge	Technological knowledge, pedagogical knowledge, content knowledge

These frameworks help educators structure digital learning outcomes to reflect higher-order thinking and effective technology use.

6.3.1 Categories of Digital Learning Outcomes

Effective digital learning hinges not only on the use of technology but on the clarity and relevance of its intended learning outcomes. To guide instruction, assessment, and learner engagement, educators increasingly use the SMART framework to develop outcomes that are Specific, Measurable, Achievable, Relevant, and Time-bound. When applied in digital learning contexts, SMART outcomes ensure that both learners and instructors remain focused, outcomes-driven, and adaptable to a variety of platforms and tools.

1. ***Specific***: A SMART learning outcome begins with specificity. In digital learning, where students may be navigating independently or asynchronously, clarity is essential. Specific outcomes define exactly what the learner will achieve and remove ambiguity. For example, instead of stating, "Students will understand website development," a specific outcome would be: "*Students will be able to create a multi-page website using HTML and CSS.*" This leaves no doubt about the task or skill being developed.
2. ***Measurable***: Digital learning outcomes must be measurable so that both learners and educators can assess progress. This is especially important in virtual environments where traditional classroom cues and informal feedback are limited. Measurable outcomes are tied to assessments, whether quizzes, digital portfolios, code submissions, or discussion contributions. For instance: "*Students will analyze and visualize sentiment trends in a dataset using Python and submit a Jupyter Notebook report.*" The product of the task provides clear evidence of achievement.
3. ***Achievable***: Outcomes should be challenging yet realistic within the time and resource constraints of the digital environment. Digital learners often vary widely in experience and access, so goals must align with their prior knowledge, tools, and support available. An unachievable outcome, such as building a full-stack

application in a two-week introductory course, can lead to frustration. A better goal might be: "*Students will design a static personal webpage using HTML, CSS, and GitHub Pages.*"

4. ***Relevant***: SMART digital outcomes should be relevant to the learners' goals, discipline, and real-world applications. In a humanities course, for example, coding should not be treated as an abstract skill but as a tool for meaningful inquiry. A relevant outcome could be: "*Students will use text-mining techniques to extract themes from a 19th-century novel and interpret the findings in a short written reflection.*" This outcome aligns the digital skill with disciplinary content and critical thinking.
5. ***Time-bound***: Finally, outcomes must be time-bound, with a clear deadline or duration for completion. This is crucial in digital learning environments where pacing can vary. Time constraints help maintain focus and ensure progression through the curriculum. For example: "*By the end of week 4, students will complete a peer-reviewed multimedia presentation using ArcGIS StoryMaps.*"

Thus, developing SMART digital learning outcomes ensures that instruction is purposeful, assessment is aligned, and students remain motivated and focused. By clearly articulating what success looks like, SMART outcomes empower learners to take ownership of their learning while helping instructors deliver meaningful, skill-based, and context-rich digital education. This structured approach is particularly valuable in online and blended environments where guidance and structure are critical to learner success.

Digital learning outcomes can be categorised into several domains [20]:

1. Cognitive Outcomes: Acquisition of digital knowledge and problem-solving skills.
2. Technical Skills: Proficiency with digital tools and platforms.
3. Collaborative Skills: Ability to work in digital teams using communication platforms.
4. Creative Output: Use of digital media to produce original content.
5. Ethical and Responsible Use: Understanding of digital citizenship and ethical technology use (Table 6.6).

Tools for Identifying and Mapping Digital Learning Outcomes

Educators can use various tools to map and identify outcomes:

1. Curriculum Mapping Software (e.g., Atlas): Aligns outcomes with standards.
2. Rubrics: Define success criteria for digital tasks.
3. Learning Management Systems (e.g., Moodle, Canvas): Track progress and align tasks with objectives.

Challenges in Identifying Digital Learning Outcomes

1. Overemphasis on Technology: Focusing on tools rather than pedagogy.
2. Lack of Clarity: Vague or generic outcomes.
3. Assessment Difficulties: Measuring digital creativity or collaboration.
4. Equity Issues: Ensuring access to digital tools for all learners.

Table 6.6 SMART criteria

SMART criteria	Explanation	Example
Specific	Clearly states what the learner will do	"Create a podcast on climate change impacts"
Measurable	Outcomes can be assessed	"Complete a quiz with 80% accuracy"
Achievable	Within the learner's capabilities	"Edit a short video using basic software"
Relevant	Linked to a course or real-world application	"Collaborate on a group project using Google Docs"
Time-bound	Expected within a certain period	"Submit a digital portfolio by the end of Week 6"

6.4 Integrating Technology into Syllabi

The integration of technology into syllabi represents a transformative shift in educational design, enabling dynamic, interactive, and student-centred learning environments. In the digital age, syllabi are no longer static documents but evolving tools that reflect modern pedagogical practices. Incorporating technology enhances engagement, accessibility, and the relevance of educational content [21] (Fig. 6.6).

6.4.1 Purpose and Benefits of Technological Integration

The integration of technology into education is not simply a matter of using digital tools, it is a purposeful transformation of teaching and learning processes to enhance engagement, accessibility, and effectiveness. As the world becomes increasingly digitized, technological integration serves as both a practical necessity and a pedagogical opportunity. When thoughtfully implemented, it enriches curriculum delivery, supports diverse learning needs, and prepares learners for a technology-driven society [23].

Purpose of Technological Integration

1. The primary purpose of technological integration is to enhance learning outcomes and educational experiences. It allows educators to deliver content in diverse, interactive, and personalized ways. For instance, digital platforms such as learning management systems (e.g., Moodle, Canvas, Google Classroom) enable blended learning models where students can access resources, submit work, and engage in discussions both in and outside the classroom.

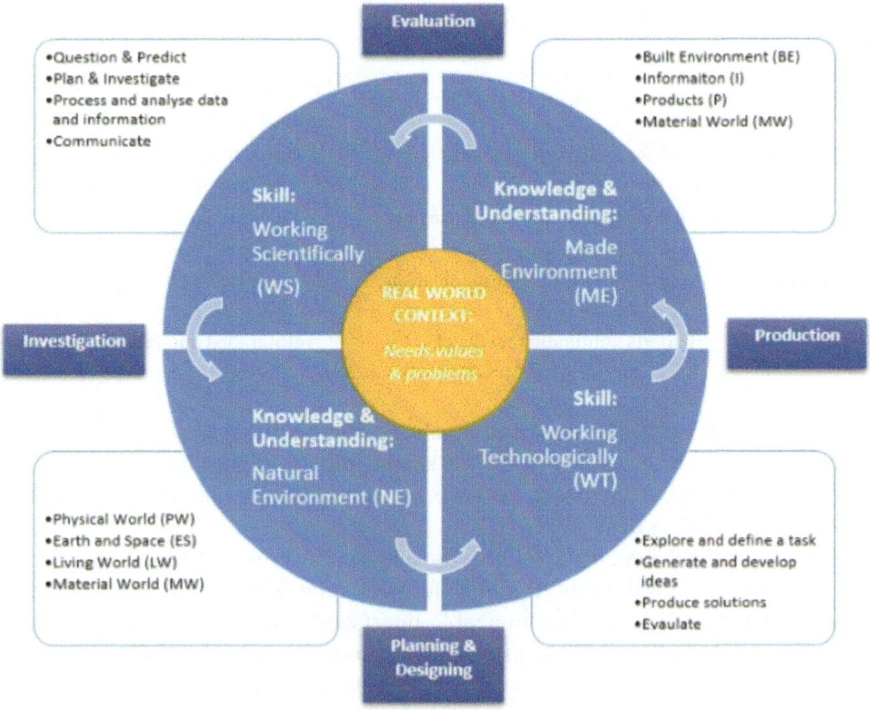

Fig. 6.6 Science and technology integration into Syllabi [22]

2. Technology also supports student-centered learning. It empowers learners to work at their own pace, revisit materials, and choose from multiple formats—videos, podcasts, texts, or interactive modules—according to their learning preferences. This shift from teacher-led instruction to learner autonomy is key to fostering independence and critical thinking.
3. Additionally, technological integration addresses the need for 21st-century skills. Skills such as digital literacy, online collaboration, coding, data interpretation, and information management are increasingly required across disciplines. Integrating technology into everyday learning ensures that students are not only consumers of content but also creators and evaluators of digital knowledge.

Benefits of Technological Integration

1. One of the most evident benefits is enhanced engagement. Interactive whiteboards, gamified learning platforms, simulations, and virtual or augmented reality can make abstract or complex topics more tangible and exciting. In literature, for example, students can explore digital maps of fictional settings or participate in interactive storytelling using platforms like Twine or StoryMapJS.
2. Technology also promotes collaboration and communication, both locally and globally. Students can work on joint projects through cloud-based tools like

Table 6.7 Benefits of technology integration in Syllabi

Benefit	Description	Example
Engagement	Use of multimedia to stimulate interest	Embedding videos and interactive quizzes
Accessibility	Reaches all students, including those with disabilities	Captioned video content, screen reader-friendly documents
Flexibility	Allows anytime-anywhere learning	Online discussion forums and recorded lectures
Interactivity	Promotes two-way communication	Live polls, chat tools during lectures
Assessment	Supports continuous feedback	Auto-graded quizzes, online rubrics

Google Docs or Trello, and participate in cross-cultural exchanges via video conferencing, discussion boards, or shared digital exhibitions. These collaborative environments mirror real-world work settings and develop essential communication skills.

3. Another critical benefit is inclusivity and accessibility. Digital tools support learners with disabilities through features like text-to-speech, captioning, adjustable text size, and customizable interfaces. Tools aligned with Universal Design for Learning (UDL) ensure that all students, regardless of learning style or physical ability, can access content and demonstrate their understanding in multiple ways.
4. From an administrative and instructional perspective, technology also enables efficient assessment and feedback. Online quizzes, interactive polls, and learning analytics help educators monitor progress in real-time and adjust instruction accordingly. Automated tools can reduce grading time and provide instant feedback, freeing up educators to focus on more meaningful engagement with students (Table 6.7).

6.4.2 Frameworks for Integration

Educational frameworks for technological integration are essential for guiding the thoughtful and effective integration of technology into education. These frameworks help educators make intentional decisions about when, how, and why to use technology to support learning. They ensure that digital tools enhance rather than distract from educational objectives. Among the most influential frameworks are TPACK (Technological Pedagogical Content Knowledge), SAMR (Substitution, Augmentation, Modification, Redefinition), and PICRAT (Creative-Transformation). TPACK emphasizes the intersection of three core knowledge areas: Content Knowledge (CK), Pedagogical Knowledge (PK), and Technical Knowledge (TK). Effective integration occurs when educators align all three domains, ensuring that technology serves learning goals rather than being an add-on or gimmick.

SAMR model provides a practical lens for evaluating the depth of technological integration, focusing on substitution, augmentation, modification, and redefinition. The goal is to move beyond basic substitution and use technology to transform learning. For example, instead of assigning a book report (Substitution), students might create a multimedia presentation or podcast interview with the author (Redefinition). PICRAT model, developed by Royce Kimmons, combines student engagement and teacher use of technology. It prompts educators to ask two key questions: What is the student's relationship to technology? (Passive, Interactive, or Creative) and what is the teacher's use of technology? (Replacement, Amplification, or Transformation). By leveraging these frameworks, schools and instructors can foster digital environments that are purposeful, inclusive, and transformative, ensuring that technology truly catalyzes deeper, more impactful learning. These frameworks help structure the integration process from planning to implementation and evaluation [24].

Components of Technology-Integrated Syllabus

A well-integrated syllabus includes [25]:

1. Course Description: Embeds digital literacy expectations
2. Learning Objectives: Includes digital competencies
3. Instructional Strategies: Details tech-enhanced learning methods
4. Assessment Methods: Includes digital assessment tools
5. Resources and Tools: Lists LMS, software, and online platforms
6. Communication Protocols: Sets expectations for online interaction

Strategies for Effective Integration

1. Use multimedia (videos, podcasts, simulations) to enhance content delivery.
2. Incorporate collaborative tools (Google Workspace, Padlet, Trello).
3. Implement formative assessments using Kahoot, Quizizz, or Socrative.
4. Provide tech tutorials and support for students.
5. Align digital tools with specific learning objectives (Table 6.8).

Table 6.8 Tools and their pedagogical applications

Tool	Application	Benefit
Google docs	Collaborative writing and feedback	Real-time editing and peer review
Edpuzzle	Interactive video lessons	Tracks understanding through embedded questions
Moodle	Course management and content delivery	Centralized access to materials
Flip (formerly Flipgrid)	Video discussion platform	Increases student voice and expression

6.4.3 Challenges and Considerations

Challenges, Assessment, and Best Practices in a Tech-Integrated Syllabus

The integration of technology into teaching and learning holds great promise for enhancing engagement, accessibility, and outcomes. However, it also introduces a unique set of **challenges and considerations** that educators must navigate carefully. To ensure that technology serves pedagogical goals effectively, instructors must also develop robust **assessment strategies** and follow established **best practices** in curriculum design and delivery.

Challenges and Considerations

1. One of the most persistent challenges is the digital divide. Despite increased access to devices and internet connectivity, significant gaps remain between students with consistent, high-quality digital access and those without. This inequality can hinder participation, performance, and student morale, especially in remote or blended learning environments.
2. Another major issue is the need for faculty training. Many educators may lack the time or support to keep up with emerging technologies. Ongoing professional development is essential to ensure instructors can use digital tools not only competently but also creatively and pedagogically.
3. The overload of tools is another common concern. Instructors often introduce multiple platforms (e.g., LMS, forums, interactive apps, video conferencing) without fully integrating them into the learning design. This can overwhelm students, leading to confusion, reduced engagement, and cognitive fatigue.
4. Privacy and data security are also critical. Instructors must be aware of legal and institutional policies such as GDPR or FERPA and ensure that tools used for teaching comply with data protection regulations.
5. Finally, there is the risk of misaligned pedagogy—using technology for technology's sake. If digital tools are not aligned with instructional goals, they may distract rather than enhance learning.

Assessment and Evaluation in Tech-Integrated Learning

1. Technology enhances both formative and summative assessment. E-portfolios, for example, allow students to document their learning journey over time, integrating artefacts such as videos, presentations, and written reflections. This supports holistic evaluation and fosters student ownership of learning.
2. Digital quizzes and polls offer immediate feedback and can be embedded into lessons for low-stakes, real-time checks on understanding. They help students monitor their progress and help instructors identify areas that require further attention.
3. Peer-assessment tools (e.g., Peergrade or within LMS platforms) encourage collaboration and reflective learning. Students develop critical thinking by evaluating others' work and receiving diverse feedback.

4. Data from LMS and analytics tools can be used to track student engagement and performance. Combined with rubrics, these tools support personalized feedback and promote continuous improvement.

Best Practices

To maximize the benefits of technology in education, educators should follow several key best practices:

1. Begin with pedagogical goals, not tools. Technology should support—not drive—the learning outcomes.
2. Ensure accessibility and usability across all devices and for all learners, including those with disabilities.
3. Provide clear instructions and expectations for how and when to use digital tools.
4. Develop contingency plans in case of technical failures, such as alternative assignments or offline resources.
5. Continuously update syllabi and instructional materials to reflect new tools, platforms, and evidence-based practices.

Integrating technology into syllabi fosters more engaging, inclusive, and effective teaching and learning experiences. By aligning technology with pedagogical intent, educators can design flexible, interactive, and outcome-driven syllabi that meet the demands of modern learners and the digital landscape.

6.5 Flipped Classrooms and Hybrid Models

The evolution of educational delivery methods has led to the development of flipped classrooms and hybrid learning models. These approaches challenge traditional paradigms by reconfiguring the time and space where learning occurs. Flipped classrooms shift content delivery outside of class time, allowing in-class sessions to focus on application and analysis. Hybrid models blend online and face-to-face interactions, offering flexibility and engagement (Fig. 6.7).

6.5.1 Flipped Classrooms: Definition and Rationale

The flipped classroom is an innovative pedagogical model that inverts the traditional structure of teaching by delivering instructional content outside of the classroom and using in-person class time for interactive, student-centered learning activities. Instead of spending face-to-face time on lectures and note-taking, students review foundational materials, such as videos, readings, or presentations—before class. Classroom sessions are then reserved for discussions, problem-solving, collaborative projects, and applied learning tasks [27].

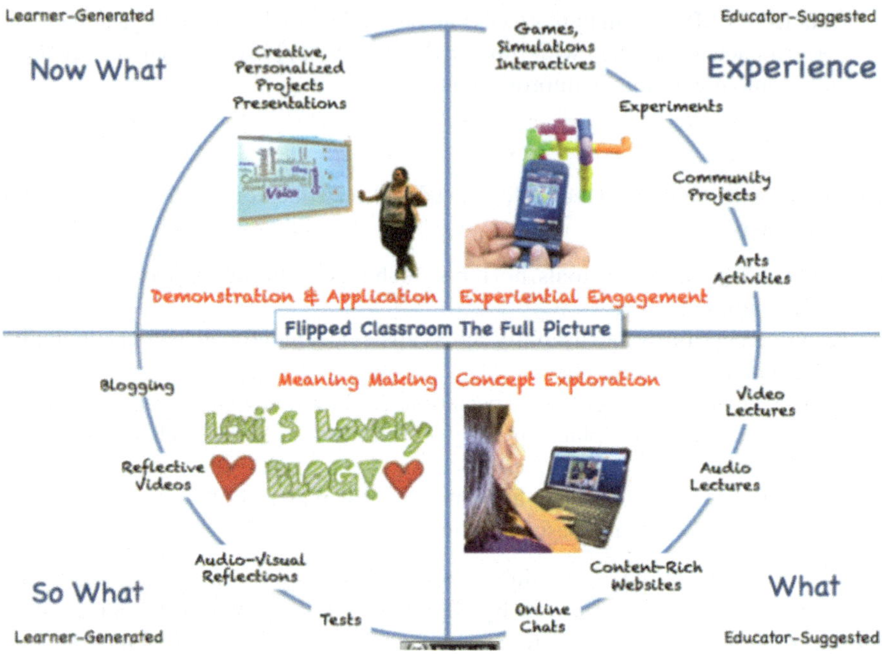

Fig. 6.7 Flipped versus traditional classroom [26]

Definition of the Flipped Classroom

In a **flipped classroom**, the sequence of instruction is reversed:

1. Before class: Students engage with core content independently through pre-recorded lectures, curated videos, readings, or multimedia modules.
2. During class: Students apply knowledge through active learning strategies—such as case studies, group discussions, labs, or peer teaching—while the teacher facilitates, supports, and provides personalized guidance.

This approach shifts the educator's role from that of a "sage on the stage" to a "guide on the side," promoting a more dynamic and engaging learning environment.

Rationale Behind Flipped Classrooms

The rationale for adopting the flipped classroom model lies in its alignment with constructivist learning theory, which posits that students learn best by constructing knowledge through experience, reflection, and social interaction. The flipped approach enables deeper engagement with content because students arrive in class prepared to analyze, synthesize, and apply what they've already reviewed.

1. **Promotes Active Learning**: By freeing up class time from passive content delivery, flipped classrooms create space for active learning. Students are

Table 6.9 Comparison between traditional and flipped classrooms

Feature	Traditional classroom	Flipped classroom
Content delivery	In-class lectures	Online videos, readings
In-class activity	Passive listening	Interactive, problem-solving
Teacher role	Lecturer	Facilitator
Student role	Note-taker	Active participant

encouraged to ask questions, engage in dialogue, and collaborate on complex problems—skills essential for critical thinking and long-term retention.

2. **Supports Differentiated Instruction**: Flipped classrooms allow students to consume learning materials at their own pace. Those who need more time can pause or rewatch videos, while advanced learners can move ahead. This flexibility fosters a more inclusive learning environment where students of varying abilities can succeed.
3. **Enhances Student Responsibility and Autonomy**: This model encourages students to take ownership of their learning. By requiring preparation before class, flipped classrooms cultivate habits of self-direction, time management, and accountability—all essential for academic and professional success.
4. **Enables Real-Time Support**: Because basic instruction happens outside of class, educators can spend class time addressing individual or group challenges. This real-time feedback and intervention are especially valuable in subjects that require cumulative understanding, such as mathematics, science, or language learning.
5. **Facilitates Technology Integration**: The flipped classroom is well-suited to modern educational technologies. LMS, video platforms, interactive quizzes, and discussion forums support content delivery, while digital collaboration tools enhance in-class engagement.

Despite its benefits, the flipped model also requires careful planning, clear communication, and student buy-in. Educators must design high-quality pre-class materials and ensure that in-class activities meaningfully build on them. Students, in turn, must adapt to a more active and participatory role in their own learning (Table 6.9).

6.5.2 Hybrid Models: Definition and Variants

As education continues to evolve in response to technological innovation and changing learner needs, hybrid learning models have emerged as powerful alternatives to traditional classroom instruction. Also referred to as blended learning, hybrid models combine face-to-face learning with online instruction, offering flexibility, personalization, and enhanced engagement. These approaches are increasingly used in secondary, tertiary, and professional education to meet the demands of diverse learners and dynamic learning environments.

Hybrid learning strategically integrates in-person interactions with digital tools and online content delivery, creating a seamless experience that leverages the strengths of both modalities. Rather than simply supplementing classroom learning with digital resources, hybrid learning reimagines the structure of instruction to foster greater learner autonomy, collaboration, and adaptability [12, 28].

Common hybrid learning models include [29, 30]:

1. **Rotation Model**: In this model, students rotate between different learning stations or modalities, such as online learning, small group discussions, and direct teacher instruction. This model is effective in structured classroom settings and allows for differentiated instruction.
2. **Flex Model**: Here, online learning forms the backbone of the course, with face-to-face sessions used only when students require additional support. This model is commonly used in self-paced or competency-based programs, allowing students to access materials on-demand and seek help as needed.
3. **Enriched Virtual Model**: This approach includes initial or periodic in-person sessions, with the bulk of learning taking place online. It is suitable for students who can work independently and benefit from occasional direct contact with instructors or peers.

Each model can be adapted to suit specific institutional goals, subject matter, or learner profiles.

Benefits of Flipped and Hybrid Models

Both **flipped classrooms** and **hybrid learning** models share several key benefits, particularly in enhancing learner engagement, flexibility, and effectiveness.

1. **Increased Flexibility**: By blending online and offline experiences, hybrid and flipped models allow students to learn anytime and anywhere. This is especially valuable for non-traditional learners, working students, or those with accessibility needs. Learners can revisit materials, pause and reflect, and proceed at their own pace.
2. **Enhanced Engagement**: Face-to-face time is no longer consumed by passive lecture. Instead, it is repurposed for active, collaborative learning such as discussions, labs, and problem-solving activities. This promotes deeper understanding and critical thinking.
3. **Personalized Learning**: These models support differentiated instruction by allowing students to access content that suits their learning styles and levels of readiness. Teachers can offer tailored support during in-person interactions while students work through digital content independently.
4. **Better Use of Instructional Time**: Educators can use classroom time more strategically, focusing on higher-order skills and one-on-one support rather than content delivery.
5. **Improved Technological Competency**: Engaging with digital tools and platforms as part of everyday learning helps students build the digital literacy and autonomy necessary for success in modern academic and professional (Table 6.10).

Table 6.10 Benefits of flipped and hybrid models

Benefit	Flipped classroom	Hybrid model
Personalized learning	✓	✓
Student engagement	✓✓	✓
Flexibility	✓	✓✓
Use of technology	✓✓	✓✓
Deeper learning	✓✓	✓

Legend: ✓ = Moderate; ✓✓ = High

6.5.3 Implementation Strategies

The success of flipped classrooms and hybrid learning models depends not only on theoretical understanding but also on effective implementation strategies. These models shift the traditional dynamics of instruction, requiring careful planning, structured design, and sustained engagement. Whether adopting a fully flipped approach or blending in-person and online experiences, educators must align pedagogy, technology, and logistics to ensure meaningful learning outcomes.

Implementation Strategies for Flipped Classrooms

A flipped classroom inverts the conventional teaching sequence by delivering instructional content outside the classroom and using in-person time for active learning. To implement this model effectively, instructors should consider the following strategies:

1. **Create or Curate Quality Pre-Class Materials**: High-quality, accessible materials are the cornerstone of a flipped classroom. These may include screencasts, short video lectures, podcasts, readings, or interactive tutorials. Materials should be concise, well-organized, and directly aligned with in-class activities. Instructors can create original content or curate existing resources (e.g., Khan Academy, YouTube EDU, open educational resources) that match their course objectives.

2. Design Active Learning Exercises for Class: With foundational knowledge covered outside class, in-person sessions should focus on deepening understanding through active engagement. Strategies include:

a. Case studies and real-world scenarios
b. Problem-based or project-based learning
c. Group discussions and debates
d. Peer teaching and collaborative activities

These methods foster critical thinking, communication, and application of concepts.

3. Use Formative Assessments to Gauge Understanding: To ensure that students come prepared, instructors can use pre-class quizzes, discussion posts, or reflections to check comprehension. These formative assessments also help identify misconceptions and guide in-class interventions. Digital tools like Google Forms, Quizlet, or LMS-based quizzes can automate feedback and data collection.

Implementation Strategies for Hybrid Models

Hybrid or blended learning combines **online and face-to-face components**, requiring clear structures and intentional integration of modalities. Effective implementation depends on balance, flexibility, and communication.

1. **Establish a Clear Schedule and Communication Plan**: One of the biggest challenges in hybrid models is managing time and expectations. Students should have a well-defined schedule outlining which components arc online or in-person, synchronous or asynchronous. Using a course calendar and regular reminders helps prevent confusion. Communication tools such as email, discussion boards, and announcements in the LMS help keep students informed and engaged.
2. **Use LMS to Organize Content**: A centralized platform such as Moodle, Canvas, or Google Classroom allows for organized content delivery, submission tracking, and student interaction. Instructors should structure content using modules or weekly folders that include objectives, instructions, materials, and assessments. Clear navigation and consistency improve learner experience and reduce cognitive load.
3. **Integrate Synchronous and Asynchronous Activities**: Hybrid models benefit from blending live sessions (e.g., webinars, Q&A, group discussions) with asynchronous work (e.g., readings, discussion forums, quizzes). Synchronous activities build community and provide real-time feedback, while asynchronous components offer flexibility and self-paced learning.

Implementing flipped and hybrid models requires intentional planning, strong communication, and alignment between content, activities, and assessment. By curating meaningful content, designing active learning opportunities, and leveraging digital platforms, educators can create flexible, engaging, and effective learning experiences. These strategies not only support diverse learning styles but also prepare students for the autonomy and adaptability required in the modern knowledge economy (Table 6.11).

Table 6.11 Tools to support flipped and hybrid learning

Tool	Use	Model
Edpuzzle	Interactive video lectures	Flipped
Zoom	Live sessions	Hybrid
Google classroom	Content and communication hub	Both
Padlet	Collaborative space	Both

6.5.4 Challenges and Solutions

Flipped and hybrid learning models have transformed the educational landscape by promoting flexibility, student-centered engagement, and technology integration. However, implementing these models effectively requires addressing several challenges while designing appropriate assessments that align with their unique structures. Thoughtful planning, institutional support, and innovative pedagogy are key to realizing their full potential.

Challenges and Solutions.

1. Student Resistance to New Learning Models: One of the most common challenges is student discomfort or resistance toward unfamiliar learning structures. Learners accustomed to passive, lecture-based instruction may initially struggle with the demands of self-directed, pre-class preparation and in-class collaboration.

Solution: Institutions and instructors should provide a clear orientation at the beginning of the course. This includes explaining the rationale behind the flipped or hybrid approach, outlining expectations, and offering examples of how it benefits learning. Introducing these models gradually and offering structured support can ease the transition.

2. Unequal Access to Technology: Flipped and hybrid models often rely on reliable internet access, digital devices, and software platforms. The digital divide can place disadvantaged students at risk of falling behind.

Solution: Institutions must work to bridge access gaps by offering loaner devices, subsidized internet, or on-campus computer labs. Courses should be designed with accessibility in mind—ensuring materials are mobile-friendly, downloadable, and usable across multiple platforms.

3. Time-Intensive Preparation for Instructors: Developing high-quality pre-class videos, online materials, and interactive activities can be time-consuming for educators.

Solution: Instructors can use templates, open educational resources (OERs), and existing digital tools to reduce workload. Collaboration among faculty or using institutionally provided content libraries can also help streamline the process. Over time, materials can be reused and refined.

4. Maintaining Engagement in Online Components: In hybrid models, students may disengage from the asynchronous or digital components if they do not perceive them as meaningful or connected to in-class work.

Solution: Online materials must be interactive, relevant, and clearly linked to assessments or class activities. Embedding quizzes, discussion forums, multimedia content, and opportunities for real-world application can sustain interest and motivation.

Table 6.12 Assessment techniques and their purpose

Technique	Purpose	Model
Pre-class quiz	Ensure content engagement	Flipped
Online discussion rubric	Evaluate participation	Hybrid
Peer review	Develop critical thinking	Both

Assessment in Flipped and Hybrid Models

Assessment in flipped and hybrid classrooms must align with the distributed and active nature of learning. It should capture both individual effort and collaborative engagement, while also leveraging the strengths of digital platforms.

1. **Formative Assessments**: These are designed to check understanding and guide instruction. In flipped models, pre-class quizzes, interactive polls, or reflection journals ensure that students engage with preparatory material and arrive in class ready for deeper learning. Digital platforms enable immediate feedback and analytics for instructors.
2. **Summative Assessments**: End-of-unit or end-of-course assessments in these models often include projects, presentations, or online exams. These allow students to demonstrate applied understanding, creativity, and synthesis of knowledge across formats.
3. **Peer and Self-Assessments**: Encouraging peer review and self-evaluation promotes accountability, collaboration, and metacognitive skills. Tools within LMS platforms or apps like Peergrade support structured and meaningful peer assessment in hybrid settings (Table 6.12).

6.6 Digital Literacy Goals

Digital literacy encompasses the ability to effectively and critically navigate, evaluate, and create information using a range of digital technologies. In contemporary education, fostering digital literacy is essential not only for academic success but also for preparing learners for the demands of the digital economy and participatory citizenship. Establishing clear digital literacy goals within the curriculum helps ensure that students acquire the competencies needed for lifelong learning and professional growth. In the twenty-first century, digital literacy has become an essential competency for education, work, and everyday life. More than just the ability to use a computer or navigate the internet, digital literacy encompasses a broad set of skills, knowledge, and ethical considerations that enable individuals to participate fully and responsibly in a digitally connected society. It is a multidimensional concept that integrates technical proficiency with critical thinking, communication, and awareness of digital ethics [31, 32].

1. **Technical Skills**: At the most fundamental level, digital literacy includes technical proficiency—the ability to use digital devices, software, and online platforms. This includes navigating operating systems, managing files, using productivity software (e.g., word processors, spreadsheets), and engaging with digital learning environments such as LMS. Technical skills also involve the ability to troubleshoot basic issues, adapt to new tools, and keep up with the fast pace of technological change. Without these foundational skills, learners and professionals risk exclusion from educational and employment opportunities.
2. **Information Literacy**: A critical component of digital literacy is information literacy—the ability to locate, evaluate, and effectively use digital information. In an era of abundant and easily accessible online content, users must develop skills to assess the credibility, relevance, and bias of digital sources. This includes understanding search engines, using academic databases, recognizing misinformation, and citing digital sources appropriately. Information literacy empowers individuals to become informed decision-makers and responsible consumers of knowledge.
3. **Media Literacy**: Media literacy involves understanding how media messages are constructed and how they influence perceptions and behavior. In digital contexts, this includes analyzing social media, news websites, advertisements, and visual content such as videos and memes. Media-literate individuals can critique the purposes, techniques, and potential biases behind digital content. This dimension is especially crucial in combating disinformation, stereotyping, and manipulation in the digital public sphere.
4. **Communication Literacy**: Digital literacy also encompasses communication literacy—the ability to engage effectively and appropriately in digital communication. This includes writing emails, participating in forums, video conferencing, and using social media for professional or academic purposes. Communication literacy also involves digital etiquette, such as tone, clarity, and cultural sensitivity, which are essential in global and virtual collaborations.
5. **Digital Ethics and Safety**: A comprehensive understanding of digital ethics and safety is vital for responsible participation in the online world. This includes respecting copyright, protecting personal data, recognizing cyberbullying, and practicing good digital citizenship. Individuals must understand privacy settings, recognize phishing attempts, and engage in respectful online behavior. Digital literacy involves not just skill, but also ethical awareness and accountability in digital spaces.

In summary, digital literacy is a multifaceted and evolving concept that goes far beyond simple computer skills. It integrates technical ability, critical analysis, effective communication, and ethical awareness. As digital tools become increasingly embedded in education, work, and social life, fostering comprehensive digital literacy is essential for equity, empowerment, and informed participation in the digital age. Educators, policymakers, and institutions must therefore prioritize digital literacy as a core component of modern education and lifelong learning (Table 6.13).

Table 6.13 Components of digital literacy

Component	Description	Example
Technical skills	Basic to advanced tool use	Operating LMS, editing videos
Information literacy	Search, validate, and apply data	Researching scholarly articles
Media literacy	Critique and interpret media	Analyzing advertisements
Communication	Effective digital interaction	Writing emails, forums, chats
Digital ethics	Online behavior and safety	Avoiding plagiarism, using secure passwords

6.6.1 Setting Digital Literacy Goals

In an increasingly digital world, digital literacy has become a foundational skill for learners across all levels of education. To ensure students are prepared for the challenges and opportunities of the digital age, educational institutions must articulate clear, measurable digital literacy goals. These goals should be closely aligned with course outcomes, institutional missions, and evolving technological trends, ensuring that digital literacy is not treated as an add-on, but as an integral part of the learning experience [31].

Purpose of Digital Literacy Goals

Digital literacy goals serve multiple purposes. They provide a roadmap for curriculum development, guide instructional strategies, and ensure students acquire the skills they need to thrive academically and professionally. These goals also allow educators to assess and track progress, helping institutions identify gaps in digital competence and provide targeted support. Furthermore, clearly defined goals communicate expectations to students and prepare them to engage responsibly, creatively, and critically in digital environments.

Sample Digital Literacy Goals

Effective digital literacy goals are specific, relevant, and assessable. Examples include:

1. Students will demonstrate proficiency in using collaborative tools such as Google Docs, Microsoft Teams, or Slack to complete group projects and engage in peer feedback.
2. Students will evaluate digital sources for credibility and bias, distinguishing between scholarly articles, opinion pieces, sponsored content, and misinformation.
3. Students will design and present multimedia content—such as videos, infographics, or digital presentations—to communicate ideas to various audiences effectively.

Table 6.14 Comparison of digital literacy frameworks

Framework	Focus	Key features
ISTE	K-12 and higher	Empowered learner, digital citizen, knowledge constructor
UNESCO	Global equity	Access, skill development, lifelong learning
JISC	UK higher	6 elements: ICT proficiency, digital creation, digital identity, etc

4. Students will demonstrate ethical behavior in digital environments, including respecting intellectual property, protecting personal information, and engaging in respectful online discourse.

These goals not only support technical skills but also promote critical thinking, creativity, and ethical awareness, which are essential in all disciplines and professions.

Frameworks and Standards for Digital Literacy

To ensure coherence and best practices, institutions often rely on **established frameworks and standards** to guide the development of digital literacy goals. These models offer structured pathways and benchmarks to support the integration of digital skills across curricula.

1. **ISTE Standards (International Society for Technology in Education)**: The ISTE Standards provide a globally recognized framework that emphasizes student empowerment, digital citizenship, computational thinking, and creative communication. These standards encourage learners to be knowledge constructors, innovative designers, and global collaborators in digital environments.
2. **UNESCO Digital Literacy Global Framework**: UNESCO's framework supports the development of inclusive, equitable digital competencies. It outlines core skills across multiple domains including information and data literacy, communication and collaboration, digital content creation, safety, and problem-solving. It is particularly relevant for shaping policy-level goals and ensuring global alignment with digital skills development.
3. **JISC Digital Capabilities Framework**: Developed in the UK, JISC's framework focuses on building digital capability for both students and educators. It emphasizes seven key elements, including digital communication, data literacy, ICT proficiency, and digital identity management. JISC offers diagnostic tools and resources to help institutions evaluate and improve digital competency levels (Table 6.14).

6.6.2 Curriculum Integration

As digital tools increasingly shape education, work, and everyday communication, digital literacy must become a foundational component of modern curricula. Rather than being treated as an isolated skillset or a single course, digital literacy should be integrated throughout the curriculum, supported by appropriate teaching strategies, and evaluated using robust assessment methods. This holistic approach ensures that learners not only acquire technical proficiency but also develop critical thinking, ethical awareness, and the adaptability needed in a dynamic digital world.

Curriculum Integration

Effective curriculum integration embeds digital literacy across disciplines, levels, and learning outcomes. It should reflect the specific needs and practices of each academic field while supporting students' progressive development.

1. **Discipline-Specific Applications**: Digital literacy must be contextualized to different fields. For instance, data analysis may be emphasized in social sciences, visual storytelling in media studies, and coding or algorithmic thinking in computer science. Embedding digital tasks relevant to each discipline fosters deeper engagement and real-world applicability.
2. **Scaffolded Progression**: Skills should be introduced incrementally, starting with foundational digital competencies (e.g., file management, online research, basic communication) in early years, then advancing to complex skills such as digital content creation, data visualization, and ethical analysis in later stages.
3. **Capstone Integration**: Capstone or final-year projects should synthesize digital skills developed throughout the program. Whether through e-portfolios, research websites, podcasts, or data dashboards, these cumulative tasks provide authentic assessments of a student's digital fluency.

Teaching Strategies

To cultivate digital literacy, instructors must adopt pedagogies that actively engage students in the creation and use of digital content. Effective strategies include:

1. **Project-Based Learning (PBL)**: Encouraging students to develop digital artefacts such as blogs, websites, videos, or infographics not only builds technical skills but also fosters creativity, research, and communication.
2. **Flipped Classrooms**: This model supports self-paced digital engagement, allowing students to explore digital resources (videos, readings, tutorials) independently before applying their knowledge in interactive class sessions.
3. **Peer Collaboration**: Leveraging collaborative tools like Google Docs, Microsoft Teams, or Padlet helps students develop teamwork and communication skills in digital environments while simulating real-world professional contexts.
4. **Workshops and Bootcamps**: Targeted sessions—either standalone or embedded within courses—can provide intensive, hands-on training in tools such as data analysis software, digital publishing platforms, or media editing tools, catering to different levels of proficiency.

Assessment of Digital Literacy

Assessment of digital literacy should align with learning goals and reflect both **technical capability** and **critical engagement**. A variety of methods can be used:

Table 6.15 Sample assessment criteria for digital literacy

Criterion	Description	Assessment Tool
Technical execution	Proficiency in digital tool use	Performance tasks
Critical analysis	Evaluation of digital sources	Annotated bibliographies
Creativity	Original expression in digital media	Multimedia presentations
Ethics	Understanding of digital responsibility	Reflective journals

1. **Digital Portfolios**: These allow students to showcase their work across different platforms and formats, illustrating growth, creativity, and breadth of skill.
2. **Rubrics**: Well-defined rubrics help evaluate technical proficiency, but also include criteria for ethical use of technology, information credibility, and critical thinking in digital contexts.
3. **Self and Peer Assessments**: These encourage students to reflect on their own practices and learn from others, promoting accountability and metacognitive awareness of digital habits.

Thus, integrating digital literacy throughout the curriculum is essential for preparing students to thrive in an interconnected, digital-first world. With thoughtful integration, engaging teaching strategies, and meaningful assessment, institutions can foster digitally fluent graduates who are not just tech-savvy, but also critical, ethical, and creative users of digital technology (Table 6.15).

6.6.3 Challenges and Recommendations

The integration of digital tools and digital literacy into educational curricula is no longer optional—it is essential. As institutions embrace technology to enrich teaching and learning, they also encounter a range of challenges that must be addressed strategically. At the same time, emerging trends such as AI, VR, and data analytics are reshaping the future of education, making it more interactive, personalized, and data-driven.

Challenges in Implementing Digital Curricula

Despite the benefits of digital transformation, institutions face several persistent barriers:

1. **Disparities in Student Access to Technology**: The digital divide continues to be a major concern. Not all students have equal access to reliable internet, devices, or conducive learning environments. This disparity affects engagement, performance, and overall equity in digital learning.

2. **Rapid Technological Change**: The speed at which digital tools and platforms evolve creates a moving target for curriculum designers. Software updates, platform changes, and the emergence of new technologies often outpace institutions' ability to adapt teaching materials and retrain staff.
3. **Faculty Preparedness**: Many educators feel underprepared to teach with or through digital tools. Without adequate support, they may default to using technology in limited or ineffective ways—undermining the potential of digital integration.

Recommendations

To address these challenges, institutions must adopt proactive and inclusive strategies:

1. **Ensure Equitable Access**: Institutions should provide reliable infrastructure, such as loaner devices, Wi-Fi hotspots, and access to digital tools through centralized platforms. Equitable access must be foundational to any digital learning strategy.
2. **Support Continuous Professional Development**: Regular and ongoing training for educators is essential. This includes not only technical skills, but also digital pedagogy, instructional design, and the ethical use of technology. Creating communities of practice and offering mentorship programs can further sustain faculty engagement and growth.
3. **Align Technology with Pedagogy**: Rather than chasing trends, institutions should invest in technologies that align with learning goals and enhance instructional effectiveness. This reduces wasted effort and ensures meaningful impact on student outcomes.

Future Trends in Digital Curricula

Looking ahead, several technological advancements are set to **reshape digital education**:

- **Artificial Intelligence (AI)**: AI is increasingly used in education for personalized learning paths, automated feedback, intelligent tutoring systems, and adaptive assessments. AI can help instructors **identify learning gaps in real-time** and offer tailored support, enhancing both student achievement and instructional efficiency.
- **Virtual and Augmented Reality (VR/AR)**: Immersive technologies offer **experiential learning opportunities** that are especially valuable in disciplines such as medicine, engineering, history, and the arts. VR simulations allow students to practice skills or explore environments in ways that traditional materials cannot provide.
- **Data Analytics and Learning Dashboards**: Institutions are using **learning analytics** to monitor student engagement and performance. Dashboards can inform real-time curriculum adjustments and targeted interventions, enabling **data-informed teaching** that responds to learner needs dynamically.

Thus, as digital education continues to evolve, addressing the current challenges while anticipating future trends is essential for building resilient, inclusive, and future-ready curricula. With the right infrastructure, faculty support, and commitment to innovation, educational institutions can harness emerging technologies to deliver transformative and equitable learning experiences for all students.

References

1. Mostert, M., Quinn, L.: Using ICTs in teaching and learning: reflections on professional development of academic staff. J. Educ. **5**(5), 72–84 (2009)
2. Kadwa, M.S., Alshenqeeti, H.: A postcolonial ecocritical reading of Zakes Mda's the whale caller and Helon Habila's oil on water. Int. J. Linguist. Lit. Transl. **3**(11), 55–67 (2023). https://doi.org/10.32996/ijllt
3. Tiron-Tudor, A., Deliu, D.: Reflections on the human-algorithm complex duality perspectives in the auditing process, vol. 19, no. 3 (2022). https://doi.org/10.1108/QRAM-04-2021-0059.
4. Coursin, A.M.: Understanding the development of alphabet knowledge in at-risk populations: The influence of pre-literacy skills. ProQuest Diss. Theses, no. August, p. 70 (2012). https://search.proquest.com/docview/1145552120?accountid=14548%0Ahttps://julac.hosted.exlibrisgroup.com/openurl/HKU_ALMA/SERVICES_PAGE??url_ver=Z39.88-2004&rft_val_fmt=info:ofi/fmt:kev:mtx:dissertation&genre=dissertations+%26+theses&sid=ProQ:ProQuest+Diss
5. Eguavoen, E.O.: ICT utilization as correlates of academic performance among students with visual impairment in Lagos State, Nigeria. Eur. Sci. J. ESJ **12**(13), 205 (2016). https://doi.org/10.19044/esj.2016.v12n13p205
6. Najafi, S.E., Nozari, H., Edalatpanah, S.A.: Artificial intelligence of things (AIoT) and industry 4.0-based supply chain (FMCG Industry). A Roadmap Enabling Ind. 4.0 by Artif. Intell. 31–42 (2022). https://doi.org/10.1002/9781119905141.ch3
7. Teizer, J. et al.: Internet of Things (IoT) for integrating environmental and localization data in Building Information Modeling (BIM). In: ISARC 2017-Proceedings of the 34th International Symposium on Automation and Robotics in Construction, no. Isarc, pp. 603–609 (2017). https://doi.org/10.22260/isarc2017/0084
8. Adebanji, B., Ojo, A., Fasina, T., Adeleye, S., Abere, J.: Integration of renewable energy with smart grid application into the Nigeria's power network: issues, challenges and opportunities. Eur. J. Eng. Technol. Res. **7**(3), 18–24 (2022). https://doi.org/10.24018/ejeng.2022.7.3.2792
9. Odeyemi, O., Awonuga, K.F., Mhlongo, N.Z., Ndubuisi, N.L., Olatoye, F.O., Daraojimba, A.I.: The role of AI in transforming auditing practices: a global perspective review. World J. Adv. Res. Rev. **21**(2), 359–370 (2023). https://doi.org/10.30574/wjarr.2024.21.2.0460
10. Aswir, Misbah, H.: Smartphones as personal digital archives? Recentering migrant authority as curating and storytelling subjects. Photosynthetica **2**(1), 1–13 (2018). http://link.springer.com/https://doi.org/10.1007/978-3-319-76887-8
11. Li, X., et al.: Blockchain-Enabled IoT-BIM Platform for Supply Chain Management in Modular Construction, vol. 148, no. 2 (2022). https://doi.org/10.1061/(asce)co.1943-7862.0002229
12. Obi, K.C., Iwuanyanwu, O.: Innovation in the assessment of primary school mathematics: an approach to revamping Nigeria education. J. Teach. Perspect. **18**(1), 51–66 (2023). http://repositorio.unan.edu.ni/2986/1/5624.pdf%0Ahttp://fiskal.kemenkeu.go.id/ejournal%0Ahttps://doi.org/10.1016/j.cirp.2016.06.001%0A10.1016/j.powtec.2016.12.055%0A10.1016/j.ijfatigue.2019.02.006%0A10.1

13. Benzidia, S., Makaoui, N., Bentahar, O.: The impact of big data analytics and artificial intelligence on green supply chain process integration and hospital environmental performance. Technol. Forecast. Soc. Change **165** (2021). https://doi.org/10.1016/j.techfore.2020.120557
14. Mikre, F.: The roles of information communication technologies in education: review article with emphasis to the computer and internet. Ethiop. J. Educ. Sci. (2011)
15. Lo Presti, A., Manuti, A., De Rosa, A., Elia, A.: Developing a sustainable career through discourse: a qualitative study on a group of Italian project managers. Int. J. Manag. Proj. Bus. **15**(8), 1–18 (2021). https://doi.org/10.1108/IJMPB-12-2020-0376
16. Don-Solomon, A., Eke, G.J.: Ontological & epistemological philosophies underlying theory building: a scholarly dilemma or axiomatic illumination- the business research perspective, vol. 6, no. 2, pp. 1–7 (2018)
17. Mursyidto, M.I.: The role of information communication technology in records creation and distribution in Uganda communications commission. Implement. Sci. **39**(1), 1–15 (2014). https://doi.org/10.4324/9781315853178
18. Aria Adibrata, J., Fikhri Khairi, N.: The impact of covid-19 blame game towards anti-Asian discrimination phenomena. J. Soc. Media **6**(1), 17–38 (2022). https://doi.org/10.26740/jsm.v6n1.p17-38
19. Bernstein, S.: The absence of great power responsibility in global environmental politics. Eur. J. Int. Relations **26**(1), 8–32 (2020). https://doi.org/10.1177/1354066119859642
20. Korot, E., et al.: Enablers and barriers to deployment of smartphone-based home vision monitoring in clinical practice settings. JAMA Ophthalmol. **140**(2), 153–160 (2022). https://doi.org/10.1001/jamaophthalmol.2021.5269
21. Kazmi, Z.A., Sodangi, M.: Modeling the constraints to the utilization of the internet of things in managing supply chains of off-site construction: an approach toward sustainable construction. Buildings **12**(3) (2022). https://doi.org/10.3390/buildings12030388
22. Cho, M., Ha, J., Park, C., Park, S.: Combinatorial feature embedding based on CNN and LSTM for biomedical named entity recognition. J. Biomed. Inform. **103**, 103381 (2020). https://doi.org/10.1016/j.jbi.2020.103381
23. Kehayov, M., Holder, L., Koch, V.: Application of artificial intelligence technology in the manufacturing process and purchasing and supply management. Procedia Comput. Sci. **200**(2019), 1209–1217 (2022). https://doi.org/10.1016/j.procs.2022.01.321
24. Chen, P., Kim, S.K.: The impact of digital transformation on innovation performance-the mediating role of innovation factors. Heliyon **9**(3), e13916 (2023). https://doi.org/10.1016/j.heliyon.2023.e13916
25. Ghosh, A., Edwards, D.J., Hosseini, M.R.: Patterns and trends in Internet of Things (IoT) research: future applications in the construction industry. Eng. Constr. Archit. Manag. **28**(2), 457–481 (2021). https://doi.org/10.1108/ECAM-04-2020-0271
26. Umek, L., Keržič, D., Tomaževič, N., Aristovnik, A.: Moodle e-learning system and students' performance in higher education: the case of public administration programmes. Proc. Int. Conf. e-Learning 2015, E-LEARNING 2015 - Part Multi Conf. Comput. Sci. Inf. Syst. 97–104 (2015)
27. Ghavifekr, S., Rosdy, W.A.W.: Teaching and learning with technology: effectiveness of ICT integration in schools. Int. J. Res. Educ. Sci. **1**(2), 175 (2015). https://doi.org/10.21890/ijres.23596
28. Liu, Z., Tang, B., Wang, X., Chen, Q.: De-identification of clinical notes via recurrent neural network and conditional random field. J. Biomed. Inform. **75**(34), S34–S42 (2017). https://doi.org/10.1016/j.jbi.2017.05.023
29. Uzuner, Ö., Solti, I., Cadag, E.: Extracting medication information from clinical text. J. Am. Med. Informatics Assoc. **17**(5), 552–556 (2010). https://doi.org/10.1136/jamia.2010.003947
30. Yang, H., Spasic, I., Keane, J.A., Nenadic, G.: A text mining approach to the prediction of disease status from clinical discharge summaries. J. Am. Med. Informatics Assoc. **16**(4), 596–600 (2009). https://doi.org/10.1197/jamia.M3096

31. Alaba, F.A., Amina, M., Ibrahim, U., Sulaimon, H.A., Zainab Aliyu, A.: Exploring the role of the internet of things (IoT) on economic growth in Nigeria: challenges and opportunities. Eng. Technol. Trends **2**(6) (2024)
32. Purcell, C., et al.: A peer-led, school-based social network intervention for young people in the UK, promoting sexual health via social media and conversations with friends: intervention development and optimisation of STASH. BMC Public Health **23**(1), 1–12 (2023). https://doi.org/10.1186/s12889-023-15541-x

Chapter 7
Interdisciplinary Collaboration and Co-teaching Models

7.1 Benefits of Cross-Disciplinary Teaching

Cross-disciplinary teaching, where educators from different academic fields collaboratively design and deliver instruction, offers a transformative approach to modern education. Students are better prepared for real-world difficulties that seldom fit neatly inside one area, and it stimulates the synthesis of information, which breaks down conventional academic silos, and it breaks down traditional academic silos [1] (Fig. 7.1 and Table 7.1). The most important advantages are as follows:

a. **Enhanced Student Engagement and Learning**: Cross-disciplinary education encourages active learning by bringing together different views. When students realise how various areas intersect. For example, how mathematics relates to art or how science influences politics, they are more likely to find meaning and purpose in their studies. This real-world application improves cognitive engagement and comprehension.
b. **Development of Critical Thinking and Problem-Solving Skills**: Students are exposed to various investigative and analytical techniques. For instance, integrating engineering with ethics in one course prompts students to consider both the practicality of technology and their ethical obligations. This fosters advanced cognitive skills, enabling students to tackle complex, multifaceted issues with a more comprehensive viewpoint.
c. **Improved Collaboration and Communication Skills**: Working across disciplines demonstrates essential collaboration abilities in today's industry. Students learn to convey ideas in a variety of intellectual languages, negotiate points of view, and combine contributions from many fields to create coherent solutions.
d. **Teacher Professional Growth**: Educators gain from acquiring new teaching methods and topics outside of their field. This cross-pollination of ideas can result in better pedagogical strategies, more creativity, and a renewed passion for teaching.

F. A. Alaba and I. G. Ajibola, *Digital Pedagogy: Bridging the Gap Between Software and the Humanities*, Information Systems Engineering and Management 85,
https://doi.org/10.1007/978-3-032-20335-9_7

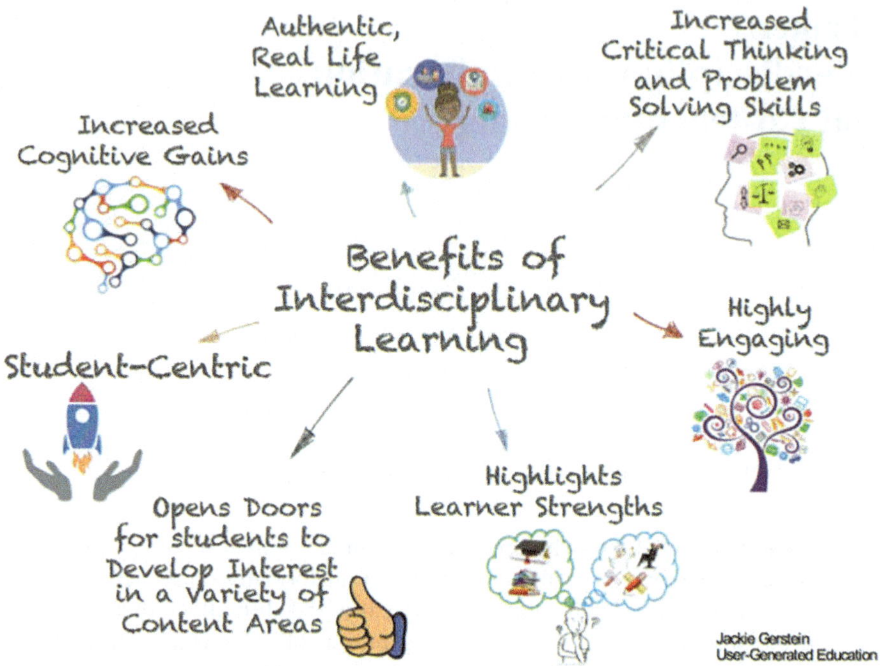

Fig. 7.1 Benefits of cross-disciplinary learning [2]

Table 7.1 Benefits of cross-disciplinary teaching for stakeholders

Stakeholder	Benefit
Students	Greater engagement, real-world application, and improved critical thinking
Educators	Broader perspectives, professional growth, and enhanced collaboration skills
Institutions	Curriculum innovation, increased student retention and satisfaction
Employers	Graduates with versatile skills, adaptability, and interdisciplinary fluency

e. **Curriculum Innovation**: Cross-disciplinary teaching fosters curriculum reform that is more integrative and experiential. It often leads to project-based or themed units that reflect real-world challenges, resulting in a more engaging educational experience.

7.2 Challenges in Interdisciplinary Work

Interdisciplinary work promises to provide innovative, creative, and comprehensive methods to education and issue resolution. Despite its revolutionary potential, it is riddled with difficulties that might jeopardise its success. These obstacles stem

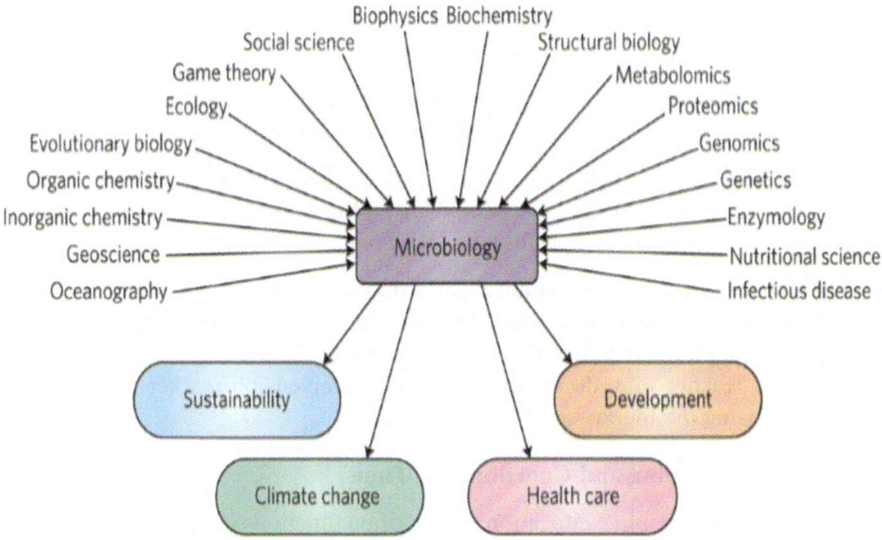

Fig. 7.2 Challenges in interdisciplinary work [4]

from long-standing academic traditions, institutional frameworks, practical considerations, and interpersonal interactions. This section delves into these problems in depth, highlighting typical mistakes and structural limits while providing graphic aids to explain and contextualise essential themes [3] (Fig. 7.2).

Differences in Disciplinary Language and Epistemology

Each discipline has a unique conceptual framework, language, beliefs, and assumptions. For example, a historian may prioritise context and story, while a physicist seeks universal rules and quantitative evidence. These epistemological distinctions determine how issues are framed, what evidence is accepted, and how knowledge is transmitted [5].

Such discrepancies may lead to:

1. Miscommunication: The word "model" might refer to a mathematical equation in engineering, a conceptual picture in social science, or a scale replica in architecture.
2. Divergent Priorities: One field may favour accuracy and objectivity, whilst another values reflexivity and criticism.
3. Difficulty in Synthesising Ideas: Attempting to combine these disparate methods without degrading either may be intellectually and pedagogically challenging.

For example, in a co-taught course on "Climate Change," the scientific professor may concentrate on data modelling and forecasts, while the humanities lecturer emphasises ethical arguments and historical reasons. Without explicit coordination, students may find it difficult to reconcile various methods.

a. **Power Imbalances and Role Ambiguity**

Collaborative education requires equality, but in fact, certain subjects may be considered more "central" or "prestigious," typically due to institutional prestige, financing, or social importance. This creates imbalanced dynamics in which:

1. One teacher may take on more decision-making authority.
2. The importance of the "secondary" discipline is underestimated.
3. Students may see one professor as the "real" expert.

Furthermore, when roles are not clearly defined, tasks such as grading, lesson preparation, and classroom leadership may become disputed or repeated.

For example, in a course that combines literature and computer science, the technical teacher may dominate owing to perceived job-market relevance, so marginalising the literary analysis component.

b. **Increased Planning and Coordination Time**

Unlike single-discipline education, which often functions within the confines of a particular subject area, interdisciplinary work necessitates frequent and persistent cooperation among teachers. This constant collaboration is vital for maintaining coherence, alignment, and depth across many disciplines. Key collaboration endeavours include the following:

1. Syllabus and teaching approaches must be aligned.
2. Agreeing on common goals and results.
3. Plan integrated evaluations and activities.

This greatly increases effort, particularly when institutional assistance (such as time release or collaborative scheduling systems) is unavailable. Even well-intentioned cooperation might be derailed by a busy academic calendar and competing commitments.

c. **Assessment Difficulties**

Evaluation is one of the most difficult topics in interdisciplinary education. Unlike typical single-subject assessments, interdisciplinary learning assesses students' ability to synthesise information, methodologies, and viewpoints from various disciplines. Students are required to combine multiple perspectives; hence the following significant challenges arise:

1. Lack of explicit rubrics: Traditional grading may not account for multidisciplinary learning outcomes.
2. Instructor disagreement: Teachers from various disciplines may favour different abilities (for example, originality vs. precision).
3. Students may be confused about how their work will be graded if expectations are unclear.

For instance, a capstone project in a cross-disciplinary sustainability course could encompass statistical analysis, policy recommendations, and visual narrative. Assessing the significance of each element might prove challenging.

d. **Institutional and Administrative Barriers**

Most universities and colleges are divided into departmental silos that manage funding, faculty recruiting, course offerings, and evaluations. This framework impedes cross-disciplinary cooperation in the following ways (Fig. 7.3 and Table 7.2).

1. Funding disputes: Which department pays for shared resources?
2. Credit hour conflicts: Which course does it count towards?
3. Concerns about promotion and tenure: Faculty members may be concerned that their multidisciplinary work will not be rewarded by standard measures.

e. **Cultural Resistance and Risk Aversion**

The benefit of interdisciplinary education is viewed with scepticism by some educators, who are concerned that it may weaken the academic rigour of their students or that it may lack consistency. When teaching outside of their area of expertise, some people may experience feelings of unpreparedness or discomfort. This resistance might be due to:

1. Philosophical: The idea that disciplines should stay independent and specialised.
2. Practical: Concerns about teacher assessment or student feedback.
3. Institutional: Absence of training, incentives, or administrative support.

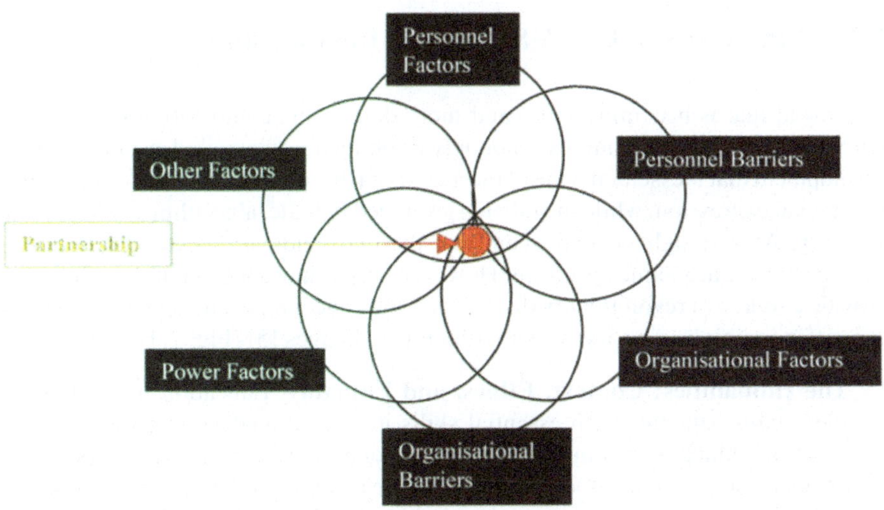

Fig. 7.3 Institutional and administrative barriers in interdisciplinary work [6]

Table 7.2 Summary of key challenges in interdisciplinary work

Challenge	Description	Impact
Disciplinary language & methodology	Varied terminologies and epistemological foundations	Communication breakdowns; integration difficulty
Power imbalance & role ambiguity	Unequal status or vague responsibilities	Teacher tension; reduced learning quality
High planning time	Need for joint planning and alignment	Faculty overload; logistical delays
Assessment complexity	Diverse criteria and expectations	Confusing grading; student anxiety
Institutional barriers	Departmental silos and funding disputes	Limited support; discouragement of collaboration
Cultural resistance	Skepticism about interdisciplinary value	Reluctance to participate; low innovation uptake
Student cognitive load	Difficulty processing multiple perspectives	Confusion; lower performance; disengagement

f. **Student Readiness and Cognitive Load**

It is not possible for all pupils to be able to think across different fields. They could be able to:

1. Expect clear boundaries between subjects.
2. Complicated the process of incorporating numerous points of view.
3. Experience "cognitive overload" as a result of juggling a variety of frameworks and content elements.

7.3 Humanities + CS: A Symbiotic Relationship

In a world that is becoming more and more digital, the combination of computer science (CS) and the humanities is not only a confluence of novel ideas; rather, it is a development that is essential. One of the areas is anchored in critical thinking, culture, and human expression, while the other is grounded in logic, algorithms, and machine accuracy. At first sight, it may seem that the two fields are located on opposing extremes of the academic spectrum. However, when they are brought together, they provide a source of revolutionary discoveries, they encourage ethical innovation, and they offer comprehensive answers to difficult challenges [5] (Fig. 7.4).

1. **The Humanities: Context, Ethics, and Empathy**: Education in the humanities equips students with essential skills in critical analysis, historical understanding, ethical reasoning, and effective communication. These competencies are increasingly vital in today's world, where technology significantly influences social behaviour, governance, privacy, and identity. Questions such as 'Should

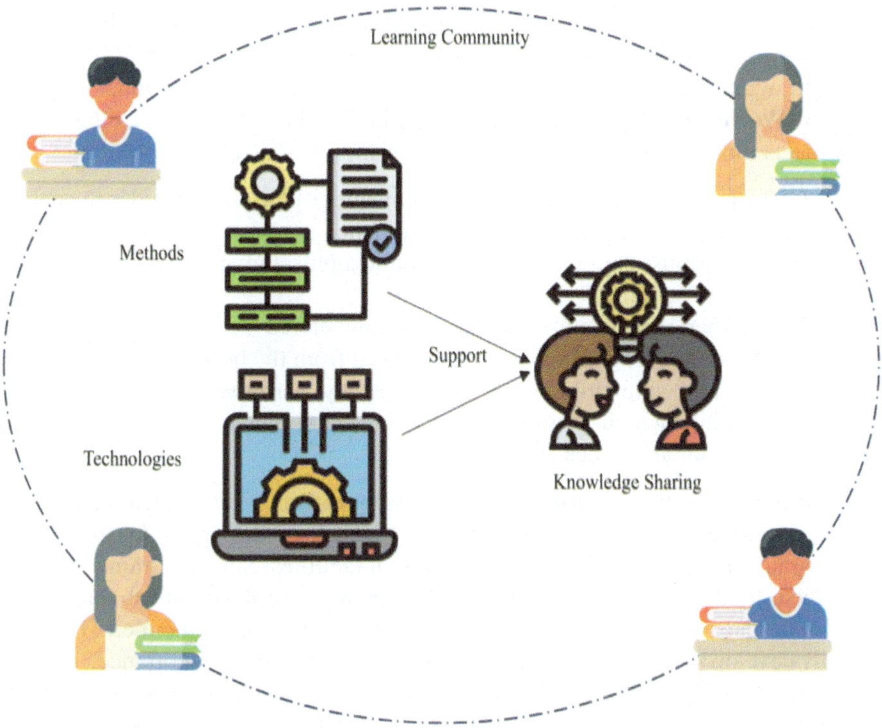

Fig. 7.4 Humanities and computer science relationship

we proceed with this development?' or 'What impact will this have on communities?' cannot be resolved solely through coding; they require the reflective insights provided by fields like philosophy, literature, and history. For example, the development of artificial intelligence necessitates humanistic reasoning to ensure justice, inclusivity, and cultural awareness. Literary theory enhances NLP by clarifying the nuances of metaphor and narrative structure. Additionally, ethics and philosophy serve as valuable frameworks in establishing ethical artificial intelligence, algorithmic transparency, and data privacy measures.

2. **Computer Science: Tools, Scale, and Innovation**: The field of computer science offers a variety of scalable technological tools and solutions. Computational linguistics enables the preservation of endangered languages in the humanities, the modelling of historical events, the visualisation of cultural data, and the analysis of vast text corpora (digital humanities). Furthermore, it allows for the modelling of historical events. Thanks to advancements in algorithms, data structures, and machine learning, humanities researchers can now pose innovative questions, analyse extensive materials, and share their findings with diverse audiences. For instance, sentiment analysis is employed to explore literary

trends across different periods, and digital archives have revolutionized historical research methodologies. Computational techniques provide both speed and depth, while humanistic insights ensure relevance and interpretation.

3. **Real-World Applications: Where Collaboration Thrives**: Collaborative projects illustrate the power of this relationship. Among the examples are:
    a. CS technologies are used by digital humanities platforms in order to analyze literature, artwork, and historical documents.
    b. AI algorithms that include moral philosophy and machine learning are referred to be ethical.
    c. User experience is informed by human-centered design in technological goods, which is based on empathy derived from the humanities.
    d. Games and simulations that are driven by narratives and make use of storytelling, which is an essential humanistic talent, and are powered by computational engines.
4. **Bridging the Divide: Educating for Integration**: Educational institutions are increasingly adopting a collaborative approach through interdisciplinary programs, joint degrees, and shared research facilities. Students trained in these diverse fields exhibit greater adaptability, possess a robust ethical framework, and are capable of fostering innovation while remaining principled. They not only develop systems but also create systems of substantial importance. To avoid isolating any discipline, it is essential to promote co-teaching strategies, collaborative research initiatives, and interdisciplinary efforts. While computer science inspires the humanities to adopt a more empathetic perspective, it simultaneously challenges the humanities to explore novel modes of thought and expression.

7.4 Planning Interdisciplinary Courses

Creating effective interdisciplinary courses that merge computer science with the humanities requires meticulous planning, the setting of shared goals, and a comprehensive understanding of the teaching methods employed in both disciplines. These courses must transcend superficial engagement with topics and promote genuine integration, ensuring that methodologies, inquiries, and concepts from each field are given equal consideration [7] (Fig. 7.5 and Table 7.3).

Principles for Course Design

1. **Shared Learning Outcomes**: Beginning with the identification of objectives that need both technical and humanistic knowledge is the first step. Take, for example:
    a. Investigate the ethical repercussions that might result from algorithmic decision-making.
    b. When investigating historical or literary datasets, you may make use of computational techniques.

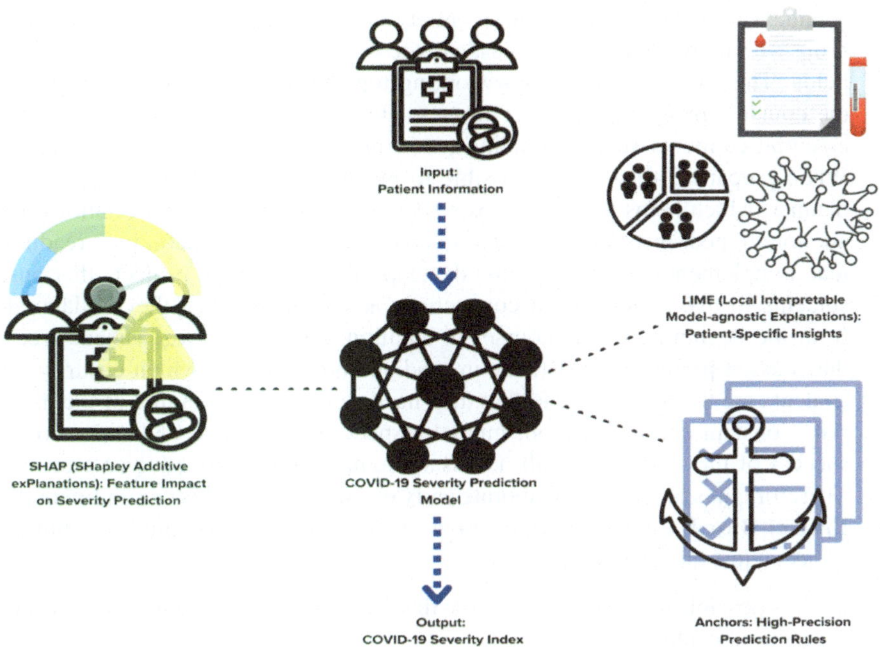

Fig. 7.5 Team principle for successful interdisciplinary research [7]

Table 7.3 Example course structures

Course title	Focus	Sample topics	Outcomes
Code & culture	Cultural studies + programming	Identity in social media, coding bias, meme cultures	Analyze and build culturally-aware software
Ethics & AI	Philosophy + AI	Moral theories, data privacy, and automation ethics	Evaluate and design ethically sound AI
Digital storytelling	Literature + CS	Narrative theory, game design, interactive media	Create interactive fiction or narrative-based games
History in data	History + Data Science	Archival methods, visualization, time-series analysis	Develop digital tools to analyze historical trends

c. It is important to do an in-depth analysis of digital depictions of culture or identity.

2. **Balanced Team Teaching**: Co-teaching by faculty from both domains ensures students benefit from diverse disciplinary expertise, gaining a more comprehensive and well-rounded understanding of the subject. The learning experience

is enhanced when teachers from a variety of areas work together because they bring with them ideas, approaches, and views that are unique to their respective fields. The faculty should begin working together early on, preferably during the course's inception, in order to guarantee that they are in agreement over essential elements such as readings, assignments, and evaluation procedures. When preparing ahead, it is possible to integrate two disciplines in a careful manner, which helps to prevent disputes at the last minute or expectations that are not aligned. Instructors are able to construct coherent syllabi, pick readings that complement one another, and develop interdisciplinary projects that push students to make meaningful connections across areas when they collaborate from the beginning of the course. In addition, coordinated assessment procedures assist preserve consistency in grading and feedback, which ensures that students are treated fairly and that they have a clear understanding of the material. Constant communication, mutual respect, and a shared dedication to the success of the students are all necessary components of effective co-teaching. Therefore, it is crucial to collaborate early on and maintain this commitment [8].

3. **Project-Based Learning**: Encourage students to work on interdisciplinary projects. just one example:

   a. It is possible to construct a chatbot that takes into account empathy, tone, and cultural context.
   b. With the help of critical theory, you may create a digital archive that includes metadata labelling.
   b. Construct a graphic representation that allows for interaction with philosophical disputes or historical contentions.

4. **Diverse Assessments**: Put together humanistic components with technical outputs, such as reflective essays and critical assessments. Examples of technological deliverables include code and data analysis. Both skill sets are given more weight in this dual evaluation system, which emphasises their significance.
5. **Scaffolded Skill Development**: The level of expertise that each student brings to the table will vary. Provide students with courses that bridge gaps, such as basic ethics and philosophy for computer science students or beginning coding for those studying the humanities.
6. **Interdisciplinary Readings**: Texts from both fields should be included. Take, for example:

   a. A book about data ethics written by Cathy O'Neil titled "Weapons of Math Destruction"
   b. The Ethical Algorithm: A Myth or Representation? (In Computer Science and Philosophy)
   c. The use of computational creativity in works of art and literature

7.5 Co-teaching Structures and Models

Successful interdisciplinary education between the Humanities and Computer Science relies on well-structured co-teaching frameworks. These models foster the integration of diverse skills, encourage mutual respect among disciplines, and enhance student engagement by providing a solid foundation. Co-teaching goes beyond merely sharing tasks; it involves collaborating to cultivate a learning atmosphere where both technical and humanistic viewpoints are valued equally [9, 10] (Fig. 7.6).

Common Co-teaching Models

1. **Team Teaching (Full Integration Model)**: Both instructors are actively involved in all aspects of the course from planning and delivery to assessment. When they are in the classroom together, they co-lead conversations, and they reply to students collaboratively. Courses that require a profound integration of abilities and views are ideal for this. As an example, a professor of philosophy and a computer scientist collaborate to teach a course titled "Ethics in Artificial Intelligence." They do this by co-facilitating discussions on case studies and critical evaluations of code.
2. **Parallel Teaching**: When instructors teach the same material to many groups of students at the same time or in rotation, they are teaching concurrently. After that,

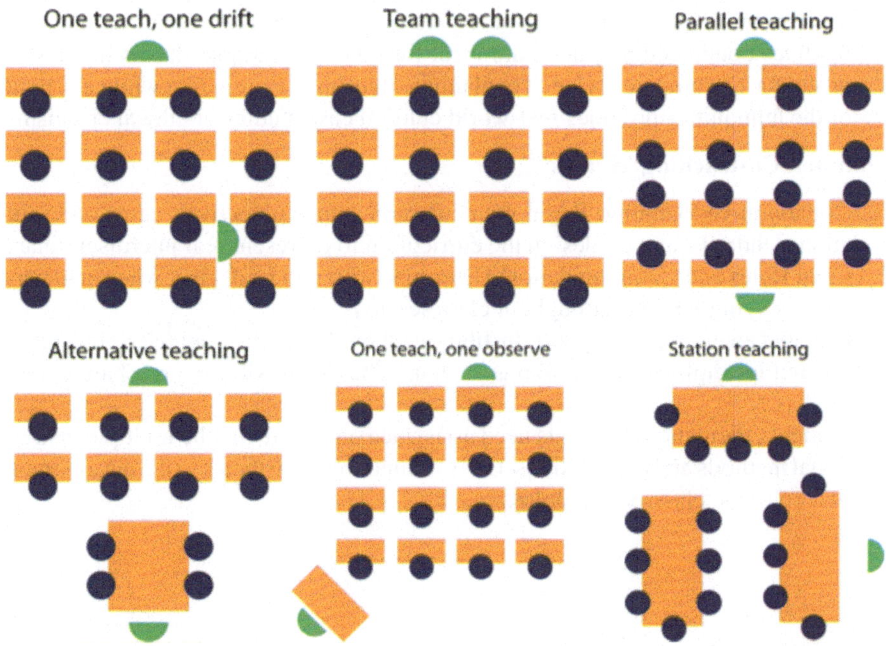

Fig. 7.6 Co-teaching models

the students get together again to discuss and integrate their newfound knowledge. Best for: Subjects that need for a solid understanding of the discipline before synthesis can be performed. In this example, students take turns participating in a Python lab for text mining and a workshop on literary analysis. After each session, they return to integrate their results.

3. **Station Teaching**: Each station is guided by a different lecturer or facilitator, and students go through them in a rotational fashion. In order to contribute to a wider interdisciplinary endeavour, each station investigates a subject from the perspective of a particular discipline. Best for: Hands-on, project-based educational environments. An example of this might be a digital storytelling course in which one station focuses on teaching narrative structure, another station discusses interactive media technologies, and a third station investigates user experience design.
4. **Sequential Teaching**: Initially, one educator teaches the fundamental principles, followed by another instructor who expands on those concepts from the perspective of a different discipline. This method is particularly effective when skills must be acquired in a specific order. This course is ideal for introductory modules or courses that draw from many disciplines. In this example, a lecturer in computer science presents data analysis techniques to a group of historians before a historian uses those approaches to the collection of documents.
5. **Complementary Teaching**: One of the instructors takes the lead, while the other teacher contributes knowledge, questions, or examples from their respective fields. In order to facilitate the seamless integration of viewpoints without overwhelming pupils, this dynamic paradigm is used. Large-scale seminars or modules conducted by guests are ideal for this. An example of this might be a professor of computer science explaining algorithmic prejudice while a scholar of the humanities provides real-world cultural case studies or ethical criticisms.

Common Co-teaching Models

1. Collaborative Planning: Begin with common course objectives, agree on evaluation standards, and co-design the curriculum to represent real interdisciplinary.
2. Clear Roles and Communication: Determine who will lead what, when, and why, while staying flexible throughout class sessions.
3. Faculty Development Support: Institutions should provide training and incentives for multidisciplinary co-teaching, such as schedule flexibility and shared credit models.
4. Student Orientation: Students need assistance on how to handle multiple instructional methods and expectations. Early orientation contributes to the development of a cohesive classroom culture.

7.5.1 Student Interdisciplinary Capstones

Capstone projects enable students to integrate their knowledge and demonstrate expertise. They offer opportunities to address real-world challenges through technological advancements and thoughtful analysis. Typically associated with interdisciplinary programs that merge computer science and the humanities, these projects exemplify the power of integrative thinking. The outcomes are often academically robust and socially significant, impacting artistic pursuits as well [7, 11] (Table 7.4).

Student Interdisciplinary Capstones

Characteristics of Strong Interdisciplinary Capstones [12]

1. Dual-Domain Integration: It is imperative that capstone projects include elements from both the humanities and computer science, going beyond mere inclusion on the surface level. It is expected of students that they will exhibit expertise in technical execution as well as critical, ethical, or cultural critique. As an example, a digital platform that incorporates human narratives and ethical data standards while simultaneously visualising patterns of refugee travel.
2. Real-World Relevance: In an ideal world, capstone projects would address social concerns or developing issues, such as algorithmic prejudice, digital preservation of culture, disinformation, or accessibility in technology. Consider a smartphone app that uses gamified storytelling and computational linguistics to encourage indigenous language acquisition.

Table 7.4 Sample capstone ideas

Project title	Description	Skills involved
AI and Ethics Simulator	An interactive platform that lets users role-play ethical dilemmas in AI deployment	Python, UX design, moral philosophy
Digital Archive of Protest Literature	A searchable repository of global protest writing with thematic analysis and historical context	NLP, metadata tagging, literary analysis
Interactive Timeline of Feminist Tech Contributions	A web-based multimedia project celebrating underrepresented innovators in technology	JavaScript, oral history, digital storytelling
Algorithmic News Bias Detector	A tool that identifies political slant in news headlines using sentiment and language analysis	Machine learning, media studies, critical theory
Virtual Museum of Indigenous Knowledge Systems	An immersive environment that preserves and shares cultural knowledge through AR	3D modelling, anthropology, narrative design

3. Research + Application: Exceptional initiatives are those that combine in-depth research with practical results. Students have the opportunity to engage in theoretical study, such as investigating power dynamics in surveillance technologies, while simultaneously developing a product, such as a browser extension that focusses on protect users' privacy.
4. Public Engagement and Impact: Encourage students to display their capstone projects at conferences, post them online, or share them with groups or organisations that are pertinent to the topic. When initiatives have meaningful audiences and stakeholders, the impact of such efforts is enhanced.

Support Structures for Capstone Success

1. **Faculty Mentorship Teams:** Assign dual mentors, one from each discipline to guide students in both theoretical depth and technical execution.
2. **Interdisciplinary Studios or Laboratories:** Create collaborative workplaces with the appropriate software, datasets, and media tools.
3. **Capstone Proposal Workshops:** Assist students in developing practical, relevant projects that address learning objectives from both disciplines.
4. **Showcase Events:** Organise public displays, demo days, or symposiums to showcase capstone accomplishments and encourage discussion.
5. **Assessment Rubrics:** Use rubrics that consider technical execution, research quality, innovation, and multidisciplinary integration.

7.5.2 Institutional Support and Frameworks

For interdisciplinary collaboration between the Humanities and Computer Science to thrive, institutions must go beyond mere encouragement; they should establish concrete structures, policies, and resources that can support and expand these initiatives. This includes re-evaluating departmental frameworks, creating incentives for collaboration, and embedding interdisciplinary into the institution's core values [13] (Fig. 7.7).

Fig. 7.7 Institutional framework

Key Elements of Institutional Support

a. Interdisciplinary Program Development

In order to combine the Humanities with Computer Science curriculum, educational institutions need to establish official programs, such as minors, dual degrees, or certificate courses. This set of programs ought to:

1. Include co-created core courses.
2. Enable cross-listing and flexible credit recognition.
3. Provide programs customised to certain subjects such as digital ethics, computational linguistics, and digital humanities.

b. Funding and Grants

In order to facilitate the development and implementation of multidisciplinary projects, dedicated financing is provided. Here are several examples:

1. Start-up incentives for new course development.
2. Research support for interdepartmental faculty cooperation.
3. Student project funding, particularly for capstones and community-engaged activities.

c. Administrative Coordination

The existence of bureaucratic silos has the potential to hinder interdisciplinary effort. It is possible to facilitate the following by appointing coordinators or creating multidisciplinary centres:

1. Schedule shared courses.
2. Managing dual faculty appointments.
3. Workload and recognition structures must be brought into harmony.

d. Faculty Incentives and Recognition

It is essential that educational institutions place a high importance on interdisciplinary teaching and research when discussing promotion, tenure, and workload. This includes the following:

1. Recognising co-teaching as full instructional credit.
2. Incorporating transdisciplinary outputs (such as public scholarship and digital initiatives) into assessment.
3. Providing release time or stipends to support cross-disciplinary course design.

e. Learning Spaces and Infrastructure

Adaptive learning spaces such as digital labs, makerspaces, and visualisation studios facilitate hands-on, interdisciplinary exploration, promoting creativity, teamwork, and problem-solving across various fields. These environments should be purposefully crafted to inspire experimentation, enabling students to translate theoretical concepts into practical, innovative applications. Crucially, they must be accessible to all students, not limited to those in STEM fields, to foster inclusivity and interdisciplinary collaboration. By inviting participants from the arts, humanities, business, and social sciences, these spaces transform into centres of diverse viewpoints, resulting in more comprehensive and well-rounded solutions. Educational institutions should actively enhance accessibility through workshops, mentorship initiatives, and interdisciplinary project opportunities, dismantling barriers between disciplines. Furthermore, incorporating these spaces into academic programs via collaborative courses or research projects can significantly boost student engagement. Ultimately, adaptive learning environments empower students to transcend conventional limits, equipping them for the intricate, interconnected challenges of the future.

f. Professional Development and Training

When it comes to the development of interdisciplinary pedagogies, faculty and staff need assistance. Those institutions ought to provide:

1. Workshops on co-teaching methods, digital humanities tools, and technological ethics.
2. Retreats or colloquia that bring together professors from different fields.
3. Access to collaborative learning and research platforms.

7.6 Institutional Support and Frameworks

While the intellectual and pedagogical advantages associated with integrating Humanities and Computer Science are clear, environmentally friendly interdisciplinary collaboration requires intentional institutional infrastructure. In the absence of sufficient support from the leadership of the academic institution, such partnerships often continue to be fragmented or rely on the passion of individual faculty members. Institutions must make investments in both formal frameworks and informal cultural adjustments in order to legitimise, incentivise, and successfully integrate these inter-disciplinary links into their fundamental academic fabric. For the purpose of promoting combined Humanities and Computer Science programs, this part investigates the institutional tactics, policy frameworks, and support structures that are necessary [14].

Institutional Support and Structural Models for Interdisciplinary Integration

When it comes to facilitating interdisciplinary teaching and research, the manner in which educational institutions organise their academic offerings is among the most important factors. Progressive educational institutions are embracing hybrid structures that promote cooperation beyond conventional academic boundaries. This is in contrast to the situation in which many universities continue to be restricted by departmental silos (Table 7.5).

These models need to be adapted to the scale of the institution, the governance structure, and the strategic aims of the organisation. Importantly, these frameworks should not just serve the CS enthusiast who is interested in the humanities or the

Table 7.5 Common structural models

Model	Description	Advantages	Challenges
Interdisciplinary institutes or centers	Dedicated hubs (e.g., Digital Humanities Centers) that host joint projects, research, and programming	Centralized space, grant-focused, high visibility	Requires sustained funding; may be seen as peripheral to mainstream departments
Joint majors or dual-degree programs	Formal undergraduate or graduate programs combining CS with Philosophy, Linguistics, Media Studies, etc	Equitable blend of disciplines, strong student appeal	Complex curricular coordination; can overload students
Thematic minors or certificates	Specialized tracks (e.g., "Tech and Society") that students can add to any major	Flexible, low-barrier interdisciplinary exposure	Limited depth without full curricular integration
Cross-listed courses	Courses that are listed in both departments and count toward either major	Encourages enrollment across disciplines	Requires ongoing faculty coordination and departmental negotiation

programmer who is passionate about ethics; rather, they should strive to achieve a balanced and meaningful interaction from both sides.

Funding, Incentives, and Resource Allocation

The administrative and time-consuming aspects of interdisciplinary work may be challenging. To this end, it is vital to provide financial support and incentives in order to promote ongoing faculty commitment and high-quality results (Table 7.6).

Table 7.6 Incentives for faculty

Incentive type	Description	Impact
Course credit sharing	Faculty from both departments receive full teaching credit for co-taught courses	Encourages sustained collaboration
Tenure & promotion recognition	Interdisciplinary work is valued equally to disciplinary work in review processes	Legitimizes cross-disciplinary scholarship
Research sabbaticals or leaves	Time off to pursue interdisciplinary teaching innovation or research	Supports intellectual exploration
Honoraria or stipends	Supplement salaries for those leading new programs or coordinating joint initiatives	Recognizes extra administrative and intellectual labor

Funding Opportunities.

1. Course Development Grants: Provide seed funds for collaboratively developed multidisciplinary courses. These awards may help fund curriculum design workshops, guest lecturers, teaching assistants, and resources.
2. Collaborative Research Funds: Encourage professors from Humanities and Computer Science to co-author papers, speak at conferences, or do community-engaged research.
3. Student Fellowships or Prizes: Encourage students to pursue ambitious capstone projects or summer internships that combine both subjects.

Also, educational institutions need to construct interdisciplinary budget lines for the purpose of purchasing course equipment, digital tools, and guest lecturers, all of which may not fit neatly under the accounts of the departments that are already in place.

Curriculum Coordination and Policy Flexibility

Interdisciplinary routes need to be accommodated by academic policy in order to provide genuine support for cooperation between the humanities and computer science. Included in this are regulations regarding the curriculum that are more adaptable, cooperation across departments, and simplified administrative procedures that are simplified.

Key Policy Reforms [15, 16]:

1. Cross-Departmental Course Recognition: Allows students to credit some computer science or humanities courses as electives or prerequisites across multiple degree programs.
2. Modular course structures allow co-taught courses to incorporate separately delivered components (for example, one module in digital ethics and one in AI design).
3. Unified Capstone Rubrics: Standardise expectations for multidisciplinary project evaluations to properly assess both technical and theoretical components.

In many educational establishments, these modifications need the formation of joint curriculum committees, coordination with the registrar, and consistent interaction among faculty members in order to align academic aims without compromising rigour.

Faculty Development and Collaborative Culture

Teaching across disciplines needs more than just knowledge in a particular field; it also necessitates the development of new educational methodologies, improvements in communication skills, and a mutual respect for different epistemologies.

Support for Faculty [17]:

1. Interdisciplinary Teaching Institutes: Organise yearly retreats or bootcamps to teach teachers on collaborative course design and project-based learning methodologies.

2. Workshops on Pedagogical Tools: Non-CS professors may learn about digital storytelling, algorithmic bias, and computational thinking.
3. Mentorship programs link experienced multidisciplinary educators with novice teachers to support them through co-teaching and collaborative research.

To establish a culture of cooperation, it is necessary to cultivate partnerships across different departments by holding frequent meetings, holding research days that include many disciplines, and using joint mailing lists or internet forums [17].

7.7 Collaboration with Libraries and Tech Units

Collaboration with university libraries and campus technology departments is one of the humanities and CS integration strategies that is often disregarded, despite the fact that it is a crucial component of effective integration. These organizations act as more than just service providers; they are essential partners in the process of enabling digital scholarship, establishing technological infrastructure, giving access to digital tools and archives, and promoting open pedagogy and the dissemination of research.

Libraries and information technology services on campus have expanded beyond their conventional functions. They serve as innovation hubs, methodological training centers, and connective tissue uniting students, professors, and developing technologies in the academic environment of today, which is characterized by multidisciplinary collaboration. The purpose of this section is to investigate the ways in which collaborative collaborations between libraries and technology departments might enhance interdisciplinary curriculum, encourage student creativity, and promote digital literacy across a variety of topic areas [18] (Fig. 7.8).

7.7.1 Libraries as Interdisciplinary Anchors

For a very long time, university libraries have been known to foster intellectual research and the accessibility of resources. Through the provision of the following services, they expand their function into the digital and computational domains within the transdisciplinary space:

1. Access to Digital Humanities Collections: Libraries provide access to large collections of digitized texts, photographs, audiovisual material, and historical records, which are critical for digital humanities and machine learning initiatives.
2. Data curation and management expertise: Librarians train students in data organization, metadata standards, FAIR principles (Findable, Accessible, Interoperable, Reusable), and long-term data preservation all of which are essential skills in both fields.

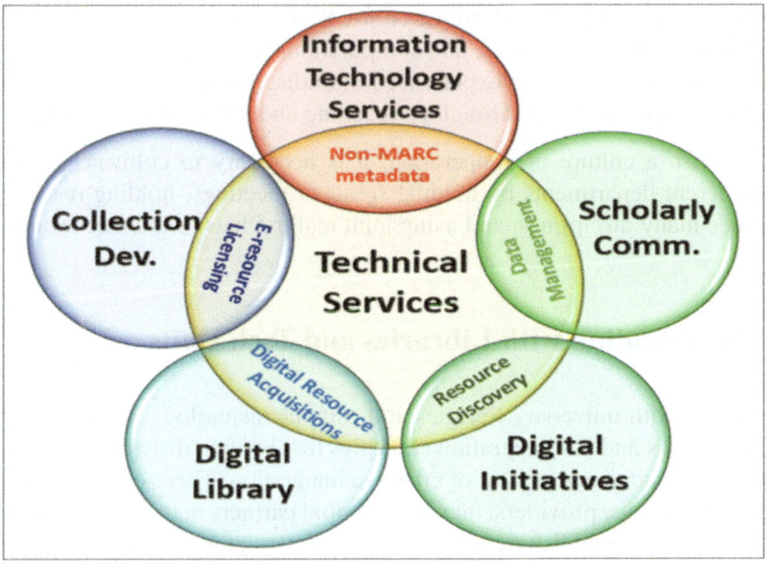

Fig. 7.8 Technical services collaboration function with different units

3. Digital Scholarship Laboratories: Many libraries now provide specialized facilities for academics and students to engage with technologies such as text mining software, GIS systems, and data visualization platforms.
4. Research Consulting and Embedded Librarianship: Subject-specialist librarians provide personalized assistance during course planning, assignment design, and collaborative research, and often co-teach or contribute to course delivery.

At Stanford University, the Centre for Interdisciplinary Digital Research (CIDR) in the library system collaborates with professors to integrate data visualisation and computational analysis into humanities courses. They provide technical help, workshops, and project consultations.

7.7.2 Campus Tech Units as Pedagogical Enablers

When implementing and providing support for the technical tools that promote transdisciplinary teaching and learning, campus IT departments are necessary. Because of the following, their relationship becomes very valuable:

1. Supporting Digital Tools in the Classroom: Tools such as Jupyter Notebooks, Omeka, Scalar, Voyant, Tableau, and ArcGIS require backend support, licensing, and training—all managed by tech teams.

2. Classroom Technology Design: Tech units play a crucial role in outfitting classrooms and labs with collaborative technologies: large-format interactive screens, high-speed internet, VR/AR systems, and digital recording tools.
3. Cybersecurity and Data Ethics: As students become more involved with real-world data and online platforms, IT departments help to ensure compliance with data protection rules (e.g., GDPR, FERPA) and ethical computing practices.
4. Cloud Computing and Research Infrastructure: Tech units handle cloud server access, compute credits, and containerised environments for large-scale projects like NLP analysis of literary corpora or AI model training utilising historical archives.

Partnership in Action: In collaboration with campus information technology and academic divisions, the Emerging Media Lab at the University of British Columbia provides financial assistance to multidisciplinary student and faculty projects that are centred on virtual reality (VR) storytelling, interactive simulations, and digital museum curation technologies.

7.7.3 Synergistic Structures: Libraries, IT, and Academics

The most successful educational institutions create collaborative ecosystems in which academic units, libraries, and IT services work together to co-design curricular experiences, co-host events, and co-invest in digital infrastructure. This allows the institutions to move beyond siloed operations [19, 20] (Table 7.7).

Integrated initiatives like this help to reinforce learning that is centered on the learner and demonstrate how cooperation across different units may be used to create learning experiences that are scalable and powerful.

Table 7.7 Synergistic structures

Collaborative initiative	Partners	Outcomes
Digital storytelling Bootcamps	Libraries, IT, Humanities Faculty	Students produce interactive narratives combining archival research with multimedia tools
Hackathons and data Jams	CS Departments, Libraries, Tech Units	Students analyse cultural datasets or address ethical tech challenges
Toolkits for computational text analysis	Libraries, DH Centres, CS Support Staff	Provide reusable codebases, guides, and annotated corpora for class use
Digital publishing platforms	Libraries + IT	Enable students to create open-access digital projects using platforms like Press books or Scalar

7.7.4 Capacity Building and Shared Training

In order to achieve long-term success in multidisciplinary work, it is essential to engage in professional growth and to share information across all groups. The faculty, librarians, and technical staff all need to make an effort to get acquainted with one other's procedures and build a common language.

Recommended Capacity-Building Strategies.

1. Interdisciplinary Tech Pedagogy Institutes: Frequent training sessions where faculty members get instruction from librarians and IT specialists on how to use tools, create syllabuses, and incorporate technology into learning objectives.
2. Train-the-Trainer Programs: Support should be scaled across several classes, and librarians and computer science graduate assistants should be given the authority to serve as in-class mentors for digital technologies.
3. Inter-unit Working Groups: Create working groups that will last for the whole semester and include members from the academics, technology, and library staff. These groups will concentrate on topics such as "Digital Archives and AI" or "Narrative and Data Visualisation."
4. Student Tech Ambassadors: Hiring student consultants who have been educated by libraries and information technology to assist their peers in utilising platforms such as Zotero, GitHub, or Python for humanities research is a great idea.

7.7.5 Challenges and Solutions

In spite of the obvious advantages, there are certain difficulties associated with institutional cooperation between academic departments, information technology, and libraries. The leadership of the administrative department has to acknowledge these obstacles and work more aggressively to create trust, shared governance, and integrated planning procedures (Table 7.8).

Table 7.8 Challenges and potential solutions

Challenge	Explanation	Potential solutions
Siloed budgets	Library and IT units often operate with different funding streams than academic units	Create interdisciplinary funding pools or grant-matching programs
Misaligned goals	Tech teams may prioritize operational reliability over experimental academic projects	Establish shared goal-setting and project planning processes
Resource constraints	Limited staff availability for custom project support	Develop scalable toolkits and asynchronous training modules
Communication gaps	Differences in language, timelines, and workflows	Use liaison roles or coordinators to bridge the units

7.7.6 Vision for the Future

The future of cooperation between the humanities and computer science is dependent not just on the inventiveness of faculty members but also on the ecosystems of infrastructure that sustain it. In their capacity as co-creators of the learning process, libraries and technology units are in a position that is unparalleled to:

- Promote digital equality by making modern technologies available to all disciplines.
- Encourage open scholarship by hosting repositories and supporting open-source projects.
- Lead in digital ethics education by encouraging transparency and ethical data use.

In order to have this vision fully realised, institutions need to formalise these collaborations by doing the following:

- Innovation centres that span many disciplines and are jointly controlled.
- Co-authored grant applications in the fields of academics, information technology, and libraries.
- Programs of shared appreciation, such as rewards for outstanding teaching innovation that span many units.

7.8 Digital Pedagogy Labs and Studios

In today's higher education system, Digital Pedagogy Labs and Studios (DPLS) are essential engines of innovation, especially in areas where multidisciplinary intersections such as the junction of computer science and the humanities are involved. The creative and technological settings that are provided by these physical and virtual venues in which ideas are converted into real and often transformational learning experiences are provided by these spaces. All of these activities are centred on digital tools, media, and approaches, and they provide support for academic experimentation, student research, collaborative creation, and community involvement programs. Unlike standard computer laboratories, which are often built for individual and technical usage, DPLS are facilities that are purposefully created for the purpose of pedagogical research, multidisciplinary cooperation, and project-based learning [21].

7.8.1 Defining Digital Pedagogy Labs and Studios

Digital Pedagogy Labs and Studios are places that combine the technical infrastructure of a computer lab, the collaborative ethos of a makerspace, and the intellectual ambition of a research centre. Their primary objective is to investigate how technology reshapes teaching, learning, and research in the 21st-century environment. The fact that these labs frequently provide support for extracurricular activities such as workshops, course-integrated projects, and faculty fellowships makes them dynamic hubs for engaged students from different fields.

Key Features [22]:

- Collaborative workspaces have adaptable layouts, writing surfaces, and digital projection.
- Multimodal equipment includes high-resolution scanners, 3D printers, AR/VR tools, and video editing suites.
- Open Access Software for digital storytelling, text mining, data visualisation, and game creation.
- Expert staff members include digital humanists, instructional designers, technologists, and librarians.

7.8.2 Purpose and Impact

Digital Pedagogy Labs and Studios are not only about giving tools; their mission extends much beyond that. Their goal is to:

- Cultivate Digital Literacy: Students learn to explore, analyse, and produce digital material critically and ethically.
- Bridge Disciplines: Students and faculty from the humanities, sciences, and technology work together to solve common intellectual challenges.
- Encourage Experimental Teaching: Faculty members try novel forms of digital storytelling, computational analysis, and interactive learning.
- Encourage assessment innovation by replacing traditional examinations with digital portfolios, podcasts, or AR/VR prototypes.

Students at the University of Michigan participate in studio-based courses that are offered by the Digital Studies Institute in collaboration with the Duderstadt Centre. These courses allow students to construct Twitter bots that interact with political rhetoric, produce critical video essays, and design digital zines. All of these activities are centred around the examination of contemporary culture through the lens of technology.

7.8.3 Organizational Models and Structures

There are a variety of governance structures, locations, and affiliations among Digital Pedagogy Labs and Studios. There are a handful that are housed inside certain departments, while others are contained within libraries, and yet others are managed by teaching and learning centres. Regardless of structure, successful DPLS prioritise

Table 7.9 Organizational models

Model	Affiliation	Strengths	Challenges
Library-based studio	Integrated within university libraries	Strong archiving and research support	May lack department-specific resources
Instructional design hub	Managed by teaching and learning centers	Aligned with pedagogical innovation goals	May be underutilized by research-focused faculty
Department-run lab	Embedded in Humanities or CS department	Highly tailored to curricular needs	Risk of siloing, less accessible to others
Interdisciplinary institute studio	Shared by multiple departments or centers	Strong cross-disciplinary engagement	Requires sustained institutional funding

openness, mentorship, and sustainability as core principles. Openness ensures accessibility to students and faculty from various disciplines, fostering collaboration and innovation beyond traditional academic boundaries. Mentorship is key, with experienced teachers and professionals helping students via hands-on projects, research, and skill development. Sustainability, both in terms of long-term institutional support and environmentally responsible design, ensures that these venues stay relevant and adaptive to changing educational demands. To survive, DPLS should include adaptable resources, such as modular workstations and digital tools, while also fostering cross-disciplinary communication. Institutional commitment through money, policy, and community engagement increases their effectiveness. By building an open, forward-thinking culture, DPLS enables students to solve complicated real-world situations while meaningfully integrating theory and practice [23] (Table 7.9).

7.8.4 Activities and Offerings

The Digital Pedagogy Labs and Studios are home to a wide variety of activities that combine teaching, research, and community responsibilities. Examples of these include:

a. Curricular Integration

- Capstone Support: Students in multidisciplinary majors utilise the lab to create digital exhibitions, text analysis tools, and games.
- Embedded Course Projects: Faculty co-teach or collaborate with lab personnel to include hands-on digital projects into course syllabuses.

b. Co-Curricular Enrichment

- Skill-Building Workshops: "Data Visualisation with Python," "Podcasting for Research," and "Feminist Hacking Practices."
- Digital Ethics Salons: Critical discussions on surveillance, algorithmic prejudice, and accessibility.

c. Faculty and Graduate Development

- Faculty fellowships provide funding and assistance for instructors who are piloting digital assignments or revising syllabi.
- Graduate assistantships provide students with research and teaching opportunities in digital pedagogy and tool creation.

Students in the Digital Competencies Program at Bryn Mawr College not only enhance their technical fluency, but they also participate in reflective projects that focus on the ways in which digital technologies influence narratives and identities. This program establishes a connection between humanistic inquiry and technical capabilities.

7.8.5 Common Tools and Technologies

A Digital Pedagogy Lab is distinctive in that it has a technological ecology that is both diversified and adaptive. The following is a selection of tools that are used frequently (Table 7.10).

Table 7.10 Common tools and technologies

Tool Type	Examples	Purpose
Text analysis	Voyant Tools, AntConc, NLTK	Analyze word frequency, sentiment, and patterns in literary corpora
Multimedia creation	Audacity, Adobe Premiere, Twine	Podcasting, video essays, interactive narratives
Visualization	Tableau, RawGraphs, Palladio	Create data-driven representations of humanistic data
Web publishing	WordPress, Omeka, Scalar	Build digital exhibits, critical editions, and e-portfolios
AR/VR & 3D	Unity, Sketchfab, HoloLens	Build immersive experiences and virtual reconstructions

7.8.6 Sustainability and Future Directions

The establishment of a Digital Pedagogy Lab is simply the first step; in order to maintain it, a long-term vision, consistent financing, and active participation from the community are required.

Strategies for Sustainability:

- Cross-Unit Collaborations: Share funds and governance across departments, libraries, and IT.
- Student Engagement: Hire students as lab fellows or technical mentors.
- Faculty incentives include course release, awards, and recognition for digital pedagogy.
- Community Outreach: Organise public events, digital exhibits, and school collaborations.

In the future, emerging tendencies indicate the following:

- AI-Powered Pedagogy: Tools such as ChatGPT are being used to teach rhetoric, programming, and research design.
- Multilingual Digital Tools: Enabling global humanistic viewpoints.
- Sustainability-Driven Design: Integrating green computing methods into lab operations.

7.9 Evaluating Interdisciplinary Success

In multidisciplinary programs, particularly those that bring together subjects such as the Humanities and CS, standard criteria of assessment often fail to match up with performance expectations. These partnerships defy simple categorisation since they span a wide range of learning goals, project kinds, and methodological methods. Therefore, in order to evaluate the effectiveness of multidisciplinary endeavours, it is necessary to have frameworks that are adaptable, multi-dimensional, and that respect both the depth of the disciplines and the breadth of the integrative approach.

At this juncture, an overview of the most important indicators, evaluation techniques, institutional tactics, and problems that are involved in determining whether or not multidisciplinary education which is beneficial will be discuss. In addition to this, it provides educators, curriculum creators, and academic leaders with real-world examples and models that may assist them in the process of designing relevant assessment systems.

7.9.1 Why Interdisciplinary Evaluation is Complex

Interdisciplinary initiatives need assessment procedures that are similar to those used in monodisciplinary programs, which have outcomes that are clearly defined and standardized:

- Context-sensitive: Different programs prioritise different types of integration (for example, applying computer science methods to humanities issues or philosophising about algorithmic prejudice).
- Flexible evidence: Success may be proved by prototypes, essays, performances, code, or community service, rather than merely exams or grades.
- Team-based: Collaborative initiatives need evaluating both individual contributions and team dynamics.
- Iterative: Because multidisciplinary integration changes with time, evaluations must reflect progress rather than static standards.

7.9.2 Defining Interdisciplinary Learning Outcomes

Institutions are required to identify what constitutes "success" for their multidisciplinary programs before assessment may take place. The following are examples of common result domains as seen in Table 7.11.

These domains guarantee that the whole student, and not just their technical output, is being developed and evaluated with regard to their performance.

Table 7.11 Common tools and technologies

Learning domain	Interdisciplinary goal	Example metrics
Knowledge integration	Combine theories/methods from two or more disciplines	Essays synthesising literature and algorithm design
collaborative problem solving	Work effectively in diverse, cross-functional teams	Peer evaluation, team reflections
Critical and ethical thinking	Reflect on the impact of tech on society and culture	Case studies, scenario analysis
Technical proficiency + Humanistic insight	Apply tools like data mining, while analysing social implications	Dual-rubric assessments (technical + reflective)
Communication skills	Communicate to both technical and non-technical audiences	Multimedia presentations, podcasts, zines

Table 7.12 Assessment tools and strategies

Criterion	Description	Score range
Code functionality	Software runs correctly, solves the intended problem	0–5
Contextual relevance	The project engages with a specific historical, social, or cultural issue	0–5
Integration quality	Tools/methods from both disciplines are used coherently	0–5
Reflective analysis	The student critically discusses process and outcomes	0–5

7.9.3 Assessment Tools and Strategies

An assessment procedure that is universally applicable does not exist. Interdisciplinary programs, on the other hand, may reap the benefits of tiered evaluation systems that mix qualitative and quantitative data (Table 7.12).

a. Rubrics for Integrated Projects

Develop grading rubrics that include both discipline criteria, such as algorithm performance, and interdisciplinary synthesis, such as historical significance or ethical concerns.

b. **Portfolios**

Students have the ability to display their development over time via the use of digital portfolios, which may include reflections, unsuccessful experiments, and iterative improvements.

c. Peer and Self-Assessment

Peer evaluation is frequently used in group projects as a means of evaluating teamwork, an equitable distribution of labour, and interpersonal communication.

d. Surveys and Interviews

- Pre/post questionnaires may assist assess confidence, perceived skill progress, and interdisciplinary awareness.
- Exit interviews give qualitative information on how students internalise interdisciplinary thinking.

e. Capstone Reviews

For climax experiences, invite multi-disciplinary panels (e.g., a literature professor + a computer scientist) to examine final projects.

7.9.4 Institutional Evaluation of Programs

From an administrative point of view, measuring the performance of interdisciplinary programs requires taking into account both the learning outcomes at the student level

Table 7.13 Key metrics

Category	Institutional metric	Purpose
Enrollment	Number of students enrolling in interdisciplinary courses	Indicates interest and growth
Retention and graduation	% of students completing interdisciplinary degrees	Reveals program sustainability
Post-graduation outcomes	Alumni employment or grad school placements	Measures real-world relevance
Faculty engagement	Cross-departmental teaching or publications	Reflects institutional collaboration
Funding and recognition	Grants, awards, or media coverage	Demonstrates program visibility

and the influence at the program level. For the purpose of determining whether or not the program satisfies the requirements of the workforce and the community, program assessments should also include comments from external advisory boards or employers (Table 7.13).

7.9.5 Student Success Stories and Narratives

Although quantitative data is essential, narrative is also very crucial. In order to better express the wider effect of interdisciplinary learning, it is helpful to record the perspectives of students who have benefited from it. It's possible that these tales emphasise:

- How a digital humanities student utilised natural language processing to examine feminist poetry.
- How a CS student learnt about colonialism by creating a decolonial game.
- Collaboration between a philosophy major and a developer resulted in an application that investigates face recognition bias.

The compelling and human-centered proof of multidisciplinary effect that these tales provide is presented here.

7.9.6 Challenges in Evaluation

In spite of the greatest efforts, measuring the effectiveness of multidisciplinary collaboration is not without impediments:

- Disciplinary bias: Reviewers may undervalue the "other" field (for example, a computer science faculty member rejects interpretative ideas as subjective).

Table 7.14 Best practice and implementation

Best practice	Implementation Tip
Align assessments with learning outcomes	Develop rubrics jointly across disciplines
Collect both qualitative and quantitative data	Use surveys, interviews, and direct assessments
Evaluate both process and product	Reward critical reflection and learning growth
Include external reviewers or advisory boards	Gain industry and civic feedback
Share findings with stakeholders	Publish evaluation reports, testimonials, and metrics

- Misaligned Timelines: Humanistic research and technical development take place on distinct time periods.
- Limited Rubric Models: Standard grading rubrics are often created for conventional essay or problem-set forms.
- Funding Prioritisation: Evaluation data may be used to support budget choices that do not accurately represent the worth of experimentation or failure.

7.9.7 Best Practices

In light of case studies conducted by educational establishments that have developed interdisciplinary programs, the following are many recommended practices for measuring success (Table 7.14).

7.9.8 The Future of Evaluation

In the future, multidisciplinary evaluation could be improved by other factors, such as:

- AI-Driven Analytics: Using machine learning to track portfolios, writing, or team dynamics over time.
- Micro-Credentialing: Designating multidisciplinary talents like "code + ethics" or "storytelling + data visualisation."
- Cross-Institutional Benchmarking: Creating consortia to exchange and compare assessment data across institutions.
- Participatory Assessment Models: Allowing students to co-create the assessment instruments that rate their work.

The improvements that have been made show that assessment itself has the potential to become a place where multidisciplinary cooperation, creativity, and progress may take place.

References

1. Canagarajah, S.: Decolonizing academic writing pedagogies for multilingual students. TESOL Q. **58**(1), 280–306 (2024). https://doi.org/10.1002/tesq.3231
2. Rakhmankulovna, A.S.: The role of education and training in professional training and the importance of foreign languages in the formation of professional competencies of students. Int. J. Adv. Sci. Res. **3**(09) 58–62 (2023). https://sciencebring.com/index.php/ijasr/article/view/374
3. Collier-sewell, F., Atherton, I., Mahoney, C., Kyle, R.G., Hughes, E., Lasater, K.: Competencies and standards in nurse education: the irresolvable tensions. Nurse Educ. Today **125**, 105782 (2023). https://doi.org/10.1016/j.nedt.2023.105782
4. Singh, V.: Digitalization, BIM ecosystem, and the future of built environment: how widely are we exploring the different possibilities? Eng. Constr. Archit. Manag. (2019). https://doi.org/10.1108/ECAM-01-2018-0004
5. Al-ababneh, M.M.: Linking ontology, epistemology and research methodology (2020). https://doi.org/10.23756/sp.v8i1.500
6. Braun, V., Clarke, V., Braun, V., Clarke, V.: Applied qualitative research in psychology. Appl. Qual. Res. Psychol. **0887**(2006) (2017). https://doi.org/10.1057/978-1-137-35913-1
7. Zhang, L., Wu, Z.: Enhancing postgraduate academic writing skills through course reform : an action research study. **5**(2), 217–233 (2024)
8. Oluwatoyin, A., Oluseun, A.: Total quality management: a test of the effect of TQM on performance and stakeholder Satisfaction. J. Clin. Nurs. **5**(1), 2007–2012 (2014)
9. Obi, K.C., Iwuanyanwu, O.: Innovation In The Assessment Of Primary School Mathematics: An Approach To Revamping Nigeria Education. J. Teach. Perspect. **18**(1), 51–66 (2023). https://repositorio.unan.edu.ni/2986/1/5624.pdf%0Ahttp://fiskal.kemenkeu.go.id/ejournal%0A10.1016/j.cirp.2016.06.001%0A10.1016/j.powtec.2016.12.055%0A10.1016/j.ijfatigue.2019.02.006%0A10.1
10. Hattie, J.: Teachers make a difference: what is the research evidence? paper presented at the building teacher quality: what does the research tell us acer research conference. Aust. Counc. Educ. Res. Teach. Make a Differ. What is Res. evidence? Disting. Expert Teach. from Novice Exp. Teach (2003). http://research.acer.edu.au/research_conference_2003
11. Weines, J., Borit, M.: Better game worlds by design: the GAS Framework for designing and analyzing games based on socio-ecological systems thinking, demonstrated on Nusfjord (2017). Games Cult. **17**(2), 262–283 (2022). https://doi.org/10.1177/15554120211027109
12. Lee, I., Mangalaraj, G.: Big data analytics in supply chain management: a systematic literature review and research directions. Big Data Cogn. Comput. **6**(1) (2022). https://doi.org/10.3390/bdcc6010017
13. Kim, H.-S.: Oil extraction and the changing dynamics of pastoral conflicts: a conjoint experiment in Turkana. Kenya. J. Mod. Afr. Stud. **61**(1), 23–47 (2023). https://doi.org/10.1017/s0022278x22000398
14. Cao, D., Li, H., Wang, G.: Impacts of isomorphic pressures on BIM adoption in construction projects. J. Constr. Eng. Manag. **140**(12), 04014056 (2014). https://doi.org/10.1061/(ASCE)CO.1943-7862.0000903
15. Mostert, M., Quinn, L.: Using ICTs in teaching and learning: reflections on professional development of academic staff. J. Educ. **5**(5), 72–84 (2009)
16. Abdulhamid, L., Yushau, B., Sambo, A.A.: Analysis of universal basic education mathematics teachers content knowledge in Nigerian public. Abacus Math. Educ. Ser. **45**(1), 50–61 (2020)
17. Bratianu, C., Stanescu, D.F., Mocanu, R.: Exploring the knowledge management impact on business education. Sustain. **13**(4), 1–16 (2021). https://doi.org/10.3390/su13042313
18. Aswir, H.M.: Smartphones as personal digital archives? recentering migrant authority as curating and storytelling subjects. Photosynthetica. **2**(1), 1–13 (2018). https://doi.org/10.1007/978-3-319-76887-8

19. Benzidia, S., Makaoui, N., Bentahar, O.: The impact of big data analytics and artificial intelligence on green supply chain process integration and hospital environmental performance. Technol. Forecast. Soc. Change. **165** (2021). https://doi.org/10.1016/j.techfore.2020.120557
20. Pham, C.T.A., Magistretti, S., Dell'Era, C.: The role of design thinking in big data innovations. Innov. Organ. Manag. **24**(2), 290–314 (2022). https://doi.org/10.1080/14479338.2021.1894942
21. Zhao, S., Pan, Q., Gao, D., Cheng, J.: Integrating internet of things and mixed reality to teach performance-based architectural design: a case study of shading devices. Educ. Inf. Technol. (2022). https://doi.org/10.1007/s10639-022-10998-6
22. Alaba, F.A., Jegede, A., Eke, C.I.: Robust data security framework for IoT. IJAMML **1**, 5–23 (2020)
23. Jamali, A.R., Bhutto, A., Khaskhely, M., Sethar, W.: Impact of leadership styles on faculty performance: moderating role of organizational culture in higher education. Manag. Sci. Lett. **12**(1), 1–20 (2022). https://doi.org/10.5267/j.msl.2021.8.005s

Chapter 8
Data, Archives, and Cultural Memory

8.1 Digital Archiving in the Humanities

Digital archiving has arisen as a transformational force in the field of humanities, transforming how academics maintain cultural and historical artefacts, get access to them, and analyse them. The shift from traditional to digital archives has not only made data more accessible, but it has also resulted in the development of new approaches to analysis, collaboration, and research that span multiple disciplines. Understanding the relevance of digital archiving, as well as the issues it faces and the potential routes it may take in the future, is becoming more important as more and more institutions, scholars, and members of the general public depend on digital platforms to interact with historical documents, artefacts, and narratives [1]. The evolution, methods, and methodology of digital archiving within the humanities are investigated in this chapter. The chapter also highlights the influence that digital archiving has had on research, pedagogy, and the preservation of cultural heritage.

8.1.1 Defining Digital Archiving

The act of gathering, digitising, storing, and conserving materials such as text, photos, audio, and video using digital technology is known as digital archiving. Within the realm of the humanities, this encompasses a diverse range of sources, ranging from manuscripts from the Middle Ages and oral histories to newspapers, artworks, and detritus. Digital archiving, in contrast to conventional archiving, offers dynamic access to collections, which enables searchability, metadata tagging, and cross-referencing across different platforms. In addition to this, it maintains the durability of delicate materials by limiting the amount of physical handling that is required and by providing alternatives for making backups and replicating data [2]. Digital archiving in the humanities is the act of turning physical or transitory cultural

F. A. Alaba and I. G. Ajibola, *Digital Pedagogy: Bridging the Gap Between Software and the Humanities*, Information Systems Engineering and Management 85,
https://doi.org/10.1007/978-3-032-20335-9_8

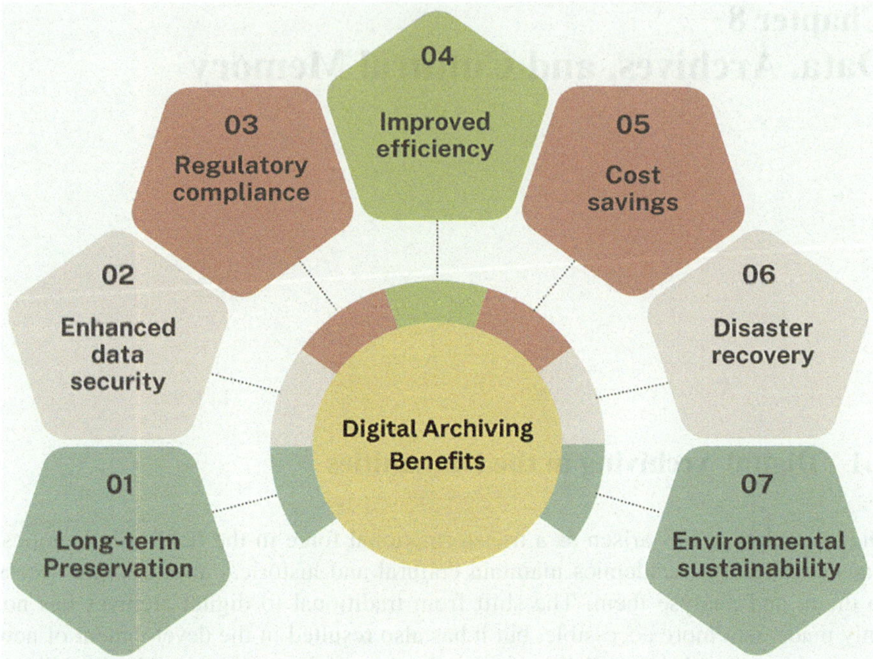

Fig. 8.1 Benefits of digital archiving [4]

assets into digital versions that are systematically maintained and made available via computer-based systems. The procedure usually entails scanning or photographing physical things, encoding texts, capturing audio or video information, and organising these resources using metadata and database systems. Unlike conventional archiving, which often needed researchers to be physically present, digital archives provide remote access, search capabilities, cross-referencing tools, and a multimedia environment [3] (Fig. 8.1).

8.2 Key Characteristics of Digital Archiving:

1. Non-invasive preservation of delicate historical items.
2. Remote and worldwide accessibility.
3. Utilisation of metadata and content management systems.
4. Scalability and long-term storage considerations.
5. Potential for multimedia integration and interactive learning.

8.2.1 The Role and Importance of Digital Archiving in the Humanities

The preservation of cultural heritage, the advancement of academic research, and the democratisation of access to information are all significantly aided by the use of digital archiving technologies. The transformation of delicate physical materials into long-lasting digital representations is one of the ways in which modern technologies protect priceless cultural artefacts from degradation while also making them accessible on a global scale. For scholars, especially those working in the humanities, where contextual interpretation is of the utmost importance, digital archives provide possibilities that have never been seen before to analyse texts, photographs, and historical records using cutting-edge techniques such as text mining, spatial analysis, and network visualisation. Examples of projects that exemplify how digital preservation may revolutionise research include the digitised manuscripts of the British Library and the huge holdings of the Internet Archive. These projects highlight how digital preservation can enable comparative studies across resources that were previously unavailable. In addition, digital archives eliminate geographical and institutional boundaries, making it possible for researchers all over the globe to search through original documents without being constrained by physical limits. Discoverability is improved via the incorporation of metadata standards and linked data technologies, and new forms of public interaction with cultural heritage are fostered through the use of interactive elements in platforms such as the Digital Public Library of America. The continued development of these solutions has the possibility of bringing about a further revolution in the ways in which we maintain, investigate, and engage with the collective knowledge of mankind [3, 5] (Fig. 8.2 and Table 8.1).

8.2.2 Tools and Technologies

To ensure the preservation, organisation, and accessibility of digital assets, digital archiving is dependent on a varied ecosystem of technologies and software applications. Each and every step of the archive lifecycle is supported by these technologies, beginning with digitisation and metadata management and continuing all the way through long-term preservation and public access. High-resolution scanners, optical character recognition (OCR) software, and three-dimensional imaging systems are examples of the types of specialised gear that are used for digitisation. These devices turn physical materials into digital forms while preserving their original quality. After the collections have been digitised, Content Management Systems (CMS) such as Omeka, DSpace, and CONTENTdm assist businesses and organisations in organising and displaying them with structured metadata. Dublin Core, METS, and MODS are examples of metadata standards that assure consistency and interoperability, which in turn enables discovery to occur without interruption across

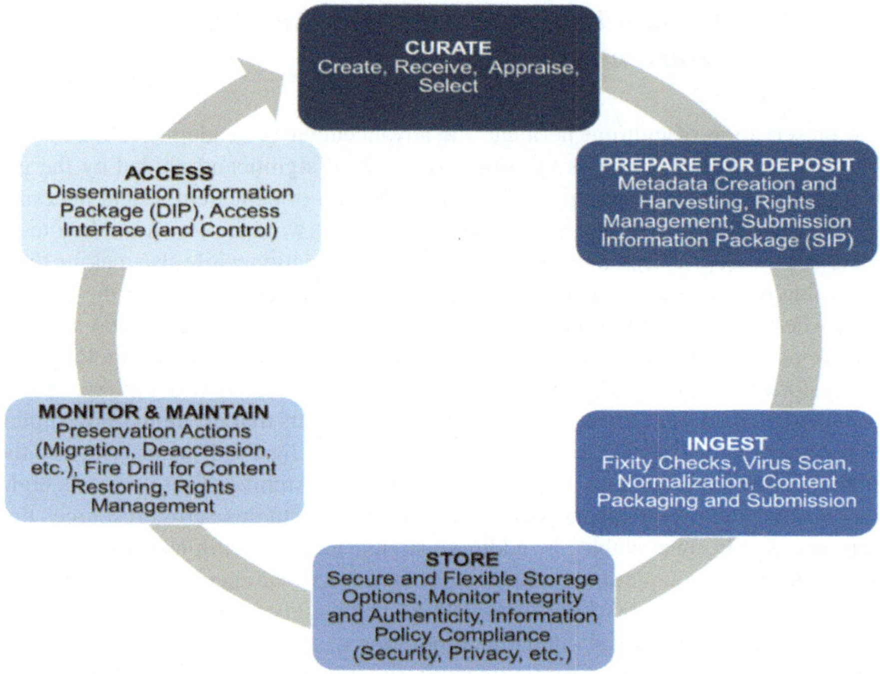

Fig. 8.2 Key stages and processes involved in digital preservation [6]

Table 8.1 Key functions of digital archiving in the humanities

Function	Description
Preservation	Protects physical materials from degradation by creating high-quality digital surrogates
Accessibility	Allows researchers, students, and the public to access content regardless of location
Interdisciplinary research	Supports collaboration between historians, linguists, technologists, and data scientists
Pedagogical tool	Enhances teaching with interactive and visual resources
Community engagement	Enables local and Indigenous communities to reclaim and share their histories

several repositories. For long-term preservation, digital preservation solutions such as Archivematica and Preservica conform to frameworks such as the Open Archival Information System (OAIS) paradigm. Additionally, sustainable file formats such as PDF/A, TIFF, and XML/TEI protect against obsolescence. Emerging technologies, such as artificial intelligence-driven metadata creation, blockchain for provenance tracking, and cloud-based archiving solutions, further improve the durability and

Table 8.2 Common tools and technologies in digital archiving

Category	Tools/Standards	Purpose
Content management	Omeka, DSpace, CONTENTdm	Store, organize, and display digital collections
Digitization	Flatbed scanners, DSLR cameras, 3D scanners	Capture images of texts, objects, and artworks
Text encoding	TEI (Text Encoding Initiative), XML	Structure and encode digitized texts
Metadata standards	Dublin Core, MODS, METS	Tag and describe items for efficient retrieval
Preservation protocols	OAIS (Open Archival Information System), LOCKSS	Ensure long-term digital storage and integrity
Visualization & analysis	GIS, Voyant Tools, Gephi, Tableau	Enable visual and data-based exploration of content

accessibility of digital archives, guaranteeing that they will continue to be used for future generations [7] (Table 8.2).

8.2.3 Notable Projects and Case Studies

The power and potential of digital archiving has been demonstrated by a number of digital humanities projects, including the following:

1. The British Library's "Endangered Archives Programme" digitises items from fragile archives worldwide, with a focus on postcolonial and marginalised groups.
2. Europeana is a digital platform that aggregates material from museums, libraries, and archives around Europe, enabling multilingual access to cultural heritage.
3. The Trans-Atlantic Slave Trade Database is an online resource that contains detailed information regarding voyages, enslaved peoples, and slave trade ports.
4. Tufts University's Perseus Digital Library provides digitised writings from ancient Greece and Rome that have been annotated and cross-referenced for scholarly study.

The capacity of digital archiving to assist social justice, historical reclamation, and inclusive research is shown by these instances, which illustrate the global and collaborative nature of digital archiving. The Digital Library of the Caribbean (dLOC), which is dedicated to the preservation of cultural documents that are in risk of being lost, and Mukurtu CMS, which is a platform built with Indigenous communities to manage holy artefacts in an ethical manner, are examples of initiatives that demonstrate how archives may empower voices that are marginalised. The Liberation records of South Africa and the documenting the Now endeavour, which records social media activity, are two examples of projects that demonstrate how digital preservation may be used to defend historical truth and oppose erasure. In the meanwhile, open-access repositories and crowdsourced transcribing programs encourage participatory research, which helps to bridge the gap between traditional academic institutions and the general public. By making access to information more accessible to the general public and putting the spotlight on narratives that are under-represented, digital archiving transforms from a merely technical activity into a transformational instrument for achieving equity and historical justice [8].

8.2.4 Ethical Considerations and Challenges

In the realm of digital archiving, there are a number of ethical considerations involved. The process of digital archiving raises a number of ethical questions that need to be taken into consideration by institutions in order to guarantee the appropriate management of historical and cultural resources. One of the most significant challenges is striking a balance between accessibility and privacy, which is especially important when preserving sensitive documents such as personal letters or medical records. In order to ensure that people' rights are respected, archivists are required to carefully evaluate what should be made accessible to the public. Another important factor to consider is cultural sensitivity, which is particularly important when dealing with groups who are marginalised or indigenous. Mukurtu CMS is an example of a project that demonstrates how archives may embrace community-based protocols. These protocols provide organisations the ability to determine how their legacy is shared and who also has access to it. Digitisation runs the potential of continuing past exploitation if such factors are not taken into account.

The preservation of digital content is further complicated by concerns pertaining to intellectual property and copyright. Although the cultural value of orphan works (materials whose owners of copyright cannot be located) cannot be overstated, the act of distributing them may constitute a violation of legal rights. Fair use requirements are often relied upon by institutions; nevertheless, these provisions differ around the world, which results in discrepancies in digital access. Additionally, ethical problems about long-term preservation are raised by the concepts of technology obsolescence and sustainability. It is possible that future access might be compromised if one relies on proprietary formats or corporate cloud services. This highlights the need of open standards and decentralised preservation solutions. Algorithmic bias in digitisation technologies, such as mistakes in OCR with non-Latin scripts, and metadata description, which may reflect obsolete or biassed language, may mistakenly exclude or misrepresent groups. The resolution of these difficulties calls for continuous engagement with a wide range of stakeholders in order to guarantee that digital archives will continue to be maintained in a manner that is fair, inclusive, and ethical for future generations [9] (Table 8.3).

Table 8.3 Ethical guidelines in digital archiving

Guideline	Description
Free, prior, and informed consent	Communities must be involved in decisions about digitizing their cultural materials
Attribution and acknowledgment	Proper credit should be given to creators, sources, and contributors
Access equity	Materials should be accessible without discrimination, considering global digital divides
Preservation of context	Metadata and annotations should include historical, cultural, and political context

8.2.5 Digital Archiving and Pedagogy

Learning is becoming more real and engaging as a result of the increased usage of digital archives in educational settings. In addition to contributing to collaborative annotation projects, students have the opportunity to investigate actual documents and transcribe oral accounts. Through the provision of educators and students with unparalleled access to original materials, the development of critical thinking, and the production of more engaging, inquiry-based learning experiences, digital archiving has introduced a revolutionary change in the field of education. Digital archives, in contrast to conventional textbooks, provide learners with the opportunity to directly engage with historical documents, multimedia artefacts, and cultural materials. This allows students to analyse raw evidence and build their interpretations of the content. This change is consistent with constructivist teaching techniques, which place an emphasis on learning that is both active and student-centred. One of the most important advantages of education is the democratisation of access. Students from all over the world are now able to examine rare manuscripts, photographs, and recordings that were previously only available in physical archives, thanks to digital platforms such as Europeana and the digital collections of the Library of Congress. The playing field in education is levelled as a result of this, particularly for institutions that have limited resources [10].

Learning that spans multiple disciplines is also supported by digital archives. A single archive collection, such as the Open Access repository maintained by the Smithsonian Institution, may serve as a resource for history, art, science, and literature classes all at the same time. In one class, students can examine letters written during the Civil War as historical documentation, while in another class, they might examine the literary elements of these letters. This demonstrates how information crosses the borders of different disciplines. In addition, digital annotation tools make it possible to collaboratively analyse data. Archival materials can be tagged, commented on, and connected by students, allowing for the creation of shared interpretations. Students, for instance, are able to annotate texts and photos via the use of the Annotation Studio, which was created by the Massachusetts Institute of Technology (MIT). Despite this, difficulties still exist. Knowledge of information is essential because students need to acquire the skills necessary to evaluate the dependability of digital sources and to contextualise them appropriately. As an additional factor, technical obstacles, such as uneven access to the internet, might prevent some students from participating [11, 12] (Table 8.4).

Students are better prepared to manage an increasingly digital world with critical awareness when digital archives are deliberately incorporated into their educational experiences. This allows students to think like historians, interact with varied viewpoints, and acquire critically important research skills. The development of fundamental digital literacy skills, the cultivation of critical thinking, and the revelation of the subtle complexity of historical sources are four ways in which interactive archives alter the learning process. The users are able to interact directly with primary sources via the usage of these dynamic platforms, which include the annotation of documents, the comparison of multimedia artefacts, and the tracking of contextual linkages. Through the use of tools such as TimelineJS, which allows for the visualisation of historical events, and StoryMapJS, which maps archive data, active analysis is encouraged rather than passive consumption. In order to simulate the procedures that are used in professional research, students are taught to assess sources, recognise

Table 8.4 Digital archiving and pedagogy

Aspect	Benefits	Examples/Tools	Challenges
Access to primary sources	Democratizes education; enables direct engagement with historical evidence	Library of Congress Digital Collections, Europeana, Smithsonian Open Access	Digital divide; uneven internet access globally
Interdisciplinary learning	Breaks down subject silos; shows real-world connections	Using Civil War letters in both history and literature courses	Requires curriculum redesign for cross-departmental collaboration
Collaborative analysis	Encourages peer learning; develops critical annotation skills	MIT's Annotation Studio, Hypothes.is for web annotations	Managing differing student interpretations; avoiding "groupthink"
Digital literacy	Prepares students for research in digital environments	Workshops on evaluating online archives, fact-checking exercises	Overcoming algorithmic bias in search tools or OCR errors
Cultural inclusivity	Centers marginalized voices; diversifies curricula	Mukurtu CMS (Indigenous archives), South Asian American Digital Archive (SAADA)	Ethical dilemmas in sharing sensitive/restricted materials
Long-term skill building	Teaches archival research methods for academia/careers	Student projects curating digital exhibits using Omeka	Ensuring sustained access to archived materials amid tech obsolescence

biases, and develop interpretations that are based on evidence. Furthermore, innovations such as crowdsourced transcription (for example, Zooniverse) or 3D artefact modification contribute to the democratisation of access while also highlighting the multifaceted character of history. Learners are equipped to question, analyse, and relate the past to contemporary concerns via the use of interactive archives, which bridge the gap between preservation and teaching. This allows learners to traverse our digital environment with both technical competency and historical knowledge [10, 13, 14].

8.3 Digitization of Historical Texts

Digitisation of historical texts is the process of transforming printed or handwritten documents, which are often uncommon, fragile, or inaccessible, into digital forms that can be saved, managed, and retrieved electronically. This process is critical to the digital humanities because it ensures the preservation, democratisation, and widespread dissemination of cultural and scholarly heritage.

8.3.1 Importance of Digitization

Historical documents, such as manuscripts, letters, newspapers, and early printed books, are prone to degradation owing to age, handling, and environmental variables. Digitisation protects these writings from future physical deterioration while also improving access for scholars, students, and the general public.

Key Objectives of Digitization:

1. Preservation: To protect original documents from wear and tear.
2. Accessibility: Making historical documents accessible internationally via the internet.
3. Searchability: To allow keyword and metadata searches using OCR.
4. Educational Applications: To facilitate digital learning and online education.
5. Interdisciplinary Research: To facilitate computational analysis and cross-disciplinary collaborations.

8.3.2 Methods of Digitization

The digitisation process varies according on the state and format of the text, but the most common approaches include scanning, photographing, and transcribing. Advanced technologies include optical character recognition (OCR) and semantic encoding [15] (Table 8.5).

Table 8.5 Common digitisation techniques

Method	Description	Use case
Flatbed scanning	Produces high-resolution images of flat documents	Printed books, pages from manuscripts
Overhead scanning	Uses a mounted camera for capturing delicate or bound texts	Ancient or fragile books
Optical character recognition (OCR)	Converts scanned images into machine-readable text	Printed texts with clear fonts
Manual transcription	Human-performed text entry, often used when OCR is ineffective	Handwritten or complex historical scripts
3D Scanning	Captures depth and texture for scrolls, tablets, or bound volumes	Clay tablets, embossed documents

8.3.3 Challenges in Digitization

Despite technical progress, numerous obstacles persist:

1. Physical Fragility: Handling ancient papers may be damaging.
2. Language and script complexity: Archaic languages, antiquated scripts, and regional dialects may be difficult to understand.
3. OCR Limitations: Low print quality and non-standard typefaces diminish OCR accuracy.
4. Metadata Creation: Creating consistent and useful metadata for historical texts demands skill.
5. Cost and Infrastructure: High-resolution scanners and digital storage demand an investment.

To achieve high-quality and sustainable digitisation initiatives, the following approaches are suggested (Table 8.6):

Table 8.6 Digitisation best practices

Practice	Benefit
Use of open file formats (e.g., TIFF, PDF/A)	Ensures long-term accessibility and interoperability
Creation of metadata during digitization	Facilitates cataloguing and user navigation
Backup and storage in multiple locations	Prevents data loss
Quality control and validation	Reduces errors and improves readability
Inclusive design for accessibility	Supports users with visual impairments or language barriers

Impact on Humanities Research

Digitisation has transformed the study of the humanities. Scholars may now use digital tools to analyse trends across enormous corpora, compare numerous manuscript copies side by side, and get access to archives that were previously inaccessible. It also facilitates inter-institutional cooperation and crowdsourcing of transcription and translation tasks. Examples of important digitalisation projects include [16]:

1. Google Books digitises millions of titles for worldwide access.
2. Project Gutenberg is a voluntary effort to digitise and preserve cultural materials.
3. Mediaeval Manuscripts from the British Library: View high-resolution photographs of rare manuscripts online.
4. HathiTrust Digital Library is a large-scale repository providing full-text search and analytics capabilities.

8.4 Metadata and Tagging Systems

Metadata and tagging systems are essential tools for digital preservation and the digital humanities. These systems organise, contextualise, and make digital resources such as writings, photographs, videos, and databases more discoverable. Proper metadata frameworks guarantee that digitised resources are not only saved but also discoverable, interpretable, and reusable.

8.4.1 Understanding Metadata

Metadata is usually referred to as "data about data." In digital archiving, metadata describes the content, structure, and administrative features of digital assets. It allows people to find, access, and interpret digital resources. Metadata can be categorized into three broad types (Table 8.7):

Table 8.7 Categories of metadata

Type	Purpose	Example fields
Descriptive	Supports discovery and identification	Title, Author, Keywords, Abstract
Structural	Indicates how components of a resource relate to each other	Table of contents, chapters, pagination
Administrative	Manages information related to file formats, rights, and preservation	File type, Copyright, Digitization date, Access rights

8.4.2 Tagging Systems in the Humanities

Tagging is the process of attaching keywords or labels to digital information in order to improve searchability and categorisation by theme. Tags may represent topic areas, genres, geographical places, dates, emotions, or rhetorical tactics. There are two main tagging approaches [17] (Table 8.8):

1. Controlled Vocabulary: A standardised collection of terminology (for example, Library of Congress Subject Headings).
2. Folksonomy (User-Generated Tags): Enables community-driven and spontaneous tagging.

Standards and Frameworks

Several metadata standards and frameworks have been created to improve interoperability and data sharing between digital archives and libraries (Table 8.9).

Metadata delivers extensive search capabilities, allows for cross-collection analysis, and ensures long-term accessibility via well-documented digital records. It also aids in the digital preservation of delicate cultural objects by including information on how, when, and by whom a digital item was generated and updated.

Key Benefits:

1. Enables sophisticated search and filtering.

Table 8.8 Tagging approaches comparison

Feature	Controlled vocabulary	Folksonomy
Source of tags	Predefined by experts or institutions	Generated by users or communities
Consistency	High	Variable
Flexibility	Low	High
Ideal use	Academic or institutional repositories	Social media, collaborative platforms

Table 8.9 Popular metadata standards in digital humanities

Standard	Description	Use case
Dublin core	Simple and generic, widely used for digital libraries	General web and academic repositories
MODS (Metadata Object Description Schema)	Rich descriptive metadata based on MARC	Library and scholarly catalogs
METS (Metadata Encoding and Transmission Standard)	Combines multiple metadata formats	Complex digital objects and archival packages
EAD (Encoded Archival Description)	Designed for encoding archival finding aids	Archival collections and manuscript inventories
TEI Header (in TEI/XML)	Metadata section for TEI-encoded texts	Literary and historical texts

2. Enables digital forensics and provenance tracing.
3. Facilitates interoperability across archives and platforms.
4. Improves user engagement with improved content discovery.

8.4.3 Challenges in Metadata Creation

Despite its importance, metadata development may be time-consuming and resource-intensive. Challenges include:

1. Standardisation versus Flexibility: Striking a balance between structural integrity and flexibility to various materials.
2. Multilingualism: Supporting diverse languages and character sets.
3. Legacy Data: Adding old information to new systems.
4. Human Labour: Manual input is often necessary for intricate or historical goods.

8.5 Cultural Preservation Through Technology

Culture refers to a group of people's beliefs, habits, rituals, languages, and artefacts, which are often handed down through generations. In an age of globalisation, climate change, violence, and displacement, maintaining cultural heritage particularly intangible forms has become more important. Technology provides novel methods for documenting, protecting, and sharing cultural practices and artefacts, guaranteeing their preservation and accessibility for future generations. Technology facilitates the digital recording and sharing of cultural information. Cultural artefacts are protected from loss, degradation, and obscurity by translating physical, oral, and intangible heritage to digital representations. This preservation serves several functions, including teaching, identity reinforcement, and cultural revitalization [18, 19]. Digital technologies provide new possibilities for (Table 8.10):

1. Document endangered languages.
2. Preserve oral traditions.
3. Digitise art, music, and rituals.
4. Create virtual museums and exhibits.
5. Encourage participative cultural heritage from marginalised communities.

8.5.1 Bridging Generations and Borders

Culture is not only preserved by technology, but interests in it are also rekindled by it. Take, for example:

1. YouTube has become a venue for indigenous groups to showcase their traditional dances, traditions, and ceremonies.

Table 8.10 Technologies for cultural preservation

Technology	Application in cultural preservation	Examples
3D Scanning & printing	Recreates artifacts and monuments in digital or physical form	Scanning Egyptian tombs, reconstructing statues
Geographic information systems (GIS)	Maps historical and cultural landscapes	Indigenous land mapping, historical geographies
Augmented reality (AR)/ Virtual reality (VR)	Immersive experiences for cultural education and tourism	Virtual museum tours, re-enactments
Artificial intelligence (AI)	Translates, classifies, and restores ancient texts/images	AI-assisted OCR of ancient manuscripts
Cloud storage & archives	Long-term preservation and collaborative sharing of digitized content	Europeana, UNESCO digital archives
Blockchain technology	Authenticates digital ownership and provenance of cultural artifacts	NFT-based indigenous artwork records

2. Mobile apps allow users to study and practise endangered languages such as Ainu and Yoruba.
3. Rare texts and recordings may be accessed globally via online archives and databases like the World Digital Library.
4. Crowdsourced systems enable communities to categorise, translate, and curate their own cultural content.

Despite its benefits, digital culture preservation presents challenges:

1. Contextual loss: A digital copy may fail to retain a cultural artifact's contextual, ceremonial, or performative context.
2. Data loss might occur in the absence of sufficient backups or sustainable systems.
3. Access disparities: Not all communities have the means to digitise or access digitised culture.
4. Cultural sensitivity: Certain cultural resources are holy or limited, and they should not be widely distributed.

8.6 TEI, XML, and Text Encoding

In the field of digital humanities, text encoding is an extremely important phenomenon, particularly when it comes to making historical, literary, and archival texts available for the sake of study, preservation, and academic cooperation. An example of one of the most well-known encoding frameworks is the Text Encoding Initiative (TEI), which use XML (Extensible Markup Language) to organise and define textual characteristics in a format that is standardised and understandable by machines. In order to guarantee that digitised texts are both legible by humans and capable of being processed by computers, these technologies are essential.

The language known as XML, which stands for Extensible Markup Language, is a markup-based language that is versatile and can format data in a manner that is legible by both humans and machines. Custom tags are used in order to specify items included inside a document.

Example of XML Markup:

```
<person>
    <name>Wole Soyinka</name>
        <birthYear>1934</birthYear>
    <nationality>Nigerian</nationality>
</person>
```

In the field of digital humanities, XML gives academics the ability to annotate and describe the structure of texts in a meaningful and standardised manner. This includes the organisation of chapters, verses, speakers, editorial comments, and other elements.

8.6.1 *Introduction to the TEI*

TEI, which stands for the Text Encoding Initiative, is a group that offers principles for the encoding of literary and historical texts in XML. Established in the 1980s, TEI provides researchers with the ability to:

1. Mark up complicated textual elements (such as footnotes, page breaks, and deletions).
2. Describe historical variations, such as other editions or spellings.
3. Include information about the authorship, date, and publication.

It is common practice for libraries, archives, and academic editing projects to make use of the standards developed by TEI (Table 8.11).

Table 8.11 Common TEI elements and their use

TEI element	Purpose	Example
< pb/ >	Indicates a page break	< pb n = “3"/ >
< hi >	Highlights or emphasizes text	< hi rend = “italic” > italic text < /hi >
< note >	Inserts a footnote or annotation	< note > See also Chap. 4. < /note >
< speaker >	Marks the speaker in dialogue	< speaker > Hamlet < /speaker >
< div >	Divides a text into chapters, acts, scenes, etc	< div type = “chapter” > … < /div >
< head >	Marks the heading of a section	< head > Chap. 1 < /head >
< lg > and < l >	Represents lines and line groups in poetry	< lg > < l > Line 1 < /l > < /lg >

Why Use TEI?

TEI provides a number of advantages in the field of digital humanities, including the following:

1. Scholarly Rigour: Enables thorough and exact text representation.
2. Interoperability: Supports a wide range of digital tools and platforms.
3. Long-Term Preservation: XML-based TEI documents are both durable and transferable.
4. Accessibility: Improves the usability of digital texts across several devices and forms.
5. Reusability: TEI-encoded texts may be analysed computationally using text-mining and visualisation software.

8.6.2 Applications of TEI in Humanities Projects

TEI is used in a wide range of initiatives that aim to preserve and investigate literary and historical materials. Among the examples are:

1. The Women authors Project (Brown University) encodes works by early modern female authors.
2. Perseus Digital Library (Tufts University) contains TEI-encoded ancient literature in Greek and Latin.
3. The Digital Pan-African Archive might utilise TEI to annotate speeches, letters, and documents from Pan-African groups.

Challenges and Considerations

TEI and XML, while their strength, are not without their drawbacks:

1. The learning curve is steep, and it is necessary to have instruction in TEI rules and XML syntax.
2. The process of detailed encoding may be time-consuming and arduous.
3. In the case of smaller projects, the overhead may be prohibitive for digital collections that are simplified.

8.7 Oral Histories and Multimedia Archives

Scholars are now able to collect, preserve, and analyse lived experiences, spoken traditions, and cultural expressions that may not be preserved in written records because to the proliferation of oral histories and multimedia archives, which have become essential components of digital humanities. These archives make use of digital resources to preserve human voices, emotions, gestures, and situations. As a result, they provide a more comprehensive picture of history and culture, with regard to populations who are sometimes excluded from conventional archives[20, 21].

Oral history is a way of collecting and preserving historical information by means of recorded conversations with persons who have firsthand knowledge of events that occurred in the past. In doing so, it acknowledges the significance of viewpoint, voice, and story while emphasizing subjective recollection rather than objective reality.

In today's digital age, oral histories are more than just voice recordings; they often include images, transcripts, metadata, geolocation, and interactive maps, which transform them into rich multimedia experiences that are easily accessible [22].

8.7.1 Importance in Cultural and Historical Research

When it comes to the preservation of cultural memory, oral histories serve as a crucial methodological and ethical basis. This is especially true in situations where dominant historical narratives have systematically ignored or distorted populations who are marginalised. As an active method of archival practice, they are necessary for the following reasons [23]:

1. Indigenous and marginalised cultures, where oral traditions are carried down.
2. Diaspora studies provide human accounts of migration and identity.
3. Postcolonial study involves questioning prevailing narratives.
4. Women's and gender studies focus on experiences that are underrepresented in official records.
5. Community memory programs focus on maintaining local knowledge, tales, and traditions.

These narratives contribute to the decolonisation of history by providing a voice to those who have been excluded from textual archives. Additionally, they provide emotional depth and contextual richness.

Multimedia Archives: Beyond Text

The modern digital archives have developed into dynamic multimedia ecosystems that have revolutionised the way historical and cultural narratives are told. This transformation has gone well beyond the traditional role of text repositories. They can accommodate a variety of media types, which enables them to create storylines that are more immersive, richer, and more accurately portray the richness of the human experience [24]:

1. Audio includes interviews, music, and oral histories.
2. Video includes performances, ceremonies, testimonials, and reenactments.
3. Photographs include family albums, events, and cultural artefacts.
4. Interactive maps and timelines: Showing trips or changes across time.
5. Transcriptions and subtitles help with accessibility and textual analysis
6. Linked metadata includes speaker biographies, topics, themes, and tags.

The combination of these elements results in the creation of immersive archives that are appealing to a diverse variety of people, including academics and students, members of the community, and policymakers (Table 8.12).

The digitisation of voices, pictures, and lived experiences via oral histories and multimedia archives brings with it significant ethical duties that go far beyond the preservation of technical aspects of the material. In the end, the moral framework that decides whose tales are saved, how they are depicted, and who ultimately controls them is formed by these concerns.

1. Informed consent requires participants to understand how their tales will be used and shared.
2. Ownership and access: Communities may want to limit who may see or reuse their resources.
3. Trauma sensitivity: Telling stories of violence or grief requires a sensitive portrayal and care.
4. Language preservation: Original languages should be preserved alongside translations if feasible.

The use of collaborative approaches, in which communities co-curate, annotate, and govern their own information, is becoming more popular as a means of maintaining archiving procedures that are both ethical and courteous.

Global Initiatives and Examples

There have been ground-breaking efforts that have emerged as a result of the digital era. These initiatives highlight the ability of oral histories and multimedia to democratise historical narratives. The following initiatives are examples of methods to preservation, access, and ethical storytelling that are at the forefront of innovation:

Table 8.12 Common tools and formats in oral and multimedia archives

Tool/Format	Functionality	Examples/Notes
MP3, WAV	Audio formats for high-quality voice preservation	WAV preferred for archival quality
MP4, MOV	Video formats for oral interviews or performances	MP4 commonly used for streaming and compatibility
ELAN	Time-aligned transcription and annotation tool	Used for linguistic and gesture annotation
OHMS (Oral History Metadata Synchronizer)	Synchronizes transcripts with audio/video	Enhances navigation and searchability
Omeka	Web publishing platform for digital archives	Supports multimedia exhibitions and metadata tagging
Transkribus	AI-powered tool for transcription and text recognition	Useful for legacy documents and handwritten notes

1. The Shoah Foundation's Visual History Archive contains more than 50,000 testimonies from genocide survivors.
2. StoryCorps (USA) is an oral history initiative that records regular American voices and broadcasts them via radio and podcasts.
3. The Pan-African Archive Project is a planned endeavour to chronicle tales of independence, migration, and resistance in African countries.
4. Voices of Rwanda preserves testimony from survivors of the 1994 massacre.

8.8 Access and Ownership in Digital Archives

The digital humanities debate revolves primarily on the topic of who has access to digital archives and who in turn owns the items that are included inside them. The transfer of archives from physical collections stored in institutions to digital repositories hosted on the cloud has resulted in an increase in the complexity of problems pertaining to accessibility, control, and intellectual property. Especially in the context of colonial histories, indigenous knowledge systems, and global disparities, it is necessary to have critical frameworks and ethical practices in order to guarantee fair and equitable access while still preserving ownership rights.

8.8.1 The Promise and Pitfalls of Open Access

The capacity of digital archives to provide more people access to information is often regarded as one of the most significant benefits of these archives. Students, researchers, and members of the general public are enabled to review rare manuscripts, oral histories, and historical documents via the use of digital archives, which can be accessed from nearly any location. This eliminates the need for physical or financial restrictions. Because of this openness, collaborative research, educational outreach, and cultural involvement are all made easier it. The open access model, on the other hand, is not a panacea. The rights of persons or groups whose materials are digitised may be ignored or overridden by it, even though it expands the scope of its reach. An open-access policy, for instance, that makes indigenous oral histories publicly accessible online might potentially contravene cultural conventions and the expectations of individuals about their privacy. It is thus necessary to strike a balance between access, ethical duty, and cultural awareness.

8.8.2 *Ownership in a Digital Context*

The ownership of digital archives is subject to a variety of legal, cultural, and technical sectors that are interrelated with one another. Each of these spheres has its own unique set of circumstances and obligations:

1. Legal ownership refers to copyright and intellectual property.
2. Custodianship refers to the institution or platform that maintains the archive.
3. Cultural or community ownership focuses on the rights of the people or descendants whose tales or artefacts are maintained.

These several levels of ownership may sometimes come into conflict with one another. It is possible that a university is legally the owner of digitised recordings of the oral history of a local community; nevertheless, the community itself may assert moral and cultural ownership of the recordings, requesting a participation in the decision-making process over how the information is acquired or distributed.

Infrastructure and the Digital Divide

Access is also influenced by technical infrastructure. Digital archives require:

1. stable Internet connections
2. enough storage space.
3. easy-to-use interfaces
4. Content choices in several languages or locales

In many regions of the Global South, inadequate internet connectivity and antiquated technology make it difficult to use digital repositories. Furthermore, archives created by Western organisations may prioritise English-language information, interface layouts, or navigation mechanisms, excluding users from non-Western backgrounds. Thus, the digital divide continues to be a barrier to both access and meaningful involvement. To close this gap, decentralise power, diversify design, and engage in community-led digitisation initiatives.

8.8.3 *Licensing and Access Models*

Digital archives depend on a complex network of licensing arrangements that govern how contents are accessed, repurposed, and shared. These frameworks represent persistent contradictions between open access goals and the restrictions required for sensitive or private material (Table 8.13).

Strategies for Equitable Access and Ownership

To achieve justice and sustainability in digital archiving, many fundamental measures may be implemented:

Table 8.13 License type

License type	Description	Use case
Creative commons (CC)	Allows creators to define usage permissions (e.g., attribution, no-commercial use)	Used in open educational resources and public archives
Institutional access only	Restricted to users affiliated with a university or library	Common in proprietary academic databases
Community licensing	Controlled by originating community; access determined by cultural protocols	Used in indigenous knowledge repositories
Public domain	No restrictions on use or distribution	Applies to out-of-copyright materials

1. Co-creation of archives: Instead of just contributing resources, encourage communities to curate and annotate them themselves.
2. Tiered access approaches allow for distinct degrees of access depending on cultural or ethical issues (e.g., public, academic, limited).
3. Localised metadata and multilingual interfaces make archives more accessible to a wide range of audiences.
4. Transparent governance arrangements include community members in decision-making and policy development.

8.9 Ethical Considerations in Curation

Curation in digital archives is not a neutral act it involves choices about what is preserved, how it is presented, and who has the right to interpret and access the materials. There is an inherent ethical nature to these judgements, particularly in situations where the information contains sensitive histories, voices from marginalised groups, or elements that are culturally valuable. The process of digital curation that adheres to ethical standards requires a method that is both thoughtful and introspective, taking into account the rights, dignity, and values of the groups and persons who are represented in the archive. The realisation that all archives are curated is the first step in ethical curation. Decisions must be taken about what things to include and what to leave out of the collection. It is possible for these judgements to either challenge or repeat preexisting prejudices. One example of a digital archive that perpetuates a distorted narrative of history is one that predominantly contains materials written by colonial authors and does not include the opinions of indigenous people. In order to properly curate, practitioners need to deal with questions such as the following:

1. Whose experiences are being conveyed, and who is left out?
2. What is the context for the material?
3. How can we portray pain or contentious histories without sensationalism?

It is the responsibility of the curator to ensure that representations are truthful, respectful, and responsible to the communities from which they originate, while also working to elevate voices that have been suppressed, criticise dominant narratives, and promote diversity.

It is becoming more important to address concerns around privacy and consent in the digital environment, where information may be accessible instantaneously and on a global scale. It is necessary for contributors to provide their informed permission, which means that they are aware of how their materials will be kept, shown, and distributed. It is especially important to keep this in mind while dealing with:

1. Personal testimonials and oral histories.
2. Images depicting holy rites or departed folks.
3. Documents concerning trauma, violence, or persecution

It is also necessary to respect the cultural norms and customs. When it comes to many indigenous and traditional civilisations, the sharing of information is not always open and accessible. There is a possibility that some information is prohibited because of gender, clan, or season. These constraints are respected by ethical curation, as opposed to being overridden in the sake of free access.

Ethical Frameworks and Practices in Digital Curation

One of the most important roles that ethical frameworks play in digital curation is to guarantee that material is maintained and distributed responsibly. Respect for credit and copyright is critical, since curators are required to adequately identify sources and comply with rules governing intellectual property. This is one of the most crucial aspects. For the purpose of ensuring that personal data and user-generated material are managed ethically, especially when dealing with sensitive information, privacy and permission are equally important. Due to the ease with which false information may be disseminated, digital curation places a significant emphasis on transparency and veracity. In order to prevent misrepresentation or distortion, curators are required to verify the sources of their knowledge comes from and display them in an honest way. Moreover, digital curation is not limited to the simple act of gathering material; rather, it is also concerned with providing consumers with valuable insights and recommendations. Instead of just gathering and presenting information without providing any context, content should be selected in a manner that increases comprehension (Table 8.14).

Ethical Challenges in the Digital Age.

The digital archives continue to encounter ethical issues, despite the fact that they have noble intentions:

1. Permanence versus Ephemerality: Once posted, things may be replicated or preserved by other parties, even if they are removed from their original location.
2. Data Colonialism: Institutions in the Global North often digitise and host content from the Global South without prior consultation or benefit sharing.

Table 8.14 Ethical frameworks and practices in digital curation

Ethical principle	Description	Applied example
Informed consent	Ensuring participants understand and agree to the use of their contributions	Oral history archives provide consent forms in the speaker's native language
Do No Harm	Avoid re-traumatizing or exposing vulnerable populations	Withholding sensitive testimonies from public view unless explicitly permitted
Cultural sensitivity	Respecting traditional knowledge systems and taboos	Limiting access to sacred songs or images based on indigenous protocols
Collaborative governance	Involving source communities in decision-making processes	Advisory boards made up of community elders and scholars
Transparency	Making curation policies and data usage clear to users and contributors	Publishing archival metadata standards and licensing terms alongside collections
Right to withdraw	Allowing contributors to retract their materials from the archive	Providing a takedown policy for oral history contributors

3. Ethics of AI and Automation: Automated metadata creation and face recognition algorithms have the potential to abuse privacy or mislabel cultural artefacts.

An ethical framework that takes into account both archival best practices and more general considerations from disciplines like digital rights, critical race studies, and indigenous epistemologies is required in light of these challenges.

Community-Centred Approaches to Ethics

One strategy that shows promise in addressing the ethical challenges that are associated with digital curation is the community-centred approach. This methodology guarantees that archives are built with direct participation from the communities whose history is being preserved with the help of the communities themselves. This method places a higher weight on local knowledge, lived experiences, and collective memory than it does on curators imposing narratives from an outside viewpoint. Communities are able to acquire agency in defining how their experiences are portrayed when active engagement is encouraged. This helps to ensure that their perspectives are heard and valued.

Because choices about information selection, organisation, and accessibility are decided with input from people who are most impacted, this collaborative technique does not just promote ethical transparency but also promotes ethical transparency. By doing so, it lessens the likelihood of misunderstandings or exploitative practices, so producing a historical record that is more precise and significant. In addition, when communities participate in the archive process, they have a feeling of ownership over their legacy, which helps to reinforce cultural identity and relationships across generations.

The community-centred approach turns archives from passive repositories into dynamic environments for storytelling, knowledge-sharing, and preservation. This is accomplished by emphasizing ethical participation and inclusion. Through the establishment of trust between curators and communities it helps to cultivate a digital curation landscape that is more responsible and is more representative. As an instance, the Mukurtu CMS, which was built in collaboration with Aboriginal tribes in Australia, gives users the ability to choose cultural permissions for content. These rights include blocking access to certain resources based on kinship or gender roles. Consequently, this demonstrates a paradigm shift away from archives as extractive institutions and towards places of shared memory and knowledge that are co-created by individuals.

Developing Ethical Guidelines and Policies

Formal ethical standards should be developed by institutions in order to guarantee sustainable and responsible curation. These rules should address the following:

1. Acquisition policies: How and from whom materials are collected.
2. Usage rights and licensing: What users may do with the resources.
3. Representation standards: How are communities and cultures portrayed and contextualised?
4. Complaint and takedown mechanisms: How to address and resolve conflicts.

8.10 Decolonizing the Archive

Decolonizing the archive is a transformative process aimed at challenging and dismantling colonial power structures embedded in traditional archival practices. Not only does it need reevaluating what is saved and how it is maintained, but it also requires reevaluating who has the authority to make decisions. History has shown that the majority of current archives, particularly those that are located in Western institutions, have traditionally marginalised the voices, knowledge systems, and narratives of people who were colonised. These systems are intended to be reoriented towards justice, inclusion, and indigenous epistemologies as part of the decolonisation process.

Colonial Legacies in Archiving

Many archives reflect a Eurocentric viewpoint, with a focus on colonial histories, Western categorisations, and state institutional views. Land records, missionary notebooks, and ethnographic pictures were often obtained by exploitative tactics and preserved without the knowledge or cooperation of the tribes affected. These collections often:

1. Ignore or distort indigenous language and knowledge systems.
2. Classify information using foreign frameworks (for example, Library of Congress Subject Headings).
3. Hide the bloodshed and dispossession that followed their construction.

Table 8.15 Decolonial archiving

Principle	Explanation
Community control	Archives should be managed or co-managed by the communities whose materials are featured
Epistemic respect	Recognize non-Western systems of knowledge, categorization, and historical understanding
Repatriation of records	Digitally or physically returning materials to their communities of origin
Contextualization	Provide narratives and metadata that reflect indigenous or local perspectives

Principles of Decolonial Archiving

When an archive is decolonised, it is not enough to merely add resources from indigenous or non-Western cultures to the collections that are already there; rather, it is necessary to make systemic changes in governance, representation, and access. Among the fundamental principles are (Table 8.15):

Tools and Models for Decolonizing Archives

These decolonial approaches have been embodied by a number of projects and platforms that have emerged:

1. Mukurtu CMS is a content management system that enables communities to define cultural rules for digital material access.
2. Local Contexts offers attribution mechanisms such as Traditional Knowledge (TK) and Biocultural (BC) labels for indigenous content.
3. Endangered Languages Archive (ELAR) collaborates with communities to archive and disseminate their linguistic history while maintaining local authority.

Digital Repatriation and Ethical Collaboration

The act of returning digitised assets to the communities that originally created them is known as digital repatriation, and it is an essential component of decolonisation. On the other hand, it must encompass more than just symbolic gestures and include:

1. Shared control over how items are represented, accessed, and interpreted
2. Capacity development for local digitisation and archive management
3. Mutual benefits include recognition, financing, and training possibilities.

8.10.1 Case Studies from Global Contexts

Case studies from across the globe demonstrate the many techniques applied to digital archiving and cultural preservation. These examples demonstrate how archives may represent local values, stimulate cooperation, and use cutting-edge technology to promote access, education, and justice. These worldwide projects, which range from indigenous protocols in Australia to community archives in Africa and digital repatriation in North America, highlight the relevance of cultural context, ethical participation, and decolonial frameworks [25–27].

Mukurtu CMS (Australia)

The Mukurtu Content Management System, created in conjunction with Aboriginal tribes in Australia, shows a culturally sensitive approach to digital preservation. Mukurtu enables communities to use cultural rules to decide who has access to certain material based on criteria such as gender, age, or kinship. It supports Traditional Knowledge (TK) labels, which ensure that the source communities retain ownership and control over the material. Mukurtu enables Indigenous people to manage their own history, ensuring cultural sensitivity is recognised and preserved in digital places.

The Endangered Languages Project (Global)

This worldwide movement seeks to chronicle and conserve the world's dying languages, many of which are associated with unique cultural identities and oral traditions. It allows linguists and community members to post, describe, and access language resources such as films, texts, and dictionaries. The initiative contributes to the preservation of linguistic variety and provides assistance for educational endeavours in areas that are more isolated and marginalised.

The Pan African Heritage Digital Museum (Ghana)

Artefacts, oral traditions, manuscripts, and creative expressions from all across Africa and the African diaspora are housed in this digital museum, which was established as a component of a wider campaign to recover African history. This project intends to rectify past misrepresentations and establish a repository that is focused on African culture and is accessible on a global scale. Through the celebration of Africa's cultural contributions and the connection of groups related to African heritage all across the world, the platform challenges colonial narratives.

The Digital Library of the Middle East (DLME)

The DLME preserves cultural heritage from conflict zones such as Iraq, Syria, and Yemen by digitising manuscripts, artifacts, and archival documents. This organisation protects endangered items by using international standards and forming relationships with local museums and academics in order to make these resources available to researchers and the general public. A digital safe haven for regional history is provided by DLME, which helps to avoid the loss of cultural memory that might occur as a result of conflict and relocation.

Plateau Peoples' Web Portal (USA)

This portal, which was developed by tribal officials and researchers, features archival materials from Native American tribes located in the Plateau area and gives digital access to those resources. There are photographs, oral histories, and historical documents included in the content, and all of them are accompanied with commentary

Table 8.16 Comparative overview of global digital archive projects

Project	Region	Focus	Community role	Unique feature
Mukurtu CMS	Australia	Indigenous heritage	Full governance by community	Cultural protocols for content access
Endangered languages project	Global	Linguistic preservation	Contributor-led	Crowd-sourced language documentation
Pan African heritage museum	Ghana/ Africa-wide	African cultural heritage	Diaspora involvement	3D virtual museum with Afrocentric lens
Digital library of the Middle East	Middle East	Conflict zone preservation	Institutional partnerships	Digital rescue of heritage under threat
Plateau peoples' web portal	USA (Northwest)	Native American history	Tribal curatorship	Tribal metadata and interpretative layers

from the tribe and interpretive metadata.Reclaiming the narrative around Native American identity, history, and culture is accomplished via the embodiment of tribal sovereignty in the sharing of information (Table 8.16).

The findings of these case studies reveal a number of significant takeaways, including the value of local agency, contextual information, technical adaptation, and fair access in the realm of digital curation. The most successful initiatives are often those in which the communities who provide the source material take the lead or co-govern the archive. This ensures that the cultural and historical settings are accurately recorded. Technology need to be aligned with traditional traditions and beliefs, rather than imposing generic tools, in order to create inclusion and authenticity. However, access alone is not sufficient to ensure fairness since in order for archives to be genuinely successful, they need to address issues such as language difficulties, cultural appropriateness, and other digital divides. In spite of these efforts, there are still obstacles that need to be addressed. These challenges include concerns of sustainability that arise from inadequate long-term financing and infrastructure, gaps in digital literacy that need community training, and political or legal impediments that make repatriation and rights management more difficult. These challenges highlight the need of continual involvement and adaptation in order to establish digital archives that are more accessible and that adhere to ethical standards.

References

1. Bhoi, N.K.: Use of information communication technology (ICT) and library operation: an overview. Int. Conf. Futur. Libr. From Promises to Pract. 445–456 (2017)

2. Agboola, O., Aremu, M.O., Eze, B.U., Wahab, N.O.: Effect of training and career development on bank employees' performance: evidence from selected banks in Nigeria. EMAJ Emerg. Mark. J. **10**(2), 56–62 (2021). https://doi.org/10.5195/emaj.2020.204
3. Pereira, C.S., Durao, N., Moreira, F., Veloso, B.: The importance of digital transformation in your business strategy. Mdpi. **14**(2), 1–26 (2022). https://blog.workana.com/en/entrepreneurship/the-importance-of-digital-transformation-in-your-business-strategy/
4. Olika, C.D.: Risk management practices among selected state governments in Southwestern Nigeria. Eur. J. Sci. Innov. Technol. **1**(2), 24–31 (2024). https://www.researchgate.net/profile/Hendra-Manurung/publication/356633046_From_Connectivity_to_Digital_Improving_Employee_Readiness_toward_Organizational_Change_in_Digital_Transformation/links/61a5b00671a23a0084c9b061/From-Connectivity-to-Digital-Improvi
5. Oyeronke, A., Fagbohun, M., Adebayo, O.: An assessment of computer and ICT skills among secondary school teachers in Ota Ogun State (2013)
6. Oyeronke, A., Fagbohun, M., Adebayo, O.: An assessment of computer and ICT skills among secondary school teachers in Ota Ogun State (2013)
7. Ruano, O., et al.: Corporate social responsibility as the organization's commitment against stakeholders. Intech, no. Tourism, p. 13, (2016). https://www.intechopen.com/books/advanced-biometric-technologies/liveness-detection-in-biometrics
8. Munawar, H.S., Ullah, F., Qayyum, S.: Big data in construction : current applications and future opportunities (2022)
9. Palacios, S., Ault, A., Krogmeier, J.V., Bhargava, B., Brinton, C.G.: AGAPECert: an auditable, generalized, automated, privacy-enabling certification framework with oblivious smart contracts. IEEE Trans. Dependable Secur. Comput. **10**, 1–18 (2022). https://doi.org/10.1109/TDSC.2022.3192852
10. Isa, A., Nte, N.D., Baba-Ahmadu, A.: Critical thinking for intelligence analysis for Nigerian students: evaluating the pedagogical challenges in real life scenarios. Indones. Educ. Adm. Leadersh. J. **5**(2), 46–60 (2023). https://doi.org/10.22437/ideal.v5i2.28639
11. Zhao, S., Pan, Q., Gao, D., Cheng, J.: Integrating internet of things and mixed reality to teach performance-based architectural design: a case study of shading devices. Educ. Inf. Technol. (2022). https://doi.org/10.1007/s10639-022-10998-6
12. Ghavifekr, S., Kunjappan, T., Ramasamy, L.: Teaching and learning with ICT tools: issues and challenges from teachers' perceptions. Malaysian Online J. Educ. Technol. **4**(2), 38–57 (2016)
13. Kozma, R.B., Andersen, R.E.: Qualitative case studies of innovative pedagogical practices using ICT. J. Comput. Assist. Learn. **18**(4), 387–394 (2002). https://doi.org/10.1046/j.0266-4909.2002.00250.doc.x
14. Adams, P.: Exploring social constructivism: theories and practicalities. Education 3–13. **34**(3), 243–257 (2006). https://doi.org/10.1080/03004270600898893
15. Krithika, L.B.: Survey on the applications of blockchain in agriculture. Agriculture **12**(9), 1333 (2022). https://doi.org/10.3390/agriculture12091333
16. Daud, I., et al.: The effect of digital marketing, digital finance and digital payment on finance performance of Indonesian smes. Int. J. Data Netw. Sci. **6**(1), 37–44 (2022). https://doi.org/10.5267/J.IJDNS.2021.10.006
17. Ghaffarianhoseini, A., et al.: Building information modelling (BIM) uptake: clear benefits, understanding its implementation, risks and challenges. Renew. Sustain. Energy Rev. **75**(2015), 1046–1053 (2017). https://doi.org/10.1016/j.rser.2016.11.083
18. Pokrovskaia, N.N.: Tax, financial and social regulatory mechanisms within the knowledge-driven economy. blockchain algorithms and fog computing for the efficient regulation. In: Proceedings. 2017 20th IEEE International Conference on Soft Computing and Measurements (SCM) 2017, pp. 709–712 (2017). https://doi.org/10.1109/SCM.2017.7970698
19. Hayakawa, S., Pan, Y., Marian, V.: Language changes medical judgments and beliefs. Int. J. Biling. **26**(1), 104–121 (2022). https://doi.org/10.1177/13670069211022851
20. Owsianiak, M., van Oers, L., Drielsma, J., Laurent, A., Hauschild, M.Z.: Identification of dissipative emissions for improved assessment of metal resources in life cycle assessment. J. Ind. Ecol. **26**(2), 406–420 (2022). https://doi.org/10.1111/jiec.13209

21. Martínez-González, J.A., Álvarez-Albelo, C.D.: Influence of site personalization and first impression on young consumers' loyalty to tourism websites. Sustainability **13**(3), 1–18 (2021). https://doi.org/10.3390/su13031425
22. Homg, M.-F., Lee, W.-T., Lee, K.-R., Kuo, Y.-H.: An adaptive approach to weighted fair queue with QoS enhanced on IP network. TENCON 2011. In: Proceedings of IEEE Region 10 International Conference on Electrical and Electronic Technology, vol. 1, no. 1, pp. 181–186 (2011). https://doi.org/10.1109/TENCON.2001.949576
23. Alonso-Bartolome, S., Segura-Bedmar, I.: Multi-modal fake news detection. Inf. Retr. Ser. **42**, 41–70 (2021). https://doi.org/10.1007/978-3-030-62696-9_3
24. Albahar, M., Almalki, J.: Deepfakes: threats and countermeasures systematic review. J. Theor. Appl. Inf. Technol. **97**(22), 3242–3250 (2019)
25. Richardson, S.: An international expansion in voluntary euthanasia/assisted dying: the implications for nursing. Int. Nurs. Rev. **70**(1), 117–126 (2023). https://doi.org/10.1111/inr.12807
26. Yong, H.H., et al.: Do social norms for Cigarette smoking and Nicotine vaping product use predict trying nicotine vaping products and attempts to quit cigarette smoking amongst adult smokers? Findings from the 2016–2020 international tobacco control four country smoking and vaping surveys. Nicotine Tob. Res. **25**(3), 505–513 (2023). https://doi.org/10.1093/ntr/ntac212
27. Hong, Y., Sepasgozar, S.M.E., Ali, A.: Key factors affecting construction organizations' acceptance of BIM: a comparative study. Modul. Offsite Constr. Summit Proc. **2006**, 67–73 (2016). https://doi.org/10.29173/mocs9

Chapter 9
Digital Storytelling and Narrative Innovation

9.1 Storytelling in the Digital Age

Digital storytelling is the use of digital resources to create and distribute tales that combine text, pictures, audio, and video to create immersive experiences. It has progressed beyond conventional narrative approaches, including interactive components and multimedia forms to engage viewers in novel ways. Narrative innovation refers to creative improvements in storytelling approaches, frequently driven by technology, that transform how tales are conveyed and perceived. Emerging technologies like artificial intelligence, virtual reality, and augmented reality are revolutionising digital storytelling by allowing for personalised, interactive, and immersive stories. AI-powered plot generators, for example, help authors create tales, while VR and AR enable viewers to enter a story and experience it firsthand. These technologies transform storytelling from passive consumption to active involvement, increasing narrative engagement and emotional resonance [1].

Storytelling is as ancient as human civilisation. From cave paintings and oral traditions to the written word and theatrical performances, narratives have played an important role in how cultures transmit information, values, and culture. However, the digital revolution has profoundly altered the techniques, media, and meanings of narrative. What was once a sequential, author-focused approach has evolved into a dynamic, multimodal, and often collaborative process. In the digital era, narrative is not limited to conventional media. It has evolved into a multimodal, interactive, and transmedia experience in which storylines play out across various platforms, influenced not just by writers but also by users, algorithms, and communities. This section investigates the historical origins, features, and ramifications of digital storytelling, demonstrating how narrative practices have changed and grown in our networked world [2, 3] (Fig. 9.1).

F. A. Alaba and I. G. Ajibola, *Digital Pedagogy: Bridging the Gap Between Software and the Humanities*, Information Systems Engineering and Management 85,
https://doi.org/10.1007/978-3-032-20335-9_9

Fig. 9.1 Storytelling in the digital age [4]

Table 9.1 Timeline of storytelling evolution

Era	Medium	Characteristics	Examples
Prehistoric	Oral, visual (cave paintings)	Communal, performative, memory-based	Folktales, legends, ritual storytelling
Ancient/Classical	Written manuscripts, scrolls	Author-driven, preserved knowledge	Epic of Gilgamesh, Homer's *Odyssey*
Medieval & Renaissance	Illuminated texts, theater	Religious, moralistic, communal theatre	Morality plays, *Divine Comedy*
Printing Revolution	Printed books, pamphlets	Mass production, democratised access to literature	Newspapers, novels, religious texts
20th Century	Film, radio, television	Visual storytelling, passive audience engagement	Hollywood cinema, radio dramas, sitcoms
Digital Age (21st Century)	Web, social media, apps, AR/VR, AI	Interactive, networked, multimodal, participatory	Web comics, video games, TikTok stories

From Oral to Digital: A Timeline of Storytelling Evolution

It is possible to get an understanding of the evolution of storytelling by examining significant historical stages. The following table provides a condensed summary on Table 9.1.

9.1.1 Defining Digital Storytelling

The term "digital storytelling" refers to the process of conveying tales via the utilisation of various digital technologies and media channels. The following are some of the formats that are included in its scope [5, 6]:

1. Short multimedia story combining pictures, text, music, and voiceovers.
2. Interactive literature and games in which user choices affect the ending.
3. Social media storytelling involves utilising consecutive posts or clips to construct episodic storylines.
4. Augmented and virtual reality experiences engage consumers in spatial tales.

Traditional narrative patterns, such as beginning, conflict, and resolution, continue to serve as the foundation for the fundamental components of digital storytelling; however, the distribution methods and audience interactions presented by digital storytelling are significantly different.

Key Features of Storytelling in the Digital Era

The transition to digital storytelling has resulted in various distinguishing qualities that have altered how tales are generated and consumed. Unlike conventional storytelling, digital platforms enable viewers to interact with material via clicks, swipes, or choices that shape the tale. Multimedia features such as films, animations, and music improve immersion and make tales livelier. Social media integration encourages real-time sharing and collaboration, transforming storytelling into a community experience. Algorithms personalise content by adapting tales to individual interests. Furthermore, non-linear tales allow readers to examine plots in unique order. These technologies democratise narrative, providing artists with a variety of tools while challenging traditional frameworks. Finally, digital storytelling combines technology and creativity, converting passive viewers into active participants in the narrative process [7, 8] (Table 9.2).

It is because of these advances that both amateur and professional storytellers have been given more power, narrative genres have been extended, and room has been made for new voices.

Table 9.2 Feature

Feature	Description
Interactivity	Audiences can influence the narrative flow, create content, or participate in live discussions
Multimodality	Integration of various media formats audio, video, images, text into a single narrative
Hyperlinking and navigation	Non-linear storytelling that allows users to explore different paths or branches of the story
Participatory culture	Encourages co-creation through comments, fan fiction, remixes, and digital art
Transmedia expansion	Stories unfold across multiple platforms (e.g., books, films, social media) to form a unified narrative
Temporal flexibility	Stories are accessible anytime, reshaping the experience of storytelling over time

9.1.2 The Role of Technology in Shaping Storytelling

Technology has fundamentally transformed the way stories are created, shared, and experienced, introducing new narrative possibilities that go beyond traditional formats. Storytellers are able to include multimedia components such as video, music, animation, and interaction into their storytelling via the use of digital technologies, which results in narratives that are more immersive and engaging for audiences. A second revolution in storytelling has been brought about by the development of artificial intelligence, virtual reality, and augmented reality. These technologies have made it possible to create experiences that are both personalised and dynamic, enabling viewers to actively engage in the process of constructing the narrative. The environment has also been altered by social media platforms, which have made storytelling more immediate and engaging. Streaming services have also provided new opportunities for producers to share their work on a worldwide scale. The progression of technology continues to push the limits of creativity, resulting in the development of novel approaches to engage with audiences and the presentation of narratives in novel forms that were previously impossible [9].

1. Internet and social media have facilitated viral storytelling, collaborative tales, and global audience reach.
2. Mobile devices have made storytelling more portable and personalised (for example, mobile journalism and digital diaries).
3. Artificial intelligence: Introduced automated or semi-autonomous storytelling technologies such as ChatGPT and AI-generated comics.
4. Augmented and virtual reality: Developed immersive experiences allowing people to explore 3D narrative settings.

These technologies not only convey tales, but they also alter the connection that exists between the audience, the narrative, and the narrator of the story.

Implications for Culture and Society

As a result of the revolution of storytelling brought about by the digital era, the manner in which tales are made, consumed, and shared has been rethought, which has resulted in many different sectors experiencing dramatic changes. The availability of digital technologies has made it possible for a variety of voices to emerge, making it possible for people and groups to communicate their experiences in ways that were previously restricted by conventional media. Storytelling has grown beyond linear formats thanks to the proliferation of social media platforms, streaming services, and interactive technology. This has encouraged experiences that are more immersive and orientated towards participation.

In addition, the advent of artificial intelligence, virtual reality, and augmented reality has opened up new avenues for the development of narrative innovation. These technologies have made it possible for audiences to interact with tales dynamically and even have some say in the results of those stories. These innovations provide a challenge to the traditional frameworks of narrative while also providing content makers with greater freedom in how they show their work. However, the expansion of digital technology also brings with it several issues, including worries about the spread of false information, ethical considerations in the use of data, and the need for digital literacy in order to traverse complicated narratives. The influence that storytelling has on culture, education, marketing, and entertainment will only continue to grow as it continues to develop. This will have a significant impact on how we engage with information and with one another in the contemporary world. The evolution of storytelling in the digital age has significant implications [8]:

1. Democratisation of Voice: Anyone with an Internet connection may become a storyteller, upending established publication barriers.
2. Fragmentation and Fluidity: Narratives are no longer stable; they change in response to audience input or algorithmic changes.
3. Cultural preservation: With digital means, endangered languages, customs, and histories may be stored and shared internationally.
4. Attention Economy: Storytellers now fight for fragmented attention, resulting in shorter forms and algorithmically optimised information.

9.2 Hypertext and Interactive Fiction

The advent of digital technology brought with it not only new platforms for storytelling but entirely new forms of narrative. These include hypertext fiction and interactive fiction (IF), which encourage readers to not only consume tales but also to explore, influence, and even co-create them. These are among the most innovative of these types of fiction. The non-sequential structures, changeable outcomes, and multimodal components that are included in these forms indicate that they go beyond linearity. A digital canvas that is always growing, hypertext and interactive fiction provide a challenge to the conventional notions of plot, authorship, and narrative closure [10]. This presents an opportunity for creative and literary creativity (Fig. 9.2).

What is Hypertext Fiction?

Hypertext fiction is a distinct genre of electronic literature that transforms narrative by enabling readers to traverse a text in a non-linear way via hyperlinks. Unlike conventional tales, which follow a set order, hypertext fiction allows readers to wander freely between related portions, resulting in a more personalised reading experience. This structure is similar to how consumers navigate websites, where clicking on links takes them to multiple pages and views rather than following a single route from start to finish. By including many paths, hypertext fiction disrupts traditional storytelling approaches, allowing readers control over their journey through the novel. The experience changes based on which links are followed, so each reading session may vary from the previous. This technique promotes involvement by immersing readers in a dynamic, web-like framework that combines literature and digital technology. Writers of hypertext fiction often employ this format to explore topics

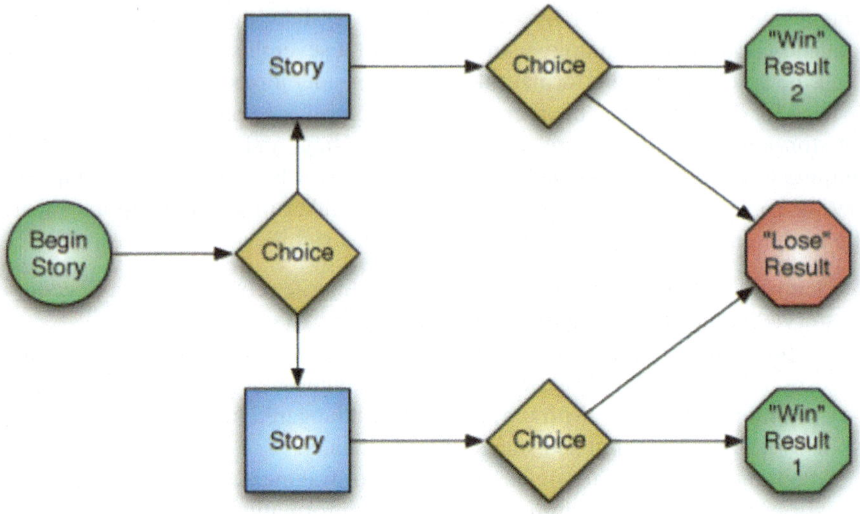

Fig. 9.2 Examples of hypertext [11]

such as fragmentation, choice, and nonlinear encounters, making it an experimental and intriguing type of narrative that will grow as digital platforms improve [12] (Table 9.3).

What is Interactive Fiction?

Interactive fiction is a more general term that encompasses a variety of different types of games, such as text-based games, narrative simulations, and choice-driven narratives. Interactive fiction, on the other hand, often places an emphasis on user agency and problem-solving inside a fictional setting, in contrast to hypertext, which frequently focusses on literary experimentation [13] (Table 9.4).

Comparing Hypertext and Interactive Fiction

When it comes to goal, interface, and design philosophy, the two genres are unique from one another. The table that follows provides an overview of some of the most important distinctions (Table 9.5):

Table 9.3 Key characteristics of hypertext fiction

Feature	Explanation
Non-linearity	The story can be read in multiple orders depending on user choices
Node-based Structure	Composed of "lexias" (text units) linked together
User Navigation	Readers click on links to move between parts of the story
Multiplicity	No fixed narrative different users may have vastly different experiences
Fragmentation	Emphasizes disjointed, poetic, or exploratory reading paths

Table 9.4 Types of interactive fiction

Type	Description	Popular examples
Parser-based IF	Users type commands to interact with the environment	Zork, Colossal Cave Adventure
Choice-based IF	Users click options to progress through the story	Twine games, Choice of Robots
Simulation IF/RPGs	Often includes branching quests and character development	Fallen London, 80 Days, Disco Elysium
Visual novels	Text-heavy games with anime-style visuals and user-driven plot lines	Doki Doki Literature Club, Steins;Gate

Table 9.5 Hypertext and interactive fiction

Aspect	Hypertext fiction	Interactive fiction
Primary medium	Written lexias with hyperlinks	Command-line input or multiple-choice menus
Focus	Literary experimentation, fragmentation	Problem-solving, immersion, player choice
Reader role	Navigator	Participant or protagonist
Structure	Loosely connected narrative fragments	Story branches based on decisions
Common platforms	Storyspace, HTML, web	Twine, Inform, Ink, ChoiceScript

9.2.1 *The Rise of Tools for Digital Narratives*

The production of digital tales has been more accessible to the general public over the course of the last twenty years because to the availability of user-friendly technologies (Table 9.6).

Table 9.6 Tools for digital narratives

Tool	Features	Audience
Twine	Visual interface for branching stories, supports HTML, CSS, JavaScript	Writers, educators, gamers
Inform 7	Natural language coding for parser-based IF	Programmers, game designers
ChoiceScript	Simple scripting language for decision-based games	Indie developers
Ink by Inkle	Narrative scripting tool for branching stories used in *80 Days*	Game studios, storytellers
Ren'Py	For creating visual novels with animation, music, and choice pathways	Visual novel creators

Table 9.7 Comparing traditional and interactive narratives

Aspect	Traditional narrative	Hypertext/Interactive fiction
Structure	Linear	Nonlinear, branching paths
Reader role	Passive	Active participant, decision-maker
Authorial control	High	Shared with the reader
Medium	Print, film, theatre	Digital platforms, web, game engines
Reusability	Limited rereading	High replayability with varied outcomes

These platforms reduce the technical constraints that users face, making it simpler for them to build complex storytelling experiences even if they do not have formal programming abilities.

Narrative Techniques in Hypertext and IF

Digital storytelling forms inspire authors to use innovative narrative strategies [14].

1. Branching Paths: Users make decisions that lead to various outcomes.
2. Multiple perspectives allow readers to see the same events through the eyes of various characters.
3. Meta-narratives are stories that recognise the user's impact or the medium itself.
4. Persistence and Memory: The system "remembers" previous user choices, which influences future content.

Educational and Cultural Applications

Beyond entertainment, hypertext and interactive fiction hold significant pedagogical and cultural value across diverse fields: literature and creative writing, they encourage experimentation with narrative structure, non-linear storytelling, and collaborative authorship; history and humanities, they enable immersive exploration of alternate historical outcomes and marginalised perspectives through choice-driven narratives; and mental health and therapy. By combining technology and narrative, these innovations reimagine engagement while also enabling education, cultural preservation, and emotional health [15] (Table 9.7).

9.2.2 *The Future of Hypertext and IF*

Emerging technologies are transforming interactive narratives by allowing AI-generated storytelling to modify storylines in real time depending on user inputs, resulting in highly personalised experiences. VR fiction immerses people in completely immersive settings while branching storylines develop all around them, whilst cross-platform storytelling effortlessly combines narratives across devices, media forms, and social networks. Together, these developments make storytelling more dynamic, inclusive, and immersive, extending narrative limits well beyond conventional print or screen-based media and altering how tales are delivered and perceived.

Games as Narrative Forms

Games are no longer seen as simple pleasures or temporary distractions; they have grown into deep storytelling mediums that compete with conventional forms of narrative expression. From the early days of text-based adventures that depended on imagination and choice-driven storytelling to the massive, immersive worlds of current role-playing games, digital games have evolved into a dynamic platform for rich, interactive tales. Unlike traditional narrative formats such as books, cinema, and theatre, video games provide unprecedented levels of player agency, enabling people to actively design their own experiences rather than passively receive them. Through branching stories, moral choices, and sophisticated character relationships, games allow users to participate in the development of a narrative in ways that conventional media frequently cannot. This emotional and intellectual involvement develops a stronger connection between players and the tales being presented, making the experience not only personal but also distinctively powerful. Furthermore, technological improvements have catapulted games into new creative realms, combining cinematic storytelling methods, sophisticated AI-driven tales, and stunning graphics to create fascinating, immersive worlds. Games now do more than simply tell tales; they elicit emotion, challenge perceptions, and create unforgettable experiences that last long after the controller is put down. Whether probing the depths of human nature, historical events, or mythical realms, video games have become a driving force in digital storytelling innovation, transforming how tales are presented and experienced in the contemporary day [16].

Understanding Ludonarrative

Ludonarrative is an important idea in understanding games as narrative forms; it combines ludus (play) and narrative to investigate the relationship between a game's mechanics and plot. When gaming mechanics coincide with a story's concepts and messages, it results in Ludonarrative harmony a seamless integration that improves the player's experience. On the other side, ludo narrative dissonance happens when the gameplay contradicts the plot, such as in a game where a character is depicted as peaceful yet must frequently participate in violent behaviors to advance. Recognising this link is critical for both game designers and analysts, as it allows for more immersive and unified narrative experiences in gaming [17] (Table 9.8).

Narrative Structures in Games

Video games, in contrast to books and movies, which adhere to a predetermined order, provide a variety of narrative frameworks that have the ability to mold the experience of the player. For example, the Uncharted series is an example of a linear narrative, which is a kind of narrative that develops in a preset order and guides the player through an organised tale. "Branching narratives" provide players the ability to make decisions that result in diverse results, so producing a more dynamic storytelling experience. "Detroit: Become Human" is a notable example of

Table 9.8 Ludonarrative alignment examples

Game	Narrative focus	Gameplay	Narrative alignment
The Last of Us	Survival and emotional loss	Stealth, combat, resource gathering	Strong harmony
Bioshock	Free will and control	First-person shooting, moral choices	Partial dissonance
Papers, Please	Bureaucracy and ethics	Document-checking under pressure	Strong harmony
Tomb Raider (2013)	Vulnerability and strength	Heavy combat	Dissonance (early gameplay)

this method. During this time, a "emergent narrative" develops organically as a result of player interactions inside the game environment. Games such as "Minecraft" and "The Sims" provide players the opportunity to create their own personal narratives. In addition, environmental storytelling is a method that is used to transmit narrative via the location, objects, and non-verbal clues. This approach is used to profoundly immerse players in the scenario, and games such as "Dark Souls" and "Gone Home" are excellent examples of applications of this strategy. A deeper level of player engagement and the ability to create deeply personalised storytelling experiences are both enabled by these diverse approaches [18].

Player Agency and Role in Storytelling

One of the most transformative aspects of modern gaming is its fundamentally interactive nature, which redefines the player's role from passive observer to active participant. Games provide players the ability to make important choices that determine the course of the tale, explore branching nonlinear routes that generate unique narrative experiences, and adopt a variety of roles and identities that directly influence the consequences. This is in contrast to conventional media, which linearly presents the story. Because of this involvement, a dynamic type of experiential storytelling is created, in which each playing becomes individually significant. The decisions that a player makes regarding character growth, moral issues, and strategic acts together weave a narrative tapestry that is unique to their trip. Not only do the tales that are produced originate from the objectives of the designers, but they also develop from the intricate interaction that exists between the game systems and the player agency. This interaction makes each experience both profoundly personal and systemically meaningful [19] (Fig. 9.3 and Table 9.9).

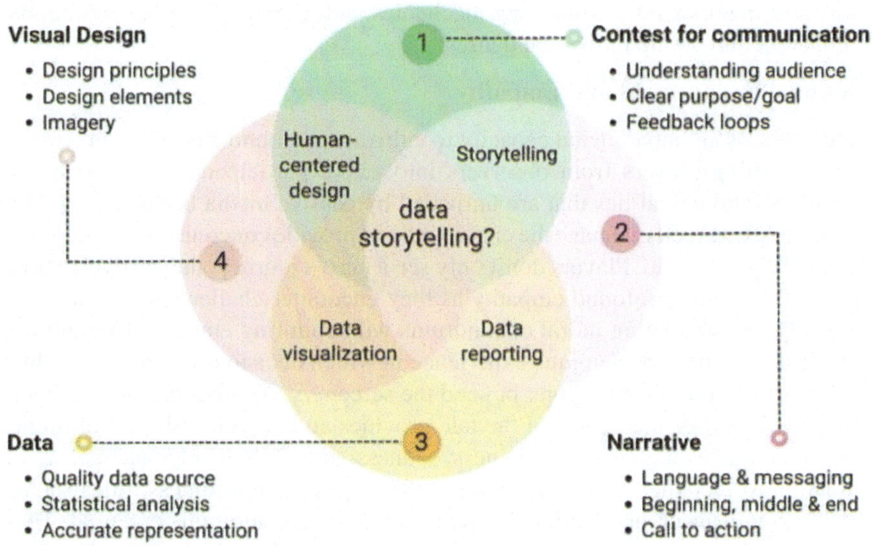

Fig. 9.3 Interactive storytelling [5]

Table 9.9 Comparison of narrative agency

Medium	Narrative control	Reader/Viewer role
Novel	Author	Passive consumer
Film	Director/Editor	Passive viewer
Game	Shared (Designer + Player)	Active participant/co-creator

9.3 Genre-Specific Storytelling

The extraordinary versatility of interactive storytelling is shown by the wide variety of game genres, with each specific style presenting a different set of narrative possibilities. The Witcher 3 and Mass Effect are two examples of role-playing games (RPGs) that place a strong emphasis on rich character development and player-driven, branching narratives. These games also include decisions that have repercussions across vast landscapes. Life is Strange and Monkey Island are two examples of adventure games that combine the elements of puzzle-solving with emotionally driven, linear tales that develop like that of fictional books. The Sims and Cities: Skylines are two examples of simulation games that produce emergent storylines via systemic interactions. These games allow for unscripted situations to evolve as a result of the player's ingenuity. In order to create suspense via the use of perspective and environmental narrative, horror games such as Amnesia: The Dark Descent and Resident Evil make use of first-person immersion or fixed camera angles. When taken as a whole, these genres demonstrate that interactivity does not limit storytelling;

rather, it reimagines it by combining mechanics and narrative into experiences that are coherent and define individual genres.

Emotional Engagement and Empathy

Games possess an unparalleled capacity to cultivate profound emotional investment by transforming players from observers into active participants. Interactive tales generate psychological ties that are unrivaled by passive media because they allow the audience to directly manage the characters and provide concrete repercussions for what they choose to do. Players don't only see a hero's journey; they actively participate in it, building profound empathy as they encounter challenges and successes personally. By simulating moral conundrums with enduring emotional weight, this media is able to provide a unique experience in which decisions on sacrifice, ethics, or relationships continue to linger beyond the screen. As a consequence, the feeling of responsibility for the results of the tale provides an unparalleled level of immersion, which makes feelings of happiness, sadness, and guilt deeply personal. Using interaction to recreate the visceral reality of loss and mental illness, titles such as That Dragon, Cancer, and Hellblade: Senua's Sacrifice exhibit this potential. These games provide a sophisticated emotional understanding that cannot be replicated by conventional narrative [20] (Fig. 9.4).

9.3.1 Narrative Innovation in Game Design

Modern game designers are pioneering innovative narrative structures that seamlessly merge traditional storytelling with interactive mechanics, creating revolutionary hybrid experiences. In order to keep the dramatic tension while still allowing the player to preserve their agency, they have developed advanced narrative pacing tools. These tools allow them to properly balance gameplay sequences with cinematic cutscenes. It is possible to see procedural generation systems in games like as Hades and No Man's Sky. These systems bring dynamic storytelling, in which material that is generated by an algorithm impacts the flow of the narrative. This ensures that no two playthroughs are similar. Outer Wilds and Twelve Minutes are two examples of games that use time loops and branching timelines to convert repetition into significant narrative exploration. Each iteration of the game uncovers new plot layers, making them perhaps the most inventive examples of this innovation. To create organic, player-driven storytelling experiences that could only be found in interactive media, these experimental approaches demonstrate how contemporary game design is fundamentally redefining the relationship between narrative and interactivity. This goes beyond the simple dichotomies that exist between cut scenes and gameplay [22] (Table 9.10).

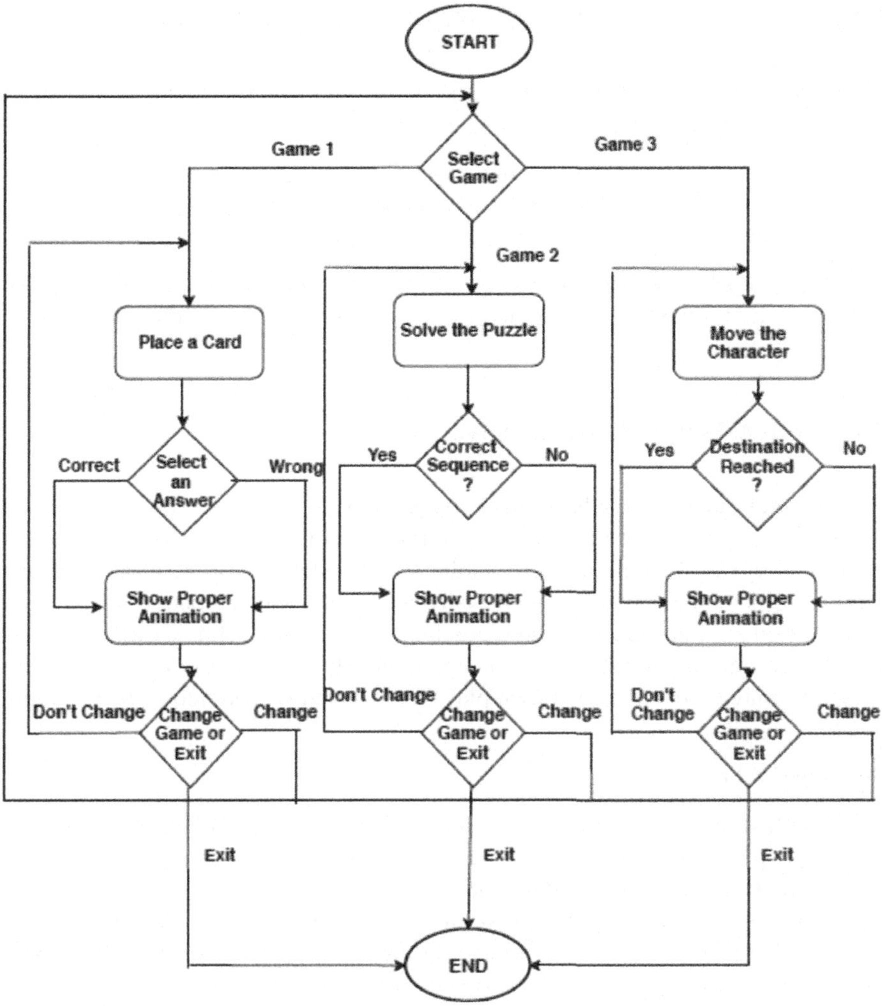

Fig. 9.4 Flowchart of game development [21]

Table 9.10 Notable narrative-driven games

Game title	Narrative feature	Unique element
Firewatch	Dialogue-driven relationship building	Story unfolds through walkie-talkie conversations
Disco Elysium	Detective narrative with deep philosophical tones	Skill-based internal monologue
Undertale	Morality system affects narrative tone	Pacifist versus genocide paths
Her story	Nonlinear story told through police archives	Player pieces together narrative via video clips

9.3.2 Educational and Social Impact of Narrative Games

Increasingly, games are being used in the fields of education, activism, and social awareness, using the narrative power of games to:

- Teach history with immersive simulations (Valiant Hearts).
- Encourage empathy and inclusivity (A Normal Lost Phone).
- Investigate ethical and social concerns (This War of Mine).

Game design has evolved into a tool for experiential learning and social interaction because it allows players to be immersed in viewpoints that are foreign to them.

9.4 Podcasts and Audio Essays

Podcasts and audio essays have arisen as prominent narrative formats in this era of digital media. These forms mix the oral storytelling traditions of the past with the technological dissemination technologies of today. These formats provide a narrative experience that is personal, easily accessible, and compatible with mobile devices. Podcasts and audio essays provide a sonic canvas for a variety of tales that engage both the mind and the imagination. These narratives may range from academic conversations and fictional plays to investigative journalism and memoirs (Table 9.11).

Defining the Forms

1. Podcasts are audio recordings that are published on a regular basis and may be streamed or downloaded. They often use episodic styles and might be narrative-driven, instructive, or conversational.
2. Audio essays are structured, thoughtful spoken-word works that are often utilised in academic, journalistic, and artistic settings. They emulate the personal tone of written writings while engaging listeners with voice, tone, and sound design.

Table 9.11 Key Differences between podcasts and audio essays

Feature	Podcasts	Audio essays
Format	Episodic, serial or standalone	Singular, structured piece
Tone	Conversational or documentary-style	Reflective, critical, often literary
Purpose	Entertainment, education, engagement	Persuasion, reflection, analysis
Typical length	20–60 min	5–20 min
Distribution	Platforms (Spotify, Apple, etc.)	Often part of academic or media outlets

9.4.1 Storytelling Techniques in Audio

One way in which audio storytelling differentiates itself from visual media is that it relies on delicate aural aspects to construct its narrative world. This allows audio storytelling to develop immersion via a fully sensory palette. The voice of the narrator serves as the emotional core of the story, and subtle shifts in tone, timbre, and inflection convey a variety of complex psychological states as well as narrative and subtextual information. The listener's imagination is used to build complete settings via the use of layered sound design, which includes carefully selected ambient noise as well as intentional musical cues. Additionally, three-dimensional audio placement may imitate real space and movement. Through the use of controlled silences to generate tension and rhythm to manage the flow of emotions, pacing becomes a potent instrument for dramatic expression. The one-of-a-kind intimacy of this medium is reminiscent of late-night radio or whispered confidences, and it helps to cultivate significant listener-narrator relationships that are both wide in their creative reach and profoundly intimate in their delivery. The resonance of audio storytelling is not achieved via the use of explicit imagery; rather, it is achieved through the activation of the listener's creative engagement in world-building as a means of making the experience uniquely collaborative [23].

9.4.2 Narrative Genres in Podcasting

Podcasting has blossomed into a sophisticated storytelling medium, cultivating diverse narrative genres that blend entertainment with intellectual engagement. Audio journalism was revolutionised by the true crime revolution that was pioneered by Serial and Criminal. This revolution enabled audio journalism to weave investigative rigor with thrilling narrative while also encouraging ethical reflection on justice systems. Personalised tales are elevated to the level of communal human records via the use of intimate memoir podcasts such as The Moth and Terrible, Thanks for Asking. These podcasts transform individual weaknesses into universal touchstones. Television programs such as Hardcore History and Throughline make use of extensive audio archives and expert commentary in order to shed light on historical patterns and political intricacies. This allows for a more profound comprehension of the background. In the meanwhile, fictional audio dramas such as "Welcome to Night Vale" and "Homecoming" are reviving the traditions of radio plays by using contemporary sound design. These dramas use just aural clues to generate complete imaginative worlds. The ability of podcasting to combine appealing narrative with critical thinking is shown across all of these media. Podcasting has the ability to transform mundane activities like commuting and housework into moments of connection, education, and wonder [24].

Table 9.12 Example applications of audio essays

Discipline	Application	Example
Media studies	Critique of digital platforms	Analysis of TikTok and media literacy
History	Reflective narration of historical events	Oral perspectives on colonial archives
Literature	Audio interpretation of poetic form	Sound-enhanced analysis of spoken word poetry
Journalism	First-person exploration of lived events	Audio documentary on migration experiences

9.4.3 Audio Essays in Academia and Media

An increasing number of educational and journalistic institutions are beginning to acknowledge audio essays as valid forms of digital research and creative nonfiction. These essays provide a dynamic and entertaining method of investigating difficult topics in language that is easily understandable. They produce an engaging narrative that connects with a variety of audiences by combining academic acumen with human storytelling. This makes complex concepts more approachable and immersive for the audience by making them more accessible. Additionally, the addition of multimodal components, including as ambient noises, interviews, and music, increases the quality of the listening experience by adding depth, emotion, and a feeling of location to the conversation. Audio essays are a great instrument for intellectual engagement and creative expression because they combine scholastic discourse with aural creativity. This allows for a more in-depth and nuanced investigation of issues, which positions audio essays as a potent tool [25] (Table 9.12).

9.4.4 Accessibility and Global Reach

Podcasts and audio essays stand out for their accessibility, requiring minimal equipment to produce while allowing listeners to engage with content even while multitasking. The fact that they are able to communicate with people all over the world via free platforms makes them an effective medium for the dissemination of information and the telling of stories. In addition to this, they function as inclusive tools, allowing those with poor literacy or visual impairments to access information and so removing obstacles to education and participation. This low barrier to entry promotes various voices and viewpoints, particularly from groups and places that are underrepresented, which in turn fosters a conversation that is richer and more varied in the digital world.

Table 9.13 Notable narrative podcasts and audio essays

Title	Narrative focus	Innovative feature
Serial	Investigative true crime	Episodic suspense built from real cases
Radiolab	Science and philosophy	Complex topics simplified through sound
The Moth	Real-life personal stories	Live storytelling without notes
99% Invisible	Design and architecture	Focus on the unnoticed with rich sound
Ear Hustle	Life inside prison	Produced by inmates, a unique insider voice

9.4.5 Voice, Identity, and Authorship

When it comes to audio forms, voice functions as both a literal and symbolic force, influencing authenticity and representation via the selection of a narrator, which may be the subject of the recording or an interpreter from a third party. This decision has an effect on who is heard, who controls the narrative, and how accents, dialects, or languages are depicted; all of these factors contribute to the formation of identity and the presence of the author. The power of voice in storytelling is highlighted by these factors, which underline its significance not only in the transmission of information but also in the construction of perception, the incorporation of cultural concerns, and the relationship that the listener has to the subject matter (Table 9.13).

9.4.6 Ethics and Authenticity in Audio Storytelling

Audio storytelling's intimate nature creates unique ethical responsibilities for creators, who must navigate complex questions of consent and representation, ensuring subjects' voices are used with permission and their stories are portrayed authentically, not reductively. While it is technically required to chop and rearrange audio, excessive manipulation runs the danger of distorting the facts or creating drama at the price of factual integrity. The editing process itself presents a number of challenges. When it comes to expressing viewpoints from disadvantaged groups, cultural sensitivity becomes of the utmost importance. Creators are required to prevent appropriation while simultaneously elevating voices from underrepresented groups. What is required for ethical audio storytelling is not only having good intentions; it also needs being transparent about the techniques used, working together with the subjects, and engaging in continuous self-reflection. In the end, the strength of audio, which is its capacity to replicate human proximity, transforms these ethical issues into not just professional recommendations, but also moral imperatives for maintaining faith in the medium [26].

Future of Audio Narratives

A renaissance for auditory narrative in digital culture is on the horizon, as seen by the fast development of audio technologies, which include everything from the streaming of audiobooks to the ecosystems of smart speakers and artificial intelligence speech synthesis. Interactive podcasts may soon provide choose-your-own-adventure tales with dynamic audio mixing that reacts to listener selections in real time. We are on the cusp of a number of innovations that will have a profound impact on the world. There is the potential for artificial intelligence narration to personalise essays by modifying the tempo, tone, and even substance dependent on the preferences of the listener, so producing individualised educational experiences. Concurrently, the use of automatic translation technologies and multilingual production pipelines has the potential to eliminate language barriers, so facilitating an unparalleled flow of stories across different cultures. Audio narratives will transcend their traditional supplementary role to become primary vehicles for knowledge transmission as these innovations converge. This will result in the reshaping of digital humanities, the reimagining of public media consumption, and the potential for the formation of new neural pathways for human storytelling cognition itself.

9.5 Video Essays and Documentary Pedagogy

In this day and age of digital media, video essays and documentary pedagogy have evolved into powerful instruments that may be used for both the telling of stories and the teaching of subjects. In order to communicate ideas, criticisms, or tales, they combine components of visual, audio, and textual communication. This provides a compelling alternative to the conventional essays and public lectures. When it comes to the digital humanities, where multimodal communication improves both comprehension and engagement, these forms are particularly effective [27] (Fig. 9.5).

Defining Video Essays and Documentary Pedagogy

A video essay is a short video that conveys an argument or story via the use of visuals, film clips, text, and narration. It is introspective, analytical, and often subjective. Documentary pedagogy is the use of documentary films and videos as a teaching tool. Visual storytelling enables learners to critically interact with real-world situations, cultures, histories, and ideologies [29] (Table 9.14).

9.5.1 The Role of Video Essays in Digital Scholarship

Video essays have achieved substantial recognition as a viable type of digital study across numerous academic disciplines, notably in topics such as cinema studies, cultural studies, and visual anthropology. When compared to typical written essays, which depend exclusively on language to express arguments and ideas, video essays use a multimodal approach by merging sound, pictures, and text in a seamless manner.

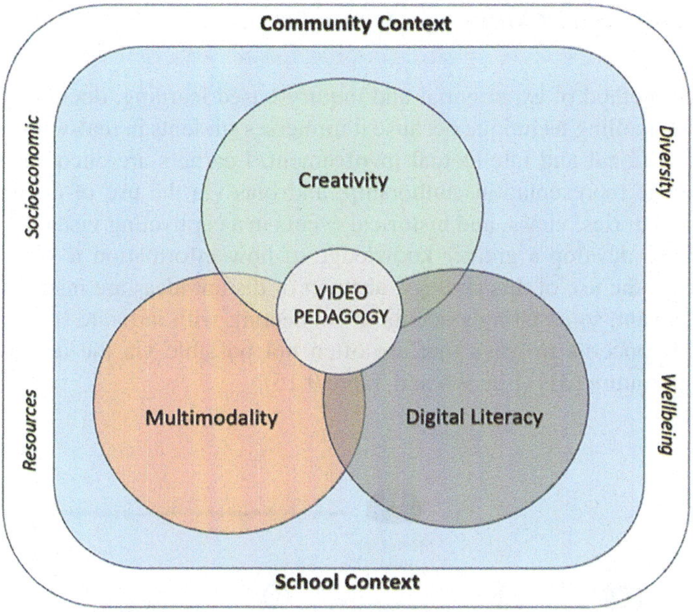

Fig. 9.5 Model of video pedagogy [28]

Table 9.14 Key characteristics

Aspect	Video essay	Documentary pedagogy
Format	Analytical or creative short videos	Full-length or short documentaries used in teaching
Purpose	Present a critical argument or idea	Educate, provoke thought, or foster discussion
Style	Artistic, subjective, essayistic	Observational, expository, participatory, reflexive
Use in Academia	Digital scholarship, film studies, media analysis	Classroom instruction, educational outreach
Tools	Editing software (Adobe Premiere, Final Cut, etc.)	Multimedia resources, discussion guides, and screenings

Scholars and producers are able to design powerful arguments that are both immersive and engaging because to this one-of-a-kind combination, which enhances the viewer's knowledge via direct visual and audio experience [30].

9.5.2 Pedagogical Value of Documentaries

As a potent method of experiential and inquiry-based learning, documentary pedagogy is a compelling technique because it immerses students in real-world situations via both emotional and intellectual involvement. Learners are encouraged to critically evaluate representation, authorship, and bias via the use of documentaries, which reveal stories, views, and historical events in a captivating visual format. This helps learners develop a greater knowledge of how information is structured and delivered. via the use of this strategy, abstract or distant ideas are made more physical and relevant, so assisting students in connecting with intricate social, political, and ethical concerns in ways that are often not possible via the use of standard academic literature [31] (Fig. 9.6 and Table 9.15).

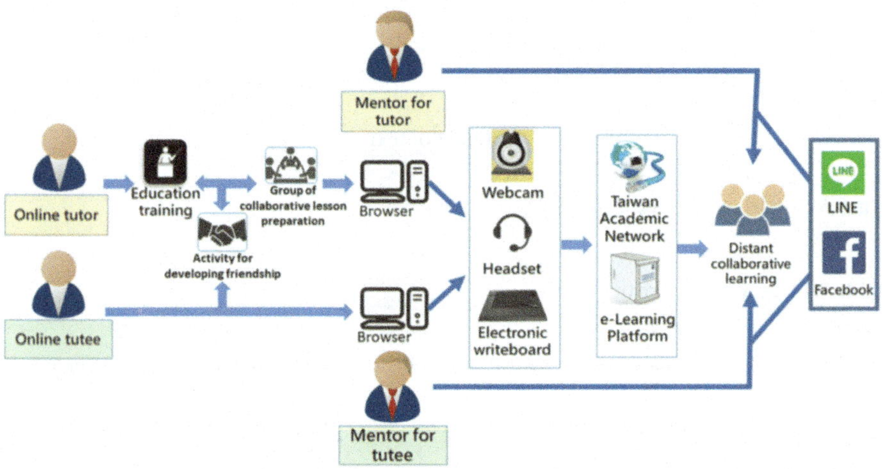

Fig. 9.6 Framework for E-tutoring [32]

Table 9.15 Educational benefits of documentary pedagogy

Learning outcome	Description
Visual literacy	Enhances ability to analyse visual and audio content
Critical thinking	Promotes questioning of sources and perspectives
Empathy and engagement	Builds emotional connections with real-world stories and characters
Multimodal understanding	Integrates learning through sight, sound, and narrative
Cultural awareness	Introduces students to diverse experiences and worldviews

9.5.3 Platforms and Tools for Creation

Through the advent of the digital revolution, the production of video essays and instructional films has been radically democratised. This has resulted in the transformation of media creation from a specialised expertise to an approachable educational tool. Traditional hurdles to dissemination have been removed thanks to the proliferation of free distribution channels such as YouTube and Vimeo, which have concurrently created venues for academic input and public participation across global learning communities. There is currently a coexistence of professional-grade editing software (such as Adobe Premiere Pro and Final Cut Pro) and user-friendly alternatives (such as iMovie and DaVinci Resolve), which makes it possible to create both advanced analyses and basic projects. Educators are now able to make sophisticated instructional films that include lecture recording, annotation capabilities, and dynamic visual explanations thanks to specialised programs such as Camtasia and Screen Flow. These technologies have further bridged the gap between academic education and the creation of digital media. This technology environment gives students and faculty the ability to generate engaging visual research regardless of the resources available to the institution. This not only promotes media literacy as a fundamental academic skill but also challenges the conventional hierarchies that have been established in the academic world about the creation and transmission of information.

9.5.4 Documentary as Activism and Pedagogy

Documentary filmmaking operates at the intersection of education and activism, using cinematic storytelling not just to inform but to mobilize. With their ability to reframe public discourse through compelling narratives and evidentiary rigour, powerful works such as Ava DuVernay's 13th (which exposes systemic racism in the United States prison system) and Laura Poitras' Citizenfour (which chronicles Edward Snowden's revelations) serve as both pedagogical resources and catalysts for social change. In educational environments, these films go beyond the realm of passive watching; they become instruments for critical participation, igniting discussions on civic duty while simultaneously bridging the gap between theoretical concepts and actual injustices in the world. Documentaries provide students with the ability to question prevailing narratives by putting the focus on viewpoints that are marginalised and by uncovering histories that are disputed. These histories range from colonial erasures to corporate misconduct. Due to the fact that it can both educate and agitate, this form is exceptionally powerful in terms of simultaneously fostering media literacy and social consciousness [33].

Table 9.16 Notable examples of video essays and educational documentaries

Title/Creator	Format	Focus	Impact
Every frame a painting (Tony Zhou)	Video essay	Film technique, editing, and directing	Widely used in film education
Hyper normalisation (Adam Curtis)	Documentary essay	Media and politics in the modern world	Popular in political theory courses
13th (Ava DuVernay)	Documentary	Mass incarceration and racial injustice	Sparks global discourse on systemic racism
The act of killing (Joshua Oppenheimer)	Documentary	Historical trauma and representation	Used in postcolonial and human rights education
Why it is almost impossible to run a Sub-2-Hour Marathon (VOX)	Video Essay	Science and sports performance	Blends scientific literacy with storytelling

9.5.5 *Ethical Considerations*

The production of video essays and documentaries demands careful ethical consideration at every stage, as these forms balance creative expression with social responsibility. Among the most important problems is the issue of representational equality, which asks who tells the tales, from whose perspective, and with what kinds of possible distortions? Particularly when photographing vulnerable groups, when real informed permission must go beyond legal formalities to assure genuine comprehension, the power imbalance that exists between the filmmaker and the subject demands ongoing attention. When collaborative tales are presented as separate works, authorship itself becomes a contentious issue. Additionally, editing decisions about inclusion, sequencing, and emphasis run the danger of propagating unconscious prejudices. When it comes to contextual framing and audience perception, more levels are revealed for instructors who make use of these resources. Therefore, the production of ethical media requires continuous reflexivity, which includes the questioning of one's own positionality, the pursuit of community-based collaborative models, and the open and honest acknowledgement of the constructed character of all nonfiction narrative, all while maintaining a commitment to journalistic integrity and cultural accuracy (Table 9.16).

9.5.6 *Student-Created Media Projects*

A substantial movement in pedagogy has occurred with the incorporation of video essay creation into the academic curriculum. This shift provides students with the ability to acquire multidimensional abilities that are necessary for the digital era.

Learners build not just conventional research skills but also essential digital literacies via the process of generating video essays. These digital literacies include activities such as evaluating sources in multimedia contexts and providing ethical citations for audiovisual content. The approach requires students to engage in higher-order cognitive activity since they are required to condense incomprehensible concepts into clear visual arguments, so combining academic study with imaginative narrative. The process of revision cultivates both analytical rigour and creative sensibility, while the hands-on experience gained via narrative writing, audio recording, and timeline-based editing contributes to the development of technical proficiency. Assignments may include doing an in-depth textual study of a film version of a book or conducting investigative documentary work on local community concerns. These assignments aim to bridge the gap between academic theory and its application in the real world. Learners who can critically consume and successfully communicate with major forms of media in the twenty-first century are eventually produced by this educational strategy, which ultimately results in more engaged learners.

9.5.7 Challenges in Implementation

Although they have many advantages, video essays and documentary pedagogy have several drawbacks:

1. Time and resources are necessary for production and evaluation.
2. Technical capabilities may vary across students and instructors.
3. Assessment standards for multimodal work are less consistent than those for textual materials.
4. Copyright difficulties with video or audio snippets used in student work.

9.6 Story Maps and Geospatial Narratives

Story maps and geographic narratives are strong digital tools that combine storytelling and spatial data. They enable users to offer tales in a location-based environment, converting abstract data into interactive experiences grounded in real-world geography. These forms are often used in digital humanities, journalism, history, environmental studies, and education to tell immersive and compelling tales (Fig. 9.7).

What are Story Maps?

The purpose of story maps is to build a narrative framework that is coherent and interactive by combining maps, multimedia information (including text, photographs, video, and audio), and narrative structure. Story maps, which were first developed by firms such as Esri (Environmental Systems Research Institute), are currently used

Fig. 9.7 Introduction to storytelling map [34]

all over the world to illustrate tales that are rooted in a particular physical location. They respond to questions like (Table 9.17):

1. Where did historic events take place?
2. How has an area evolved throughout time?
3. What are the links between various locations?

The Role of Geospatial Narratives

Geospatial narratives are tales that are inextricably tied to particular locales or travel across space. These tales provide events with geographical context and provide insights that text-based descriptions do not. They are particularly helpful in:

1. Historical reconstructions (such as tracing slave trade routes)
2. Migration studies (such as refugee trips)
3. Ecological narrative (such as deforestation or species movement)
4. Cultural mapping (e.g., indigenous place-names or holy geography)

Table 9.17 Components of a typical story map

Component	Description
Interactive map	A geospatial layer that displays key locations or movements
Narrative panel	Textual description that guides the user through the story
Multimedia elements	Images, videos, and audio that enhance emotional and factual engagement
Navigation tools	User-friendly interfaces (scrolling, clicking, zooming) to explore content

9.6.1 Applications in Humanities and Education

Digital researchers and educators utilise story maps and geographic narratives for a variety of objectives.

1. History and archaeology include mapping old trade routes, wars, and archeological sites.
2. Literature: Visualizing characters' voyages in books (for example, Odysseus' trip in The Odyssey).
3. Sociology and Anthropology: Documenting cultural customs from different places.
4. Environmental studies demonstrate the influence of climate change on particular regions.

Students, for example, may develop a narrative map that traces the Underground Railroad, integrating historical documents with modern geography and cultural memory (Table 9.18).

Tools for Creating Story Maps

Many digital tools and platforms enable narrative map generation. User technical expertise, intent, and resources determine the option (Table 9.19).

Table 9.18 Use cases in the humanities

Discipline	Story map example
History	Mapping Civil Rights Movement sites across the American South
Literature	Following Leopold Bloom's route in *Ulysses* by James Joyce
Anthropology	Documenting oral histories with geotagged interviews in native communities
Environmental studies	Charting climate-induced displacement in coastal cities

Table 9.19 Popular story map tools

Platform	Features	Ideal for
Esri StoryMaps	Professional-grade, integrates with ArcGIS; supports maps, media, and text	Academic research, environmental projects
Google earth web	User-friendly interface, simple tours and pins	Classroom activities, basic visualizations
StoryMapJS	Free, open-source tool from Knight Lab; timeline-style map navigation	Journalism, historical storytelling
Mapbox	Custom, interactive maps with full coding flexibility	Developers, data-heavy projects

Pedagogical and Communicative Power

Story maps promote active learning and student participation. They:

1. Encourage critical thinking on space, place, and viewpoint.
2. Enable multimodal learning by merging visual, linguistic, and spatial elements.
3. Develop digital literacy and research skills.
4. Encourage collaborative and project-based methods in education.

Educators believe that mapping projects help students perceive complicated material, reveal hidden spatial links, and creatively combine disparate sources. Scholars and practitioners must maintain cultural awareness, especially when mapping indigenous or contentious regions.

9.7 AR/VR in Storytelling

AR and VR have transformed storytelling from a linear, passive experience into one that is interactive, spatial, and immersive. The line between reality and fiction becomes blurrier as these technologies immerse users in the narrative universe. Augmented and virtual reality provide potent resources for investigating multi-layered stories via the medium of embodied experience in the fields of digital humanities, journalism, education, and the creative arts [35] (Fig. 9.8).

Fig. 9.8 Leveraging AR/VR for brand storytelling

Table 9.20 AR versus VR in storytelling

Aspect	Augmented reality (AR)	Virtual reality (VR)
Environment	Enhances the real world with digital overlays	Fully immersive virtual environment
Devices	Smartphones, tablets, AR glasses	VR headsets (e.g., Oculus, HTC Vive)
User movement	Limited; depends on camera orientation	Full-body navigation or head-tracking
Narrative engagement	Layered on physical spaces	Deep immersion in virtual settings
Ideal use cases	Museum guides, site-specific storytelling	Historical reconstructions, fictional narratives

Understanding AR and VR

AR is the application of digital content, such as pictures, sounds, or text, to the actual environment via devices such as smartphones, tablets, or AR glasses. VR uses headgear to immerse users in a completely simulated environment, enabling them to explore 3D environments apart from the actual world. These technologies serve distinct storytelling purposes: AR often enriches reality and context in real time, while VR transfers people to a new narrative experience (Table 9.20).

Immersion and Presence in Storytelling

AR/VR storytelling's effectiveness resides in its capacity to establish presence—the psychological feeling of "being there." Unlike text or film, which are fundamentally representational, AR/VR tales engage the audience as active participants rather than passive observers. This yields:

1. Emotional connection to characters or events
2. Increased empathy in educational or humanitarian settings.
3. Interactive learning, particularly for complicated topics like history, anatomy, and architecture.

For example, the UN's VR documentary Clouds Over Sidra immerses viewers in the life of a 12-year-old Syrian refugee. The immersive experience creates a strong emotional connection with displaced populations.

9.7.1 AR/VR in Humanities and Cultural Heritage

AR/VR technologies have transformed how we perceive and retain cultural narratives. These technologies are applied by museums, archives, and scholars to:

1. Restore ancient sites and ruins (e.g., Ancient Rome, Egyptian tombs)
2. Visualise archaeological strata in real time.
3. Present complex tales using location-based storytelling

4. Translate oral histories into interactive scenes.

In AR, for example, a tourist standing in front of a historic ruin may view a digital recreation of its original appearance on their mobile device. This puts historical imagination and knowledge to the forefront.

9.8 Student-Created Digital Narratives

Digital storytelling is becoming a student-led, interactive process that fosters creativity, critical thinking, cooperation, and digital literacy. Digital narratives help students connect with knowledge, improve communication, and learn technological skills. Students may create multimodal, multimedia tales using podcasts, video essays, interactive websites, and AR experiences.

9.8.1 Educational Value of Student-Driven Storytelling

Encouraging students to develop their own digital narratives turns them from passive receivers of information to active knowledge producers. This participatory paradigm coincides with constructivist learning theories, in which learners acquire understanding through experience and reflection.

Formats of Student-Created Digital Narratives

Student narratives may take a number of forms based on the project's aims, audience, and accessible resources.

1. Video essays are popular for analysing literature, history, and art using narrative and visual media.
2. Digital Timelines–Tools such as Timeline JS allow students to plot events chronologically using photos, audio, and links.
3. StoryMapJS and Google Earth allow for geographical storytelling.
4. Podcasts are ideal for facilitating discussions, interviews, and storytelling with minimal equipment.
5. Web-Based Storytelling–Blogs, wikis, and platforms such as Adobe Express or Wix enable multimedia story creation.
6. AR Projects–Students use tools such as Zappar or HP Reveal to create AR experiences based on objects or locations.
7. Social media campaigns–Narratives delivered on Instagram, TikTok, or Twitter may address current issues in novel ways.

Process and Pedagogy

To support both storytelling and technical development, instructors should scaffold the process of assigning digital narrative projects. A typical workflow may

Table 9.21 Sample rubric snippet

Criteria	Excellent (A)	Satisfactory (B–C)	Needs improvement (D–F)
Narrative coherence	Clear, compelling, and well-structured	Mostly coherent, with minor gaps	Disjointed or unclear storytelling
Use of media	Seamless integration of multimedia	Basic use of media elements	Poor or distracting media usage
Research & accuracy	Insightful, well-supported claims	Some support with sources	Lacks credible or sufficient sources
Technical execution	High-quality production	Functional but uneven quality	Technical issues hinder experience

include: Topic selection and storyboarding, Research and scriptwriting, Tool training, Production, and Students create the digital artefact.

Assessment and Evaluation

The evaluation of student-created digital narratives must strike a balance between creative expression and academic rigour. The rubrics should assess: Narrative structure and coherence, Research and factual correctness, Creativity and inventiveness, Technical quality (audio and video clarity, editing), Engaging with the audience, and Reflective comment or written reasoning (Table 9.21).

9.8.2 Challenges and Considerations

While digital storytelling empowers, it also poses challenges:

1. The digital divide means that not all pupils have access to the same gadgets and software.
2. Skill disparity: Different levels of familiarity with media technologies may influence results.
3. Time-consuming: Digital storytelling projects can take more time than conventional essays.
4. Assessment Complexity: Creativity and multimedia aspects may make standardised grading difficult.
5. Copyright Issues: Students must learn about fair use, Creative Commons, and intellectual property.

9.8.3 Publishing Platforms (Medium, Scalar)

Digital storytelling needs narrative creation, presentation, and dissemination. Medium and Scalar are vital for publishing digital narratives in academic, artistic, and educational settings. Images, videos, hyperlinks, metadata, and interactive media may be seamlessly integrated into these platforms to enable multimodal expression and creativity beyond text. Scalar is a non-linear, academic publishing platform for complicated digital humanities projects, whereas Medium is easy to use for linear narrative writing and audience participation. Choose a platform based on narrative aims, interaction, and audience [36].

9.8.4 Medium: Simplicity and Social Sharing

Medium is a web-based publishing platform noted for its clean design, simplicity of use, and large audience. Originally developed for long-form blog entries and essays, it has now become a popular platform for journalists, authors, educators, and students looking to share their story with a large audience [37].

1. Medium offers a clean, minimalist editor for distraction-free writing.
2. Easily embed pictures, videos, tweets, and links.
3. Built-in statistics and audience participation (likes, highlights, and comments).
4. Increasing exposure via topic-specific tagging and curation.
5. Collaborative editing and publishing using organisational funds.

Use in Education and Storytelling: Instructors and students write reflective essays, opinion articles, digital storytelling, and multimedia journalism on Medium. Medium helps students write for genuine, public audiences by supporting simple sharing and mobile readability.

9.8.5 Scalar: Multimodal Scholarly Storytelling

Scalar, created by the Alliance for Networking Visual Culture, is a free, open-source platform for digital humanities academics. Multiple reading pathways, deep metadata, and semantic labelling are possible with non-linear content structure.

Key Features of Scalar:

1. Non-linear routes (users may go through the material in various orders).
2. Multimedia integration (video, music, maps, photos, and galleries).
3. Semantic labelling and ontology-based linkages.
4. Custom annotations and remarks.
5. Import tools for material from YouTube, Vimeo, SoundCloud, the Internet Archive, and other platforms.

Table 9.22 Comparative summary medium versus scalar

Feature	Medium	Scalar
Ease of use	Very high ideal for beginners	Moderate some learning curve
Narrative structure	Linear	Non-linear / modular
Multimedia support	Images, video, GIFs	Extensive supports video, audio, maps, and interactive media
Metadata and annotation	Basic tagging	Advanced semantic metadata and annotations
Academic use	Good for reflections, general audience	Ideal for scholarly publishing and digital humanities projects
Collaboration	Supports shared publications	Supports multiple contributors with version control
Hosting	Medium.com (proprietary)	User-hosted or Scalar server (open source)
Audience	Public, wide social reach	Academic, specialised audiences

Why Scalar is Ideal for Academic Narratives: Scalar is ideal for complicated digital storytelling that requires layered meaning, such as historiographies, annotated texts, and projects with significant visual components. Its scholarly design facilitates citations, peer review, and metadata development, which are all required for academic publication (Table 9.22).

9.8.6 Pedagogical Implementation

Integrating publishing platforms into course assignments increases student engagement by providing real-world authoring opportunities. Instructors could create assignments like this:

1. Medium Assignment: Create a personal tale based on a topic from a literature or ethics class. Use hyperlinks and photos to provide context.
2. Scalar Assignment: Create a multimedia digital book on a local oral history project that includes interviews, timelines, and geographical maps.

Instructors should help students with:

1. Workshops in digital literacy and content structure.
2. Examples of well-created digital storytelling.
3. Rubrics that value innovation, coherence, and technical proficiency.

9.8.7 Challenges and Considerations

Each platform has constraints, which should be considered during project planning:

1. Medium inhibits advanced customisation and may establish paywalls for readers.
2. Scalar demands technical knowledge and may present accessibility issues.
3. Ownership and Archiving: The medium is proprietary, and the material may vary depending on the platform.
4. Scalar enables institutional archiving but needs digital infrastructure.

To respond to these issues, teachers can:

1. Emphasise exportable formats and backup plans.
2. Encourage students to think critically about platform selection depending on their target audience and purpose.
3. Provide alternate tools or analogue assignments to students who have restricted digital access.

9.8.8 Assessment of Narrative Innovation

Narrative innovation assessment is crucial and complicated as digital storytelling grows via hypertext fiction, AR/VR, podcasting, and interactive mapping. Digital tales are multimodal, nonlinear, and immersive, yet essay rubrics typically fail to capture them. Form, content, audience engagement, innovation, technology skill, and narrative coherence must be assessed in this area. Educational and creative professionals may learn what makes a digital tale useful, original, and successful using a comprehensive assessment methodology.

9.8.9 Key Dimensions of Narrative Innovation

Through a number of different aspects, innovative digital storytelling goes beyond the boundaries of traditional storytelling (Table 9.23).

9.8.10 Developing an Assessment Rubric

It is important for instructors and reviewers to employ rubrics that take into consideration both creative and technical execution capabilities in order to guarantee fairness and consistency. An example of an evaluation rubric that may be used in a humanities classroom to evaluate digital tales that were developed by students is shown in Table 9.24.

Table 9.23 Dimension of story telling

Dimension	Description
Narrative structure	Use of non-linear, branching, or interactive storytelling techniques
Multimodality	Integration of various media types (text, image, video, audio, maps)
Audience interaction	Engagement through choices, feedback, comments, or adaptive content
Thematic depth	Complexity and originality of the narrative's message or inquiry
Technological integration	Use of platforms, tools, or code that enhances rather than distracts from the story

Table 9.24 Sample assessment rubric for narrative innovation

Criteria	Points	Indicators of excellence
Originality and creativity	20	Unique concept, original use of media, innovative storytelling style
Narrative coherence	15	Clear structure or purpose; logical progression even in nonlinear formats
Technical proficiency	15	Effective use of tools/platforms; minimal glitches; proper formatting and embedding
Multimodal integration	15	Balanced and purposeful use of images, audio, video, and text
User engagement & interactivity	10	Allows user choice, feedback, exploration, or immersive interaction
Thematic and intellectual depth	15	The story reflects critical thinking, cultural awareness, or scholarly depth
Accessibility & usability	10	Mobile-friendly, readable, accessible to diverse users (e.g., captions, alt text)

9.8.11 Assessment Methods

There are numerous evaluation methodologies suitable for digital narratives, particularly in academic and project-based settings:

1. Instructor/Expert Evaluation: Evaluators evaluate using standardised rubrics and checklists. Suitable for summative grading or publication review.
2. Peer Review: Students provide comments, frequently emphasising audience reaction, intelligibility, and involvement. Encourages collaborative learning.
3. Self-reflection: Students engage in metacognition by writing about their choices, problems, and learning results.
4. Analytic-Based Evaluation: Platforms like as Medium and Scalar provide user data (clicks, views, and interaction pathways) that may be used to analyse engagement.

9.8.12 *Inclusive and Ethical Evaluation*

Narrative innovation often relies on varied cultural experiences and storytelling traditions, which may not conform to established literary rules. So, evaluators should:

1. Avoid using linearity as a criteria of coherence.
2. Consider cultural storytelling forms such as oral traditions or visual symbols.
3. Ensure that rubrics are adaptive to accessibility needs (for example, screen readers or multilingual storytelling).
4. Instead than focussing just on technical polish, consider the creator's aim and the context of the story.

9.8.13 *Challenges in Assessing Innovation*

Despite the greatest attempts, judging digital story innovation poses substantial obstacles.

1. Subjectivity in creativity: Innovation is hard to measure.
2. Tool and platform limitations: Student access and familiarity with platforms may have an impact on output.
3. Time-consuming evaluation: Multimedia initiatives need more careful examination than printed texts.
4. Rapidly shifting standards: New forms develop regularly, necessitating continuous rubric updating.
5. Educators should set clear standards, give many rounds of formative feedback, and allow for experimentation.

9.8.14 *Case Example: Classroom Implementation*

Students in a University of Washington digital humanities course created nonlinear multimedia biographies using Scalar, an open-source publication platform for academic multimedia work. Participants have to create networked narratives using digitised archive resources, oral history footage, interactive timelines, and critical comments to go beyond biographical essays. One notable study studied Northwest poet Richard Hugo's life via wartime letters, audio recordings of readings, and GIS-mapped places from his poems, showing how digital technologies may expand literary history. The project developed metadata tagging and hyperlinking abilities, nonlinear narrative theory, and ethical awareness while managing live persons' digital legacies. Such assignments are at the forefront of humanities instruction, as students actively contribute to expanding academic communication while struggling with the epistemological issues of depicting life via networked media.

References

1. Korot, E., et al.: Enablers and barriers to deployment of smartphone-based home vision monitoring in clinical practice settings. JAMA Ophthalmol. **140**(2), 153–160 (2022). https://doi.org/10.1001/jamaophthalmol.2021.5269
2. Ghosh, A., Edwards, D.J., Hosseini, M.R.: Patterns and trends in Internet of Things (IoT) research: future applications in the construction industry. Eng. Constr. Archit. Manag. **28**(2), 457–481 (2021). https://doi.org/10.1108/ECAM-04-2020-0271
3. Francis, T., Hoefel, F.: True Gen': Generation Z and its Implications for Companies. McKinsey Co., p. 10 (2018). https://www.mckinsey.com/~/media/McKinsey/Industries/Consumer Packaged Goods/Our Insights/True Gen Generation Z and its implications for companies/Generation-Z-and-its-implication-for-companies.ashx
4. Del Vecchio, P., Mele, G., Ndou, V., Secundo, G.: Open innovation and social big data for sustainability: evidence from the tourism industry. Sustain **10**(9) (2018). https://doi.org/10.3390/su10093215
5. Pujiastuti, E.E., Soeprapto, A., Susanta, S., Utomo, H.S., Maharaniputri, A.: The role of perceived value in understanding tourist experience and post experience at heritage destinations. J. Siasat Bisnis **26**(1), 35–56 (2022). https://doi.org/10.20885/jsb.vol26.iss1.art3
6. Yang, C.M., Liu, T.H., Zheng, Y.Y.: A study on design thinking based creative product design process in a design project. Int. J. Syst. Innov. **5**(3), 62–71 (2019)
7. Yin, C., Pereira, P., Hua, T., Liu, Y., Zhu, J., Zhao, W.: Recover the food-energy-water nexus from COVID-19 under sustainable development goals acceleration actions. Sci. Total Environ **817**, 153013 (2022). https://doi.org/10.1016/j.scitotenv.2022.153013
8. Wan, L., Bai, Y.: Application research on the BIM and internet of things technology in construction logistics management in the period of big data BT. In: Proceedings of the Fourteenth International Conference on Management Science and Engineering Management. In: Xu, J., Duca, G., Ahmed, S.E., García Márquez, F.P., Hajiyev, A (eds.), pp. 704–716. Cham: Springer International Publishing (2021)
9. Odonkor, B., Kaggwa, S., Uwaoma, P.U., Hassan, A.O., Farayola, O.A.: The impact of AI on accounting practices: a review: exploring how artificial intelligence is transforming traditional accounting methods and financial reporting. World J. Adv. Res. Rev. **21**(1), 172–188 (2024). https://doi.org/10.30574/wjarr.2024.21.1.2721
10. Hussain, A., et al.: Security framework for IoT based real-time health applications. Electron. **10**(6), 1–15 (2021). https://doi.org/10.3390/electronics10060719
11. Zhang, L.: Smartphone app security: vulnerabilities and implementations. Biomass Chem Eng. **3**(2), 1–123 (2019). http://journal.stainkudus.ac.id/index.php/equilibrium/article/view/1268/1127
12. Porsani, G.B., de Lersundi, K.D.V., Gutiérrez, A.S.O., Bandera, C.F.: Interoperability between building information modelling (BIM) and building energy model (BEM). Appl. Sci. **11**(5), 1–20 (2021). https://doi.org/10.3390/app11052167
13. Hong, T., Sun, H., Chen, Y., Taylor-Lange, S.C., Yan, D.: An occupant behavior modeling tool for co-simulation. Energy Build. **117**, 272–281 (2016). https://doi.org/10.1016/j.enbuild.2015.10.033
14. Tavana, M., Shaabani, A., Vanani, I.R., Gangadhari, R.K.: A review of digital transformation on supply chain process management using text mining. Processes **10**(5), 1–19 (2022). https://doi.org/10.3390/pr10050842
15. Ofori-Kuragu, J.K., Osei-Kyei, R.: Mainstreaming pre-manufactured offsite processes in construction—are we nearly there? Constr. Innov. **21**(4), 743–760 (2021). https://doi.org/10.1108/CI-06-2020-0092
16. Soriano, F.R., Samper-Zapater, J.J., Martinez-Dura, J.J., Cirilo-Gimeno, R.V., Martinez Plume, J.: Smart mobility trends: open data and other tools. IEEE Intell. Transp. Syst. Mag. **10**(2), 6–16 (2018). https://doi.org/10.1109/MITS.2017.2743203

17. Khan, A., Sepasgozar, S., Liu, T., Yu, R.: Integration of BIM and immersive technologies for AEC: a scientometric-SWOT analysis and critical content review. Buildings **11**(3), 1–34 (2021). https://doi.org/10.3390/buildings11030126
18. Kumar, N.M., et al.: Distributed energy resources and the application of AI, IoT, and blockchain in smart grids. Energies. **13**(21) (2020). https://doi.org/10.3390/en13215739
19. Fadele, A.A., Othman, M., Hashem, I.A.T., Alotaibi, F.: Internet of things security: a survey. J. Netw. Comput. Appl. **88**(April), 10–28 (2017). https://doi.org/10.1016/j.jnca.2017.04.002
20. Xie, L., Pinto, J., Zhong, B.: Building community resilience on social media to help recover from the COVID-19 pandemic. Comput. Human Behav. **134**, 107294 (2022). https://doi.org/10.1016/j.chb.2022.107294
21. Mahmoud, I.M.: Analysis of oil spill impacts along pipelines and the fate of sensitive environments in Nigeria. Int. Oil Spill Conf. Proc. **1**, 2021 (2021). https://doi.org/10.7901/2169-3358-2021.1.629863
22. Shah, S.A.A., Ahmed, E., Rodrigues, J.J.P.C., Ali, I., Noor, R.M.: Shapely value perspective on adapting transmit power for periodic vehicular communications. IEEE Trans. Intell. Transp. Syst. **19**(3), 977–986 (2018). https://doi.org/10.1109/TITS.2017.2775965
23. Wu, Y., Jiang, M., Lei, J., Xu, H.: Named entity recognition in Chinese clinical text using deep neural network. Stud. Health Technol. Inform. **216**(2), 624–628 (2019). https://doi.org/10.3233/978-1-61499-564-7-624
24. Arije, O., Madan, J., Hlungwani, T.: Quality of sexual and reproductive health services for adolescents and young people in public health facilities in Southwest Nigeria: a mystery client study. Glob. Health Action **15**(1) (2022). https://doi.org/10.1080/16549716.2022.2145690
25. Lawrence, D.O.: Aligning strategic intelligence and graduate employability: a conceptual review. J. Strateg. Manag. **7**(5), 61–80 (2023). https://doi.org/10.53819/81018102t4178
26. Dauvergne, P.: Is artificial intelligence greening global supply chains? Exposing the political economy of environmental costs. Rev. Int. Polit. Econ. **29**(3), 696–718 (2022). https://doi.org/10.1080/09692290.2020.1814381
27. Isa, A., Nte, N.D., Baba-Ahmadu, A.: Critical thinking for intelligence analysis for Nigerian students: evaluating the pedagogical challenges in real life scenarios. Indones. Educ. Adm. Leadersh. J. **5**(2), 46–60 (2023). https://doi.org/10.22437/ideal.v5i2.28639
28. Zhao, S., Pan, Q., Gao, D., Cheng, J.: Integrating internet of things and mixed reality to teach performance-based architectural design: a case study of shading devices. Educ. Inf. Technol (2022). https://doi.org/10.1007/s10639-022-10998-6.
29. Kozma, R.B., Andersen, R.E.: Qualitative case studies of innovative pedagogical practices using ICT. J. Comput. Assist. Learn. **18**(4), 387–394 (2002). https://doi.org/10.1046/j.0266-4909.2002.00250.doc.x
30. Zhang, L., Wu, Z.: Enhancing postgraduate academic writing skills through course reform : an Action research study. **5**(2), 217–233 (2024)
31. Tella, A., Tella, A., Toyobo, O.M., Adika, L.O., Adeyinka, A.A.: An assessment of secondary school teachers uses of ICT's: implications for further development of ICT's use in Nigerian secondary schools. Turkish Online J. Educ. Technol. **6**(3), 13 (2007)
32. Alharbi, E.: A Study on the Use of ICT in Teaching in Secondary Schools in Kuwait Eid Alharbi Thesis Fulfilment for the Degree of Doctor of Education (PhD) Cardiff School of Education Cardiff Metropolitan University, p. 279 (2014). https://repository.cardiffmet.ac.uk/dspace/bitstream/10369/5675/1/Eid Al-Harbi Thesis.pdf%5Cn
33. Mukama, E., Andersson, S.B.: Coping with change in ICT-based learning environments: newly qualified Rwandan teachers' reflections. J. Comput. Assist. Learn. **24**(2), 156–166 (2008). https://doi.org/10.1111/j.1365-2729.2007.00249.x
34. Atiq, M., Abid, G., Anwar, A., Ijaz, M.F.: Influencer marketing on Instagram: a sequential mediation model of storytelling content and audience engagement via relatability and trust. Information **13**(7) (2022). https://doi.org/10.3390/info13070345
35. Orunbon, N.O, Ibikunle, G.A., Badmus, A.A.: Principals' digital transformational leadership, teachers' organisational commitment and job satisfaction during Covid-19 in Lagos state education district V, Nigeria. Int. J. Humanit. Technol. Civiliz. **8**(1), 68–73 (2023). https://doi.org/10.15282/ijhtc.v8i1.9424

36. Sokolova, K., Kefi, H., Dutot, V.: Beyond the shallows of physical attractiveness: perfection and objectifying gaze on Instagram. Int. J. Inf. Manage. **67** (2022). https://doi.org/10.1016/j.ijinfomgt.2022.102546
37. de Beer, D., Matthee, M.: Approaches to identify fake news: a systematic literature review. **136**, no. Macaulay 2018. Springer International Publishing (2021). https://doi.org/10.1007/978-3-030-49264-9_2

Chapter 10
Assessment and Evaluation in Digital Pedagogy

10.1 Reimagining Assessment for Digital Work

The digital age has changed learning from static books and closed-book exams to multimedia presentations, blogs, interactive simulations, podcasts, digital storytelling, and more. Thus, linear essays and multiple-choice assessments no longer measure students' depth, complexity, and inventiveness. Reimagining assessment entails creating tools and frameworks that effectively measure student learning in digital contexts while fostering fairness, inclusion, and creativity. What does digital learning look like? In multimodal projects, how can we quantify creativity, cooperation, and critical engagement?

10.1.1 Principles of Reimagined Assessment

To match assessment techniques with the capabilities of digital learning, educators must adopt a framework that prioritises process, creativity, and context. The key concepts include: (see Table 10.1, Figs. 10.1 and 10.2)

10.1.2 Moving from Products to Performances

Traditional evaluations concentrate on finished output, such as essays and tests. In digital pedagogy, there is a growing focus on performance, demonstrating learning via action, iteration, and digital production (see Table 2).

F. A. Alaba and I. G. Ajibola, *Digital Pedagogy: Bridging the Gap Between Software and the Humanities*, Information Systems Engineering and Management 85,
https://doi.org/10.1007/978-3-032-20335-9_10

Table 1 Digital assessment

Principle	Implications for digital assessment
Authenticity	Use real-world tasks, such as blogs, podcasts, or infographics.
Transparency	Share criteria and expectations clearly through rubrics and exemplars.
Process-oriented evaluation	Evaluate planning, drafting, revision, and reflection not just the final product.
Flexibility	Allow students to choose formats (e.g., video essay vs. written report).
Equity and accessibility	Ensure all learners have access to tools and support for success.

Fig. 1 Principles of authentic assessment [1]

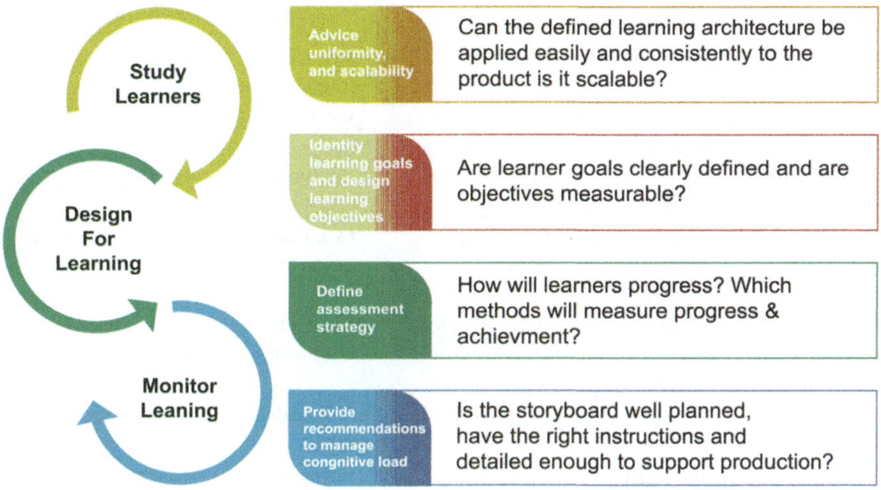

Fig. 2 Introduction to digital learning design [2]

Table 2 Assessment types

Assessment type	Traditional approach	Reimagined digital equivalent
Essay on a historical topic	Typed paper graded on thesis and grammar	Podcast series exploring historical narratives through interviews
Science experiment report	Lab write-up with predefined sections	Interactive website with video logs, charts, and simulations
Art interpretation	Text-based analysis of a painting	Multimedia essay with voice-over and visual annotations

10.1.3 Multimodal Rubrics: A New Toolset

Reimagined assessment also entails creating new rubrics that can meet the particular requirements of digital work. Rather of considering a "thesis statement," we may evaluate:

1. Narrative coherence in a nonlinear framework.
2. Efficient integration of media elements.
3. User Engagement and Interface Design Technical competency with selected digital tools (see Table 3).

This rubric may be customised based on the nature of the digital output.

Table 3 Rubric for a multimedia project

Criteria	Excellent (4)	Good (3)	Needs improvement (2)	Poor (1)
Media integration	Seamless, enhances message	Mostly clear and relevant	Some media disconnected	Media lacks purpose or clarity
Creativity and originality	Highly innovative approach	Some unique elements	Common or predictable	Minimal creativity
Narrative coherence	Clear, engaging structure	Generally clear	Somewhat confusing	Lacks coherence or direction
Technical execution	Flawless and professional	Minor issues	Noticeable errors	Poorly executed
Reflection on process	Insightful, detailed	Clear reflection	Surface-level	No reflection or incomplete

10.1.4 The Role of Reflection in Digital Assessment

In digital pedagogy, student reflection is an important source of evidence. Students demonstrate their thinking by expressing their design decisions, problem-solving tactics, and learning experiences. Reflection may be incorporated as follows:

1. Embedded blog posts or videos
2. Process journals
3. Design Statements
4. Post-project criticism essays

These tools may be used to measure both intellectual depth and metacognitive improvement.

10.1.5 Inclusive and Culturally Responsive Assessment

Reimagining assessment also requires inclusive practices that recognise:

1. Various modes of knowledge and expression (e.g., oral storytelling, visual arts)
2. Language diversity and accessibility requirements.
3. Diverse technological access.

Assessment should highlight rather than penalise diversity. Educators must develop criteria that take into consideration context, effort, and resource limits, particularly in global or under-resourced contexts [3].

Table 4 Challenges in reimagining assessment

Challenge	Implication
Educator training	Teachers may lack familiarity with digital tools or new media literacies
Time and resource intensity	Digital projects often require more time to evaluate
Subjectivity in creativity	Defining standards for innovation can be difficult
Equity of access	Students may face tech-related disadvantages

10.1.6 Challenges in Reimagining Assessment

Reimagining assessment comes with several challenges, especially as education evolves to be more student-centred and digitally integrated. Some key obstacles can be seen on Table 4.

10.1.7 Limitations of Traditional Assessment Models

Essays, multiple-choice tests, and standardised examinations are industrial-age evaluation methods. They value result, linear thinking, and memorisation above process. Digital projects are iterative, multi-media, and collaborative, making these methods unsuitable for evaluation. Traditional tests typically neglect learning. Digital labour values learning as much as production. For instance, iterative coding requires debugging and recurrent testing, which are not obvious in a single program review. The last aspect of digital narrative requires scriptwriting, audio mixing, and editing, which static grading ignores. Second, typical evaluations ignore digital format variety. Podcasts and video essays may not be graded using essay rubrics. This demeans multimodal literacy and modern communication abilities. Third, conventional models typically ignore student agency. Students have little say in defining success or evaluating their work. This gap may demotivate and disengage pupils, especially those who love digital output [4].

10.1.8 Principles for Reimagining Assessment

To effectively evaluate digital work, educators must embrace new concepts that prioritise relevance, flexibility, and diversity [5] (see Table 5).

Table 5 Five foundational principles for reimagining assessment

Principle	Description	Application in digital work
Authenticity	Emphasize real-world relevance and application	Assess students on projects like creating digital campaigns, simulations, or community-based multimedia stories
Flexibility	Adapt to varied digital outputs and formats	Develop rubrics that assess skills such as creativity, critical thinking, and technical proficiency rather than specific tools or platforms
Process-oriented	Value learning as a journey	Use version control systems, logs, and reflections to evaluate the stages of development in a digital project
Co-construction	Involve students in assessment design	Collaborate with learners to define rubrics, select artifacts, and determine success criteria
Transparency	Clearly communicate expectations and feedback	Provide annotated exemplars, continuous feedback, and clear criteria for each stage of project development

10.1.9 Emerging Assessment Models

To fulfil the demands of digital pedagogy, a number of unique evaluation methods have arisen. Each of these approaches is consistent with the concepts outlined above and provides distinct benefits to both educators and students [6].

1. **Digital Portfolios**: Digital portfolios are carefully maintained compilations of student work that demonstrate progress, accomplishments, and reflections throughout time. They promote process-oriented learning and may use a variety of media formats, including text, graphics, audio, and video. Benefits: encourage introspection and metacognition, highlight progress on a timeline, and provide real proof of learning.
2. **Peer and Self-Assessment**: Involving students in the review process encourages critical thinking and personal responsibility. Google Docs, Padlet, and Hypothes.is are examples of digital systems that allow for peer feedback, collaborative annotation, and shared learning. Benefits: encourage active participation in others' work, improve evaluation and communication abilities, and develop a feeling of ownership over learning.
3. **Badge and Micro-Credentialing Systems**: Digital badges, which are often modular or gamified, recognise certain abilities or accomplishments. These qualifications may be shown on portfolios, resumes, and social media sites. Benefits: Motivate learners with visible milestones, acknowledge gradual learning, and enable flexible and personalised routes.
4. **Narrative and Reflective Assessments**: Reflections, learning notebooks, and project narratives let students express their thoughts, problems, and strategies. These evaluations emphasise voice and individuality. Benefits: centre learner

experiences, make unseen learning apparent, and enhance critical and creative thinking.

5. **Performance-Based Assessments**: Public-facing initiatives, such as blogging, podcasting, or establishing an instructive YouTube series, enable students to exhibit real-world skills. Benefits: emphasise audience awareness, and encourage high standards of responsibility.

10.1.10 Alternative Grading Models

As digital teaching advances, so should our grading systems. Traditional mathematical or letter-based methods often fall short of conveying the depth, inventiveness, and process-oriented character of digital production. Alternative grading methods provide more egalitarian, transparent, and student-centered approaches to assess learning. This section examines many major methods and how they relate to digital education. Contract grading is a mutual agreement between the teacher and the student on the quantity and quality of work necessary to get a certain grade [7]. Rather of making subjective judgements, pupils follow specific expectations to achieve their intended objective.

Key features:

1. Students pick their workload depending on the desired grade.
2. Contracts specify assignment number, kind, and quality.
3. Promotes autonomy and minimises grade anxiety.

10.1.11 Ungrading

Ungrading eliminates letter and numerical grades, focusing instead on formative, narrative feedback. Students often reflect on their progress and offer a final grade. Benefits: promotes intrinsic motivation, prioritises learning above performance, and reduces exam anxiety while encouraging open-ended inquiry.

10.1.12 Specifications Grading

Specifications grading, sometimes known as "specs" grading, offers unambiguous, binary standards for assessing student work: assignments either completely match the established requirements or do not. Unlike conventional grading systems, which depend on partial credit or subjective judgements, specifications grading emphasises mastery by demanding students to create work that exactly matches the specified learning goals. This method enables students to engage fully with the topic, reflect on comments, and deepen their comprehension across numerous iterations. By

Table 6 Portfolio-based assessment

Component	Description	Evaluation focus
Introduction	Overview and goals of the portfolio	Clarity, purpose
Selected works	Representative projects	Creativity, execution
Reflective commentary	Personal insight into the learning process	Depth, growth
Technical artifacts	Code, prototypes, media	Functionality and innovation

Table 7 Peer and self-assessment

Assessment type	Purpose	Tool examples
Peer review	Evaluate contributions and teamwork	Google forms, Peergrade
Self-reflection	Reflect on challenges and growth	Journals, video diaries

allowing several efforts to satisfy standards, specs grading turns the emphasis away from gaining points and towards acquiring competence, promoting a growth-oriented learning environment in which students take responsibility of their development and build resilience in the face of obstacles [8, 9].

10.1.13 Portfolio-Based Assessment

Digital portfolios enable students to present a selected selection of their work, such as reflections, process documents, and multimodal outputs. Advantages: shows progress over time, promotes self-assessment and metacognition, and supports multimodal artefacts, such as blogs, videos, and infographics (see Table 6).

10.1.14 Peer and Self-Assessment

In digital learning, where cooperation is ubiquitous, peer and self-assessment promote responsibility and critical thinking. Implementation Strategy: Rubrics co-created with students, anonymous peer evaluations may help decrease prejudice, and self-assessment notebooks describe the process (see Table 7).

10.1.15 Digital Rubrics and Peer Review

A rubric is a grading instrument that clearly defines the expectations for an assignment by listing the criteria and identifying the degrees of excellence for each one. A digital

Table 8 Traditional versus digital rubrics

Feature	Traditional rubrics	Digital rubrics
Accessibility	Paper-based, often lost or misplaced	Stored in LMS or cloud; easily accessible
Feedback delivery	Delayed or limited to numeric scores	Immediate, with detailed comments
Revision/reuse	Requires reprinting	Easily editable and shareable
Integration with grades	Manual grade entry	Automated scoring and LMS integration
Student interaction	Passive (view only)	Interactive (students can self-assess)

rubric extends the evaluation framework into LMS or digital assessment platforms. Digital rubrics have numerous major benefits:

1. Transparency and Clarity: Students understand exactly how their work will be judged.
2. Consistency: Instructors and teaching helpers use the same standards, eliminating subjective grading disparities.
3. Efficiency: Automated scoring and comment features save time on grading.
4. Feedback Richness: Comments that are linked to particular criteria assist students recognise their strengths and areas for growth (see Table 8).

10.1.16 Designing Effective Digital Rubrics

Effective digital rubrics should include [10, 11]:

1. Criteria are clearly specified components of performance (for example, argumentation, evidence, and clarity).
2. Performance Levels: Descriptions that range from bad to exceptional performance (for example, Unsatisfactory to Exemplary).
3. Descriptors are qualitative descriptions of each performance level under each criterion.
4. Scoring Mechanism: Holistic (one total score) or analytic (scores for each criteria).

Educators may embed rubrics directly into assignments using digital systems like Canvas, Moodle, Google Classroom, and Turnitin. This integration simplifies grading and increases transparency, especially in big classrooms where grading uniformity is critical.

Table 9 Key features of digital peer review platforms

Platform	Anonymity	Rubric integration	Feedback analytics	Instructor moderation
Peergrade	Yes	Yes	Yes	Yes
Eli review	Optional	Yes	Yes	Yes
Moodle workshop	Yes	Yes	Limited	Yes
Google docs	No	Manual rubrics	No	Partial (comment control)

10.1.17 Peer Review in Digital Pedagogy

Peer review is the procedure in which students assess one another's work using predetermined criteria. In digital learning situations, this might take the shape of asynchronous feedback via digital platforms or real-time evaluations during synchronous sessions. Advantages of peer review include:

1. Metacognitive Development: Students learn how to evaluate quality and use criteria critically.
2. Collaboration and Communication: Reviewing others' work fosters a sense of community among learners.
3. Ownership of Learning: Students take an active role in the evaluation process.

Peergrade, Eli Review, Moodle's Workshop Module, and Google Docs are examples of digital technologies that offer organised peer evaluations. These tools enable instructors to assign anonymous reviews, monitor student participation, and manage feedback quality (see Table 9).

10.1.18 Challenges and Considerations

While digital rubrics and peer review systems provide significant benefits, their implementation is not without problems.

1. Technical Literacy: Both students and instructors must be comfortable utilising digital platforms.
2. Bias in Peer Feedback: Despite being anonymous, peer feedback may be influenced by social dynamics or a lack of experience.
3. Time Investment: Creating thorough rubrics and teaching students on appropriate feedback techniques may be time-consuming.
4. Concerns about Plagiarism: Without controls, work in peer-reviewed assignments may be duplicated or disseminated unethically.
5. Educators must support the peer review process by modelling positive comments, holding training sessions, and providing incentives for high-quality submissions.

10.1.19 Strategies for Implementation

To successfully include digital rubrics and peer review into digital education, institutions and educators should consider the following strategies:

1. Begin small by testing rubrics on one or two tasks before scaling them up.
2. Train students by providing examples of constructive comments and rubric interpretation.
3. Iterate and Improve: Use data from LMS systems to fine-tune rubrics and discover frequent misconceptions.
4. Incorporate Reflection: Ask students to think about the comments they received and how they will utilise them.
5. Balance Instructor Oversight: Make sure that peer input is combined with professional assessment.

10.1.20 The Pedagogical Foundation of Formative Feedback

Formative feedback is based on constructivist learning theory, which holds that learners actively build knowledge rather than passively receive information. According to Nicol and Macfarlane-Dick (2006), formative assessment empowers students by clarifying objectives, promoting self-evaluation, stimulating discourse, and inspiring effort via constructive feedback.

Key features of successful formative feedback:

1. Timely: Delivered quickly so that students may act on it.
2. Specific: Emphasises certain facets of performance.
3. Actionable: Offers clear suggestions on how to improve.
4. Dialogic: Promotes thought and discourse between student and instructor.

Formative feedback must be included in digital platforms that enable interaction, multimedia integration, and learner monitoring.

10.1.21 Categories of Digital Formative Feedback Tools

Digital instruments for formative feedback are divided into three main groups, each having its unique pedagogical affordance [12]:

1. Real-time response tools: Used during synchronous meetings to assess comprehension immediately. Examples include Kahoot!, Mentimeter, Slido, and Poll Everywhere.
2. Asynchronous Quiz and Polling Tools: Self-paced tools allow for several tries and explanations. Examples include Google Forms, Microsoft Forms, Quizizz, and EdPuzzle.

Table 10 Types of digital formative feedback tools and their functions

Tool type	Example tools	Main function	Best use case
Live polling	Mentimeter, Kahoot!	Check immediate understanding	Lectures, workshops
Asynchronous quizzing	Google forms, Quizizz	Self-assessment and retrieval practice	Homework, revision activities
Inline annotation	Hypothesis, Google docs	Targeted feedback on student writing	Essay drafts, peer review
Learning analytics	Canvas, Moodle	Track engagement and performance trends	Early intervention, instructor planning
Automated writing feedback	Grammarly, Turnitin	Instant feedback on grammar, structure	Writing-intensive courses

3. Annotation and Commenting Platforms: Allow professors and peers to offer in-text comments on students' work. Examples include Hypothesis, Google Docs, and Kami.
4. Learning Analytics Dashboards: Offer insights based on student performance and engagement indicators. Examples include Canvas Analytics, Moodle Gradebook, and Hapara.
5. Automated Feedback Systems: Use artificial intelligence or programmed logic to provide instant feedback on workouts. Examples include Grammarly, Turnitin Draft Coach, AutoTutor, and Codio (for programming) (see Table 10).

10.1.22 Benefits of Formative Feedback Tools

Digital formative feedback tools can enhance education by promoting active learning, improving retention through frequent low-stakes quizzes, and fostering student reflection through inline comments and automated suggestions. Digital formative feedback technologies provide a number of educational benefits, particularly when connected with well stated learning goals.

1. Encourages Active Learning: Tools such as polling platforms transform passive lecture formats into participatory ones.
2. Improves Retention: Frequent low-stakes quizzes promote long-term memory consolidation (the "testing effect").
3. Encourages Student Reflection: Inline comments and automated suggestions help students evaluate their own work.
4. Improves Instructor Decision-Making: Dashboards identify areas that need more explanation or reteaching.
5. Supports Diverse Learners: Multimedia and personalised feedback forms accommodate a variety of learning methods and demands.

10.1.23 Case Example: Using EdPuzzle for Video-Based Formative Feedback

In a flipped classroom style, an instructor assigns pre-recorded lessons using EdPuzzle. Each movie has embedded multiple-choice and open-ended questions. Students must answer before proceeding. Observed benefits:

1. 92% of students finished videos on time.
2. The average quiz score increased by 18% from pre-lesson to post-lesson.
3. Students expressed increased confidence in engaging in synchronous conversations.

The Instructor's Perspective: "EdPuzzle allowed me to spot misconceptions before class, so I could tailor my live sessions more effectively."

10.1.24 Designing Formative Feedback for Engagement and Equity

To maximise their effectiveness, formative feedback instruments should be created with engagement and equality in mind:

1. Anonymous Response Options: Allow timid or under-represented students to express their ideas without fear.
2. Multimodality: To accommodate varied learners, use text, voice, video, and pictures.
3. Feedback Timing: To improve relevance, offer feedback near to the learning activity.

Culturally Responsive Approaches: When creating feedback language, keep in consideration the diversity of experiences and perceptions (see Table 11).

Table 11 Challenges and mitigation strategies

Challenge	Explanation	Suggested solution
Overuse of superficial feedback	Automated tools may not provide deep, contextual critique	Supplement with teacher comments
Technological barriers	Access issues or platform unfamiliarity	Offer training, ensure mobile compatibility
Feedback fatigue	Too much feedback can be overwhelming	Prioritize key issues, stagger activities
Interpretation gaps	Students may misread automated or peer feedback	Provide models and scaffolded responses

10.1.25 Best Practices for Implementation

For the best use of formative feedback tools in digital pedagogy:

1. Align with learning outcomes: Select tools that directly support the lesson's goals.
2. Integrate Seamlessly: Incorporate feedback opportunities into the flow of learning rather than as add-ons.
3. Foster a Growth Mindset: Consider feedback as part of an ongoing learning process.
4. Encourage Student Feedback Literacy: Show students how to understand and use feedback successfully.
5. Use a variety of tools: Combine real-time and asynchronous techniques to provide maximum coverage and flexibility.

10.1.26 Public-Facing and Open Assignments

In the digital era, learning is no longer limited to the classroom walls or the confines of a conventional curriculum. Public-facing and open assignments, often known as "open pedagogy" or "renewable assignments," call into question the closed, disposable character of conventional academic labour by encouraging students to develop knowledge for realistic, real-world audiences. This strategy enables students to participate actively in public conversation, contribute to global knowledge networks, and see their work as significant beyond evaluation. The open education movement emphasises access, equality, and knowledge co-creation, which matches public-facing tasks. Constructivist and connectivist learning theories emphasise student agency, cooperation, and social networks in knowledge generation.

10.1.27 Types and Formats of Public-Facing Assignments

Public-facing assignments may take a variety of forms, depending on the course setting and learning objectives. They often use digital tools and platforms for publishing, collaboration, and engagement (see Table 12).

10.1.28 Pedagogical Benefits

Public-facing assignments have several advantages, including not just student learning but also community participation and information dissemination [13]:

Table 12 Common types of public-facing assignments

Assignment type	Description	Typical platform / Tool
Blog posts / commentaries	Short, reflective or analytical posts for public view	WordPress, medium, blogger
Wikipedia edits	Contribution to public encyclopedia entries	Wikipedia, WikiEdu
Podcasts	Audio episodes exploring course-related topics	SoundCloud, Spotify, anchor
Infographics/posters	Visual representations of research or data	Canva, Piktochart, Google slides
Open textbooks/guides	Student-authored chapters or resources	Pressbooks, GitBook, manifold
Digital exhibitions	Curated digital collections or storytelling projects	Omeka, scalar, StoryMapJS
Social media campaigns	Content designed for advocacy or education	Instagram, twitter (X), TikTok
Public presentations	Recorded or streamed talks, lectures, or panels	YouTube, zoom, Panopto

1. Increased Motivation and Ownership: Knowing that their work has a genuine audience motivates students to put out greater effort and take pleasure in their results.
2. Real-World Skills Development: Students acquire practical experience in digital communication, media creation, and public involvement.
3. Civic and social engagement assignments might focus on community concerns, such as advocacy, awareness, or social justice.
4. Knowledge democratisation: Outputs may help to bridge gaps in publicly accessible knowledge, especially on under-represented themes or viewpoints.
5. Portfolio Development: Students will depart with tangible, shareable artefacts that may be utilised in career or academic applications.

10.1.29 Case Study: Wikipedia Editing in a History Course

In a university history course, students were tasked with editing or creating Wikipedia pages on lesser-known African historical events. Students used WikiEdu's dashboard to learn how to cite sources, adhere to editorial standards, and write objectively for a worldwide audience. Results:

1. There were 14 new articles created and 26 old ones updated.
2. Over 50,000 people saw the articles in six months.
3. Students indicated that their work "mattered" and that they improved their research, writing, and digital literacy abilities.

Instructor observations: "Students engaged more deeply with historical sources because they knew their work would be public and held to Wikipedia's standards."

10.1.30 Designing Effective Open Assignments

Effective public-facing tasks need careful preparation and scaffolding. The instructors should:

1. Clarify Purpose and Audience: Define who the assignment is for and why it is important.
2. Include Ethical Considerations: Discuss privacy, representation, intellectual property, and misinformation.
3. Provide Technical Support: Choose platforms that are accessible to all students and provide tutorials or walkthroughs.
4. Integrate Peer Feedback and Revision: Ensure quality by iteratively draughting, reviewing, and reflecting.
5. Provide Opt-Out Options: Offer private options to students who are uncomfortable publicising their work.

10.1.31 Challenges and Mitigation Strategies

Public-facing assignments, which require students to express their work beyond the classroom, whether through blogs, presentations, social media, or collaborations with external audiences, introduce unique complexities that instructors must carefully consider. Unlike standard assignments in academic contexts, public-facing projects need more attention to ethical issues, audience awareness, and possible real-world effect. Instructors should prepare for problems linked to privacy, intellectual property, and changing audience expectations. Furthermore, when their work is submitted to public inspection, students may feel more pressure or anxiety, thus clear instructions, support mechanisms, and revision chances are crucial. By proactively addressing these complexities, educators can create meaningful learning experiences that help students develop communication skills, digital literacy, and the ability to engage thoughtfully with broader communities, ensuring that public-facing assignments are both enriching and pedagogically effective [14, 15] (see Table 13).

10.1.32 Ethical and Inclusive Practices

When students develop public-facing projects, they face new ethical challenges that define the integrity and effect of their work. Instructors have a critical role in leading students through these concerns, ensuring that their participation in public debate is

Table 13 Public-facing assignments

Challenge	Explanation	Mitigation strategy
Privacy and safety concerns	Students may be uncomfortable sharing work publicly	Allow pseudonyms, opt-outs, or private alternatives
Quality control	Public errors can reflect poorly on students/institution	Emphasize peer review, revision cycles
Platform unfamiliarity	Students may struggle with new digital tools	Offer training, use institution-supported platforms
Time demands	Public projects often require more time and effort	Extend timelines, scaffold in phases
Equity and access	Not all students have the same digital access or literacy	Ensure tech support and device compatibility

intelligent, responsible, and informed. Educators may give guidance in the following key areas [16, 17]:

1. Use inclusive language to prevent damaging representations: Students should be taught to recognise and combat stereotypes, ensuring that their work represents multiple viewpoints and values many identities and experiences. Thoughtful communication promotes a more egalitarian and courteous public debate.
2. Accurate citation of sources and grasp of digital authorship norms: To ensure academic integrity and protect intellectual property rights, public works must be properly attributed. To avoid misinformation or plagiarism, instructors should emphasise the necessity of properly crediting sources and adhering to copyright and fair use rules.
3. Interacting appropriately with audience feedback: Public assignments often elicit comments, conversations, and criticisms. Students should be prepared to react constructively, accepting opposing opinions while being professional and respectful in their interactions.
4. Understanding digital footprints and long-term visibility: Once shared publicly, material may stay available permanently. Instructors may assist students in comprehending the consequences of online permanence by encouraging them to think about how their work will be seen in the future and how it fits into their personal and professional identities.

10.1.33 Assessing Collaboration and Process

In the ever-changing world of digital pedagogy, collaborative learning has evolved as a critical component of knowledge generation and community participation. While conventional evaluations sometimes concentrate on individual outputs, measuring cooperation and process is just as important, especially when learning goals emphasise communication, teamwork, and iterative development. In digital settings, where

Table 14 Key elements in assessing collaboration

Dimension	Indicators	Assessment tools
Contribution	Frequency, quality, and relevance of participation	Peer evaluations, participation logs
Communication	Clarity, tone, responsiveness	Discussion forum analytics, chat transcripts
Coordination	Task management, role division, workflow	Project plans, version histories
Problem-solving	Strategies used to address challenges	Group reflections, process journals
Reflection	Critical appraisal of the collaborative experience	Individual reports, self-assessments
Equity and inclusion	Fairness in workload, respect for diverse viewpoints	Anonymous peer feedback, facilitator notes

group work may occur asynchronously and across cultures, examining the how of learning (rather than merely the what) becomes a pedagogical necessity. Collaborative learning is founded on socio-constructivist theory, which views learning as a social, dialogic, and interactive process. According to [18], learners get greater comprehension when they collaborate with more knowledgeable peers. Assessing collaborative processes is consistent with:

1. Formative learning objectives include promoting development, reflection, and metacognition.
2. Process-oriented teaching focuses on how students engage, problem solve, and negotiate.
3. Twenty-First-century skill development, such as collaboration, digital communication, and dispute resolution.

10.1.34 Dimensions of Collaborative Work to Assess

To evaluate cooperation holistically, educators must consider more than just the end product and take into account numerous aspects of the group process (see Table 14).

10.1.35 Digital Tools for Tracking and Assessing Process

Digital learning platforms provide integrated monitoring tools that allow instructors and students to monitor interactions in real time or retrospectively (see Table 15).

Table 15 Digital tools for process assessment

Tool/Platform	Feature/Use
Google docs/MS 365	Version history, comment threads, real-time collaboration
Slack/Discord	Threaded discussions, file sharing, asynchronous dialogue
Trello/Asana	Task assignment, progress tracking, time logs
LMS (e.g., Moodle)	Group spaces, forum activity tracking, submission monitoring
Miro/Padlet	Visual collaboration, brainstorm tracking
Peergrade/FeedbackFruits	Peer review workflows with rubrics and feedback analytics

Table 16 Process-focused rubric criteria

Criteria	Excellent (4)	Good (3)	Fair (2)	Poor (1)
Communication	Proactive, inclusive, timely	Generally responsive	Inconsistent	Minimal or disruptive
Contribution	Regular, high-quality ideas	Contributes fairly	Uneven participation	Rarely contributes
Responsibility	Takes initiative and meets deadlines	Usually dependable	Sometimes unreliable	Often unaccountable
Conflict resolution	Resolves conflict constructively	Attempts compromise	Struggles to resolve	Escalates tension
Reflection	Insightful, connects theory to practice	Some insight	Minimal reflection	No reflection evident

10.1.36 Rubrics for Assessing Collaboration

Clear, accessible rubrics are required for evaluating group projects. Process-oriented rubrics are concerned with how the group accomplishes its goals, not just what they accomplish (see Table 16).

10.1.37 Balancing Creativity and Academic Rigor

Digital pedagogy provides a unique chance to reinvent conventional academic limits by including creative means of expression like multimedia storytelling, data visualisation, podcasting, virtual exhibits, and interactive timelines. These approaches inspire students to connect with course information in ways that are both personally meaningful and socially relevant. However, incorporating creative projects into academic environments presents an important question: how can we retain academic rigour while still encouraging creativity and artistic freedom? The case for incorporating creativity into evaluation depends on many instructional traditions:

Table 17 Examples of rigor in creative projects

Project type	Academic indicators
Podcast	Script with citations, interview protocol, thematic focus
Digital story	Narrative arc, historical context, critical commentary
Infographic	Data accuracy, source reliability, and visual hierarchy
Video essay	Integration of visuals with scholarly argument
Interactive map	Historical/geographic accuracy, annotated sources

1. Constructivist learning theory holds that learners create meaning via active participation and personal interpretation.
2. Critical pedagogy argues for giving students the opportunity to utilise their voices and living experiences in academic work.
3. Multiliteracies theory (New London Group, 1996) broadens literacy beyond textual analysis to encompass visual, spatial, and digital forms of meaning construction.

When combined with rigorous investigation, creative outputs may exemplify critical thinking, concept synthesis, and deep engagement with course issues.

10.1.38 Defining Academic Rigor in Creative Work

Academic rigour should not be mistaken for traditionalism or strict adherence to traditional forms (such as essays or examinations). Instead, it refers to: depth of analysis, use of reputable sources, application of theoretical frameworks, logical coherence, and originality rooted in scholarship (see Table 17).

10.1.39 Designing for Both Creativity and Scholarship

Effective scaffolding for creative academic endeavours requires a systematic strategy that preserves scholarly depth while encouraging originality. Instructors should start by demanding extensive research proposals in which students define precise questions, explain their preferred medium (video essay, podcast, or digital exhibit), identify important source materials, and make specific linkages to course topics. An annotated bibliography component promotes interaction with peer-reviewed research and appropriate vetting of original materials, so rooting creative work in academic debate. The storyboard or outline step next connects research and execution, requiring students to carefully organise information and incorporate academic analysis into their preferred style. Finally, a reflective metacommentary that details editorial choices, technical challenges, and thematic decisions turns the creative process into a subject of critical analysis, ensuring that students articulate how their

Table 18 Balanced assessment rubric

Criteria	Excellent (4)	Good (3)	Fair (2)	Needs improvement (1)
Conceptual depth	Strong engagement with theory and research	Demonstrates insight	Limited conceptual basis	Superficial ideas
Creative expression	Original, compelling, well-executed	Clear and thoughtful	Uneven or clichéd	Lacks creativity
Research & Sources	Extensive, well-integrated references	Adequate use of sources	Minimal citation	Inaccurate or missing
Technical execution	Polished, professional-quality	Mostly clear and functional	Technically flawed	Poor or incomplete
Reflective analysis	Deep reflection on process and choices	Some insight into process	Vague or descriptive only	Missing reflection

work contributes to disciplinary conversations while developing transferable digital literacy skills.

10.1.40 Rubrics for Creative Assessment

Rubrics should be transparent and flexible, allowing for artistic risk while upholding scholarly standards (see Table 18).

10.1.41 Student Voice and Autonomy

Creativity flourishes in settings where pupils have agency. Allowing pupils to choose their medium or platform (within certain restrictions) promotes ownership and deeper commitment. However, setting explicit limits helps to avoid scope creep and misunderstanding. Strategies to support student autonomy: provide choice boards with various project categories, provide examples that indicate academic excellence, create milestones and check-ins to ensure feedback and alignment, and encourage risk-taking with thoughtful reasoning.

10.1.42 Challenges and Considerations

Balancing creativity and academic rigour necessitates solving a number of obstacles to guarantee that new undertakings remain intellectually grounded. One major

Table 19 Challenges and considerations

Challenge	Description	Strategy
Subjectivity in evaluation	Creative work may seem harder to grade fairly	Use clear rubrics and multiple evaluators if possible
Technical skill disparities	Students may have different access or experience	Provide tutorials, scaffolded tasks, or alternative formats
Misalignment with outcomes	Creativity may overshadow academic learning	Connect each element explicitly to learning outcomes
Time-intensive feedback	Creative projects often require more individualized guidance	Use formative checkpoints and peer review

problem is maintaining academic depth, creative work must be supported by research, theoretical frameworks, and explicit links to course goals. Another challenge is structuring open-ended assignments, in which students may struggle with ambiguity or lack clear standards for success. Evaluating creative endeavours fairly may also be difficult, since typical grading rubrics may not always capture the subtleties of creativity and intellectual engagement. Furthermore, giving direction without strangling creativity is critical; too much structure might restrict inquiry, while too little can result in superficial work. Instructors may assist students in combining creative expression with rigorous academic inquiry by carefully developing scaffolded support, open assessment standards, and revision chances. This fosters both creativity and deep intellectual engagement (see Table 19).

10.1.43 Quantitative Versus Qualitative Metrics

Assessments of digital pedagogy can highlight basic problems about how learning should be assessed. Traditional educational institutions have traditionally relied on quantitative metrics test scores, attendance rates, assignment completion, and grade distributions as the key markers of success. While these measurements provide comparability and efficiency, they often fail to convey the complexity, subtlety, and transformational power of digital and multimodal learning experiences. In contrast, qualitative measurements concentrate on the quality of student involvement, critical thinking, creativity, and collaboration aspects that are especially important in digital contexts [19] (see Table 20).

Table 20 Defining the metrics

Type	Description	Common examples
Quantitative	Numerical indicators of performance or participation	Grades, test scores, number of posts, log-in frequency
Qualitative	Descriptive and interpretive assessment	Reflective journals, project narratives, instructor feedback

10.1.44 The Role of Quantitative Metrics in Digital Pedagogy

Quantitative indicators are valuable in digital situations, particularly for measuring involvement and identifying disengaged learners. For example: Learning management systems may report: Number of logins, Time spent doing activities, Quiz scores and Completion rates. Analytics dashboards can display: Engagement patterns throughout weeks, and Correlation of involvement and results.

10.1.45 The Value of Qualitative Metrics

Qualitative metrics better reflect digital pedagogy's process-oriented, student-centred attitude. They evaluate how students perceive, apply, and reflect on information. Examples of Qualitative Data In Practice:

- Discussion forum posts were evaluated for insightfulness, creativity, and relevance to readings.
- Digital portfolios: Demonstrating learning progression over time.
- Peer reviews and instructor commentary reflect the depth of engagement with others' work.
- Reflective essays present introspective accounts of learning experiences.

10.1.46 Mixed-Methods Assessment

The most successful evaluation methods incorporate both measures to provide a more complete picture of learning (see Table 21).

10.1.47 Future Directions: AI, Sentiment Analysis, and Learning Analytics

Emerging technologies are opening up new prospects for automated hybrid assessment:

Table 21 Mixed-method assessment design example

Activity	Quantitative metric	Qualitative metric
Weekly discussion posts	Number of posts	Depth of analysis and use of course materials
Multimedia project	Submission on time	Narrative quality, research depth, creative execution
Group work	Peer rating (1–5 scale)	Reflective statement on the collaboration process
Self-assessment	Likert-scale confidence ratings	Descriptive reflections on learning gains and challenges

1. Natural Language Processing (NLP) systems may evaluate conversation threads for complexity, sentiment, and subject coverage.
2. AI-powered feedback systems may provide quick formative feedback on student work.
3. Learning analytics solutions include qualitative labelling (e.g., "critical thinking," "collaboration") into dashboards.

While promising, these technologies must be utilised with caution in order to prevent monitoring and provide ethical learning settings.

10.1.48 Assessing Media Literacy

Media literacy has evolved into a fundamental ability as pupils interact more with digital surroundings. It covers the capacity for access, analysis, assessment, creation, and action based on many kinds of communication. Not only does media literacy in educational settings guarantee technological competency, but it also helps to produce critical, ethical, and reflective digital citizens. According to the Center for Media Literacy, media literacy is "the ability to access, analyze, evaluate, create, and act using all forms of communication." It has the following essential dimensions:

1. Critical engagement involves questioning and understanding media material.
2. Technical proficiency: Navigating and creating material using digital tools.
3. Civic and ethical awareness: Recognising prejudice, ownership, and consequences
4. Multimodality: Reading and generating texts that mix picture, sound, text, and design.

Table 22 National association for media literacy education

Competency	Example in educational context
Access	Navigating a digital archive or media repository
Analyze	Evaluating bias in a news article or social media thread
Evaluate	Assessing the credibility of digital sources
Create	Designing an infographic or video narrative
Act	Engaging in civic discourse through media (e.g., blog post)

10.1.49 Core Competencies in Media Literacy

There are a number of skills that may be evaluated for media literacy. Several fundamental elements are identified by the National Association for Media Literacy Education (NAMLE) (see Table 22).

10.1.50 Assessment Frameworks for Media Literacy

There are a few different frameworks that are used to guide the evaluation of media literacy in formal education. Among the notable ones are:

1. The SAMR Model (Substitution, Augmentation, Modification, and Redefinition) assesses the extent to which technological advancements cause significant changes in educational activities.
2. The acronym TPACK, which stands for "Technological Pedagogical Content Knowledge," places more emphasis on the interaction of technology, pedagogy, and content.
3. There are performance measures that include access, comprehension, and involvement that are provided by the National Association of Media Literacy (NAMLE) (see Table 23).

10.1.51 Types of Media Literacy Assessments

1. Media Artefact Projects: Students develop a product, such as a podcast, video essay, or meme series, and accompany it with a written justification or storyboard that explains their decisions and how they relate to the course objectives.
2. Critical Media Analysis Essays: Students analyse a piece of media (such as an advertisement, news broadcast, or Twitter thread) to determine its aim, audience, bias, and rhetorical methods.
3. Digital Literacy Diaries: A weekly reflective journal of students' media consumption and output, including platform use, interaction patterns, and observations.

Table 23 Example rubric for media literacy assessment

Criteria	Excellent (4)	Good (3)	Fair (2)	Needs improvement (1)
Source evaluation	Demonstrates advanced source critique	Uses credible sources with minor gaps	Inconsistent source reliability	Relies on unverified sources
Message analysis	Deep understanding of media framing and bias	Identifies bias with some depth	Limited or vague analysis	Little to no analysis
Technical production	Creative, polished, well-executed media artifact	Functional with minor flaws	Basic execution, lacks clarity	Poor technical quality
Multimodal integration	Seamless combination of modes (text, image, audio)	Mostly integrated	Some disjointed elements	Modes used without purpose
Reflective commentary	Strong metacognitive insight into media choices	General reflection provided	Superficial or incomplete	Missing or off-topic

4. Participatory Media Projects: Activism or civic involvement is carried out using digital channels, such as generating awareness campaigns or producing online exhibits (see Table 24).

Table 24 Challenges in media literacy assessment

Challenge	Description	Strategy
Variability in tech skills	Students may lack prior experience with tools or editing software	Provide scaffolded workshops or alternative assignment formats
Overemphasis on aesthetics	Flashy projects may overshadow weak analysis	Require reflective commentary and source transparency
Implicit bias in evaluation	Instructors may conflate style with substance	Use clearly defined rubrics with peer calibration
Plagiarism in remix culture	Reuse of media without citation	Educate about fair use and require attribution practices

Table 25 Short-term versus. Long-term outcomes

Aspect	Short-term outcome	Long-term outcome
Focus	Specific task performance	Transferable understanding and behaviour change
Duration	Within the course timeframe	Persists after the course ends
Measurement	Grades, submissions, participation logs	Alumni surveys, portfolio reviews, longitudinal studies
Example	Scoring 90% on a quiz	Applying digital analysis skills in a professional setting

10.1.52 Long-Term Learning Outcomes

Ultimately, the purpose of education is to cultivate long-term comprehension, marketable abilities, and a thirst for knowledge, even if digital pedagogy places a heavy emphasis on instant performance evaluation (e.g., grades, submissions, participation). Long-term learning outcomes relate to how educational experiences affect a student's cognitive, emotional, professional, and civic growth. Long-term learning outcomes extend beyond material retention to assess:

1. Knowledge and abilities are transferable across settings.
2. Metacognition and Self-directed Learning Habits
3. Civic and ethical responsibilities
4. Technological proficiency and adaptation
5. Critical and creative thinking.

10.1.53 Theoretical Foundations

Several educational theories guide the assessment of long-term learning:

1. Constructivism emphasises profound, conceptual learning that pupils develop over time.
2. Situated Cognition proposes that meaningful learning happens when information is anchored in real-world situations.
3. Bloom's Taxonomy: Higher-order outcomes, including evaluation, creation, and reflection, are important predictors of long-term retention (see Table 25).

10.1.54 Assessing Long-Term Impact in Digital Pedagogy

Evaluating long-term learning outcomes in digital contexts requires a multidimensional approach that captures steady progression beyond current courses. E-portfolios

Table 26 Digital tools for tracking long-term outcomes

Tool / Method	Function	Example use
LMS-integrated e-portfolios	Store student artifacts over the years	Canvas or Moodle portfolios track learning paths
Alumni follow-up platforms	Survey graduates on skills use	Google forms, Qualtrics, Microsoft forms
Learning analytics dashboards	Monitor longitudinal engagement or progression trends	Digication, Portfolium, PebblePad
Social media/digital networks	Observe professional and civic engagement post-course	LinkedIn groups, alumni discord servers

serve as dynamic stores for students' changing capacities; comparing early and later work demonstrates improvements in digital output quality, depth of research integration, and feedback application over semesters. Structured alumni surveys (conducted 6–24 months after the course) offer career-relevant data points by monitoring how graduates use digital literacy and critical thinking abilities in work situations. Capstone projects provide focused proof of longitudinal progress, since their multidisciplinary complexity demonstrates years of accumulated analytical and technical skills. Most innovatively, built-in longitudinal feedback mechanisms like scheduled digital check-ins or alumni discussion forums transform assessment into an ongoing dialogue that fosters lifelong learning communities while providing instructors with rich qualitative data about educational impact over time [20] (see Table 26).

References

1. Wu, X.W., Yang, E.H., Wang, J.: Lightweight security protocols for the internet of things. In: IEEE international symposium on personal. Indoor and Mobile Radio Communications (PIMRC). **2017**, 1–7 (2018). https://doi.org/10.1109/PIMRC.2017.8292779
2. Sanchez-lite, A., Zulueta, P., Sampaio, A.Z., Gonzalez-gaya, C.: BIM for the realization of sustainable digital models in a uni- versity-business collaborative learning environment: assessment of use and students'perception, 1–14 (2022)
3. Benzidia, S., Makaoui, N., Bentahar, O.: The impact of big data analytics and artificial intelligence on green supply chain process integration and hospital environmental performance. Technol. Forecast. Soc. Change. **165** (2021). https://doi.org/10.1016/j.techfore.2020.120557
4. Pereira, C.S., Durao, N., Moreira, F., Veloso, B.: The importance of digital transformation in your business strategy. Mdpi. **14**(2), 1–26 (2022) https://blog.workana.com/en/entrepreneurship/the-importance-of-digital-transformation-in-your-business-strategy/
5. Olika, C.D.: Risk management practices among selected state governments in southwestern Nigeria. Eur. J. Sci. Innov. Technol. **1**(2), 24–31 (2024) https://www.researchgate.net/profile/Hendra-Manurung/publication/356633046_From_Connectivity_to_Digital_Improving_Employee_Readiness_toward_Organizational_Change_in_Digital_Transformation/links/61a5b00671a23a0084c9b061/From-Connectivity-to-Digital-Improvi
6. Oyeronke, A., Fagbohun, M., Adebayo, O.: An Assessment of Computer and ICT Skills among Secondary School Teachers in Ota Ogun State (2013). http://digitalcommons.unl.edu/libphilprac/

7. Paay, J., Kjeldskov, J., Aaen, I., Bank, M: User-centred iterative design of a smartwatch system supporting spontaneous reminiscence therapy for people living with dementia. Health Informatics J. **28**(2) (2022). https://doi.org/10.1177/14604582221106002
8. Purba, A.K., Thomson, R.M., Henery, P.M., Pearce, A., Henderson, M., Katikireddi, S.V.: Social media use and health risk behaviours in young people: systematic review and meta-analysis. BMJ. **1–16** (2023). https://doi.org/10.1136/bmj-2022-073552
9. Shapiro, G., et al.: SARS-CoV-2 vaccine acceptance disparity among israeli defense forces personnel. Mil. Med. **188**(7–8), e2592–e2597 (2023). https://doi.org/10.1093/milmed/usac122
10. Agboola, O., Aremu, M.O., Eze, B.U., Wahab, NO.: Effect of training and career development on bank employees' performance: evidence from selected banks in Nigeria. EMAJ Emerg. Mark. J. 10(2), 56–62 (2021). https://doi.org/10.5195/emaj.2020.204
11. Korpi, H., Sjögren, T., Mikkonen, K., Piirainen, A.: A systematic review and metasynthesis of qualitative studies on the competencies of health and rehabilitation science educators. NERP. **64**(2), 270–282 (2020) https://www.academia.edu/download/88209414/cr_20ltec_202019_20paper_2021.pdf
12. Derges, J., et al.: Mental health practitioners' and young people's experiences of talking about social media during mental health consultations: qualitative focus group and interview study. JMIR Form. Res. **7** (2023). https://doi.org/10.2196/43115
13. König, J., Jäger-biela, D.J., Glutsch, N.: Adapting to online teaching during COVID-19 school closure: teacher education and teacher competence effects among early career teachers in Germany. Eur. J. Teach. Educ. **43**(4), 608–622 (2020). https://doi.org/10.1080/02619768.2020.1809650
14. Goulart, C.M.A.: O conceito de letramento em questão: por uma perspectiva discursiva da alfabetização. Bakhtiniana Rev. Estud. do Discurso. **9**(2), 35–51 (2014). https://doi.org/10.1590/s2176-45732014000200004
15. Assar, S.: Information and communications technology in education. Int. Encycl. Soc. Behav. Sci. Second Ed. **12**, 66–71 (2015). https://doi.org/10.1016/B978-0-08-097086-8.92104-4
16. Noubissi Domguia, E.: Taxing for a better life? The impact of environmental taxes on income distribution and inclusive education. Heliyon. **9**(11), e21443 (2023). https://doi.org/10.1016/j.heliyon.2023.e21443
17. Mohammed, M.: A comparative analysis of environmental justice between urban neighbourhoods. J. Incl. Cities Built Environ. **1**(2), 27–40 (2021). https://doi.org/10.54030/2788-564X/2021/v1i2a4
18. Uwuigbe, U., Jafaru, J., et al.: Dividend policy and firm performance. J. Real Estate Portf. Manag. **7**(1), 79–88 (2012) http://www.jstor.org/stable/24882251
19. Naz, F., Agrawal, R., Luthra, S., Kumar, A.: Reviewing the applications of artificial intelligence in sustainable supply chains: exploring research propositions for future directions. December. **2022**, 2400–2423 (2021). https://doi.org/10.1002/bse.3034
20. Zhao, S., Pan, Q., Gao, D., Cheng, J.: Integrating internet of things and mixed reality to teach performance-based architectural design: a case study of shading devices. Educ. Inf. Technol. (2022). https://doi.org/10.1007/s10639-022-10998-6

Chapter 11
Barriers, Challenges, and Institutional Realities

11.1 Infrastructure and Technological Gaps

The promise of digital pedagogy is dependent on reliable, accessible, and equitable technological infrastructure. However, gaps in device availability, internet speed, software, and institutional support persist, resulting in unequal learning environments, especially in underserved or rural areas. These infrastructural and technical deficiencies create substantial challenges to the effective adoption and evaluation of digital learning systems. The phrase "infrastructure gap" refers to disparities in the availability and quality of infrastructure, including hardware such as computers, tablets, and cellphones. Connectivity (reliable internet and data coverage). Electricity supply (particularly in rural or poor areas). Software and platforms (Licensed versus Open-Source Tools). Technical help and maintenance. And digital training and assistance for students and teachers [1] (see Fig. 11.1).

11.1.1 Global and Regional Disparities

In Sub-Saharan Africa, frequent power outages and poor internet penetration interrupt online course delivery, while rural India confronts additional problems owing to restricted hardware availability and low-tech awareness, which limit participation in blended learning. The U.S. urban-rural split exposes substantial broadband disparities, giving rural pupils less options for real-time or multimedia-rich learning. Meanwhile, Southeast Asia's dependence on mobile devices, despite its strong connection, limits learning depth owing to smartphone platform restrictions. These instances

F. A. Alaba and I. G. Ajibola, *Digital Pedagogy: Bridging the Gap Between Software and the Humanities*, Information Systems Engineering and Management 85,
https://doi.org/10.1007/978-3-032-20335-9_11

Fig. 11.1 AI as an infrastructure gap [2]

demonstrate that digital education is not necessarily inclusive; its success is dependent on local infrastructure, resources, and literacy. Without addressing these discrepancies, technology solutions have the potential to exacerbate rather than alleviate current educational inequality.

Technological deficiencies have a direct impact [3]:

1. Assessment validity: Students are unable to access and engage with assessments equitably.
2. Digital literacy development: inconsistent exposure to digital instruments impedes skill acquisition.
3. Equity of experience: Disadvantaged pupils risk exclusion or decreased involvement.
4. Instructional design requires educators to adjust curriculum to the lowest common technological denominator.

For example, in a multimodal assignment that requires video editing, a student who does not have access to editing software or a laptop is effectively barred from equal participation.

Institutional and Policy Responses

To reduce technical gaps in education, institutions and authorities have implemented specific solutions that address regional restrictions. Device lending schemes, such as the United Kingdom's COVID-era Digital Access Fund, deliver hardware directly to impoverished students, and zero-rating efforts (e.g., South African institutions' ISP agreements) erase data costs for critical learning platforms. Content adaptation tactics are also important—educators are already creating hybrid materials like offline-readable PDFs and audio lectures for low-bandwidth environments. Infrastructure development is still crucial, with national efforts such as Nigeria's Broadband Plan attempting to improve rural access gradually. UNESCO's educator training frameworks complement these technology solutions by providing instructors with low-tech pedagogical abilities, ensuring that learning continues despite technological limits. These multilayered methods demonstrate how context-specific solutions can gradually reduce digital disparities while maintaining educational quality across diverse technology environments [4].

11.1.2 Open-Source and Offline Solutions

A growing number of schools and organisations are turning to open-source and offline-capable digital solutions to overcome technical constraints. LibreOffice is a free productivity suite for organisations that cannot buy licenses. These technologies promote the democratisation of digital education by providing alternatives in areas where infrastructure is limited.

Digital Pedagogy for Low-Tech Contexts

Designers of infrastructure gaps need new, context-sensitive instructional methods that prioritise accessibility and adaptability. Mobile-first learning solutions work with the most popular devices, while SMS- or WhatsApp-based systems distribute material and promote conversations in low-bandwidth contexts. Offline printable copies of digital materials bridge connection gaps, while peer-to-peer learning networks let students assist one other when institutional resources are short. A northern Kenyan community college created a solar-powered learning facility to solve energy and internet constraints. The institution created a small Wi-Fi network using Raspberry Pi servers with instructional content to let students access resources wirelessly. These adaptable solutions show how to use resources creatively and create community-driven learning environments to keep education excellent and inclusive despite technology constraints [5] (see Fig. 11.2).

Technology gaps in schooling create structural inequalities with serious ethical consequences. Digital redlining the systemic exclusion of marginalised populations from connectivity, creates educational apartheid based on location and economics. This reinforces privilege cycles since pupils with stable high-speed internet and contemporary equipment outperform others with antiquated technology or sporadic connections. The "invisible learners" situation is most concerning when disengaged pupils discreetly withdraw, concealed by aggregate participation measures. These realities need a change towards rights-based digital pedagogy that sees strong technology infrastructure as essential to education. Schools must offer equal device

Fig. 11.2 Online Teaching [6]

access, connection, and digital literacy assistance to avoid technology from becoming another educational stratification axis, much like electricity and textbooks.

11.1.3 Faculty Training and Support Deficits

The successful integration of digital pedagogy is dependent not just on infrastructure and student preparation, but also on faculty ability to develop, administer, and analyse digital learning experiences. However, many instructors lack enough training and support, resulting in unequal teaching quality and underutilised digital resources. Faculty training and support shortages emerge in a variety of areas [7]:

1. Technical proficiency: Difficulty utilising LMS, digital grading systems, or interactive media.
2. Pedagogical adaptation: Challenges in redesigning courses for online or mixed delivery.
3. Assessment literacy: little experience to multimodal or formative digital assessments.
4. Ongoing support: No access to IT help desks, instructional designers, or peer mentorship.

Faculty training and support shortages are often caused by institutional, budgetary, and cultural reasons. A lack of institutional commitment might cause colleges to prioritise research above teaching quality, resulting in insufficient investment in faculty development. Budget constraints further restrict access to professional training opportunities, and outdated training models may overlook critical aspects

Table 11.1 Root causes

Root cause	Description
Insufficient institutional investment	Professional development is not prioritized or funded
Rapid tech evolution	Constant updates outpace faculty learning
Resistance to change	Cultural or generational reluctance to adopt new technologies
Time constraints	Faculty are already overburdened with teaching, research, and service
Lack of incentives	No reward for digital innovation in tenure or promotion criteria

such as student mental health and educational technology. Furthermore, academic members sometimes have excessive responsibilities, making it difficult to devote time to professional development. Limited knowledge and inadequate incentives hinder participation in accessible training programs, while organisational impediments such as bureaucratic hurdles and tight institutional norms might further limit access to needed assistance. Addressing these difficulties needs a diversified strategy that improves institutional commitment, finance, and training program accessibility [8] (see Table 11.1).

11.1.4 Impact on Digital Assessment

Digital assessment has changed how educators evaluate student performance, with benefits and drawbacks. Grading automation, immediate feedback, and reduced administrative effort boost efficiency. Digital evaluations may also interest students with interactive forms and gamification. Academic integrity problems, the requirement for trustworthy technical infrastructure, and automated grading system biases are difficulties. Digital exams motivate students and peer learning by encouraging involvement and knowledge via interactive online evaluations. Innovation, fairness, and accessibility must be balanced when educational institutions use digital evaluation methods [9] (see Table 11.2).

Table 11.2 Impact on digital assessment

Area of Impact	Consequences for students and courses
Assessment design	Use of outdated or inappropriate evaluation tools
Feedback quality	Delayed, generic, or misaligned feedback due to unfamiliarity with tools
Accessibility	Lack of inclusive design practices in digital assignments
Engagement	Reduced student motivation due to poor digital interaction
Data-driven improvement	Minimal use of learning analytics to inform instructional strategies

11.1.5 Training Needs across Faculty Roles

Faculty members across different roles have distinct training needs to enhance their effectiveness in teaching, research, and administrative responsibilities. Pedagogical tactics, educational technology, and student involvement training help professors and lecturers increase learning. To enhance scholarship, researchers need grant writing, data analysis, and academic publication help. Administrative faculty like department heads and deans need leadership, policy, and institutional management training to run smoothly. Faculty engaging in community outreach get public interaction, teamwork, and social impact evaluation training [10] (see Table 11.3).

Institutional Practices and Recommendations

Best practices in research quality, cooperation, and professional development may boost institutional effectiveness. Create a supportive research culture, ensure fairness and inclusion, and adopt clear faculty and student involvement procedures. Institutions should also prioritise ethical research training, offer enough resources for scholarly activity, and recognise efforts with rewards. Strong leadership and communication may also help institutions innovate and overcome problems [11] (see Table 11.4).

Table 11.3 Training needs across faculty roles

Role	Specific training needs
Adjunct instructors	Orientation to institutional tools and policies
Senior faculty	Pedagogical strategies for blended and online learning
Early-career academics	Building digital confidence and integrating assessment software
Technologists/designers	Collaboration and communication with teaching staff

Table 11.4 Institutional practices and recommendations

Institutional strategy	Description	Example
Center for teaching and learning (CTL)	Central hub offering ongoing workshops, drop-ins, and consultations	CTLs at MIT, Indiana University, university of cape town
Faculty workload credits	Incentivizing professional development through release time	One course release for major digital course redesign
Annual digital teaching symposiums	Sharing innovations and failures within the academic community	"Digital futures" conference at Australian National University
Faculty digital portfolios	Reflective practice and documentation of pedagogical growth	Encouraged as part of promotion review at progressive institutions

11.1.6 Equity and Support for all Faculty Types

Despite the fact that contingent and adjunct instructors make up the bulk of teaching personnel at many universities, they are often underpaid. Addressing this imbalance includes: providing equitable access to training and technology, compensating for engagement in faculty development, integrating adjuncts into digital planning and decision-making, and providing mobile-first training resources for time-constrained educators [12].

Digital Divide and Equity Concerns

Despite the expanding use of digital tools in pedagogy and evaluation, the digital gap remains a persistent and multidimensional impediment to egalitarian education. This gap includes not just access to technology and connection, but also disparities in digital literacy, cultural capital, and institutional support. In digital assessment settings, these differences may considerably distort learning results, perpetuate existing inequities, and marginalise already disadvantaged communities. The digital divide is typically divided into three layers [13] (see Table 11.5).

Dimensions of Digital Inequity

Digital disparity takes many forms, influencing access, skills, and opportunities in the digital environment. Access to technology, where disparities in internet connectivity and device availability create barriers; digital literacy, which influences how effectively individuals can navigate and use digital tools; affordability, as high device and internet service costs disproportionately affect lower-income groups; and content relevance, ensuring that digital resources cater to diverse cultural and linguistic needs. Institutional assistance also helps to bridge gaps, since policies and actions may either reduce or increase digital inequality [14] (see Table 11.6).

Impact on Assessment Validity and Fairness

1. Biassed outcomes: Students with less resources often do worse not owing to a lack of knowledge, but due to technological limitations.
2. Participation gaps: Asynchronous or real-time evaluations exclude those who do not have consistent internet connections.
3. Stress and cognitive overload: Technological problems might detract from learning and assessment performance.

Table 11.5 Understanding the digital divide

Layer	Description
Access divide	Unequal availability of digital devices and internet connectivity
Skills divide	Variation in digital literacy and the ability to use tools effectively
Usage divide	Differences in how individuals use technology to create, learn, or assess

Table 11.6 Dimensions of digital inequity

Equity concern	Manifestation in digital assessment
Socioeconomic status	Limited access to devices/software; stress due to unstable connectivity
Rural versus urban disparities	Poor broadband access in rural areas affects submission and real-time exams
Disability	Inaccessible platforms or non-compliant content (e.g., no captions)
Language barriers	Interfaces and instructions not available in students' first language
Institutional support gaps	Smaller institutions lack infrastructure or tech support for students

4. Self-censorship: Marginalised students may avoid multimedia tasks if they are inexperienced with the appropriate instruments.

Intersectionality and the Digital Divide

Equity problems are heightened at the intersections of race, gender, disability, and geographic location. A female student from a remote region with inadequate English ability may encounter many challenges. Understanding the platform instructions (language).

11.1.7 Inclusive Design and Assessment Strategies

Educators have the ability to implement the following inclusive strategies in order to reduce inequalities (see Table 11.7).

Table 11.7 Inclusive design and assessment strategies

Strategy	Description
Universal Design for Learning (UDL)	Design assessments that provide multiple means of engagement and response
Asynchronous flexibility	Allow submissions over wider time windows
Multimodal options	Let students choose between video, audio, text, or visual responses
Mobile-first design	Ensure assessments are functional on smartphones
Use of open educational resources (OER)	Reduce cost barriers and ensure equal access
Language and accessibility support	Offer instructions in multiple languages; use screen-reader friendly formats

Role of Institutional Policy
Institutions have an important role in resolving the digital gap.

1. Digital equity audits: Identify and document gaps in student access.
2. Equitable technical provisioning: Provide devices, internet stipends, or on-campus access.
3. Bias-aware grading policies: Recognise non-technical barriers to quality assessment.
4. Monitor equity indicators by using disaggregated data to monitor assessment discrepancies.

Ethical and Pedagogical Imperatives
Ignoring digital disparities in assessment violates core educational ideals.

1. Justice: Every student deserves an equal chance to succeed.
2. Care: Teachers must react to their pupils' daily experiences.
3. Reflection: Faculty should constantly review how their decisions impact inclusivity.

A pedagogy of equality considers not just who succeeds, but also who is excluded and why.

Budgeting and Resource Allocation

Effective digital assessment strategies are heavily reliant on institutional investments in infrastructure, training, tools, and human resources. However, disparities in budgeting and resource allocation across institutions and departments can severely limit the successful implementation of digital assessments. Implementing and sustaining digital evaluation systems necessitates investment in multiple interconnected areas (see Table 11.8).

Table 11.8 Categories of financial investment in digital assessment

Budget Category	Examples
Hardware and infrastructure	Computers, servers, routers, mobile devices, smart classrooms
Software licensing	Learning management systems (LMS), plagiarism checkers, e-portfolio platforms
Training and professional development	Faculty workshops, instructional design support, certifications
Support personnel	IT staff, instructional technologists, accessibility coordinators
Maintenance and upgrades	System updates, cloud storage fees, cybersecurity measures
Content development	Multimedia assessment materials, OER creation, language localization

11.1.8 Cost-Benefit Considerations

Institutions often confront tough choices when weighing expenses and possible advantages.

1. Short-term versus long-term investment: A full LMS may need a considerable initial investment but will save money over time compared to fragmentary solutions.
2. Open-source versus proprietary software: Open platforms have lower license prices but may need more maintenance and customisation.
3. Cloud versus on-premise solutions: Cloud-based exams provide scalability but come with ongoing subscription fees and data privacy issues.

Resource Allocation and Equity

Uneven resource allocation across departments or campuses might result in assessment inequalities. For example:

1. A humanities department may not have the same multimedia laboratories as engineering students, restricting their capacity to conduct digital storytelling exams.
2. Urban campuses may get more bandwidth investment than rural branches, which affects the practicality of real-time assessments.

For example, a university's central allocation strategy may promote high-enrollment departments while leaving low-enrollment schools with inadequate digital assessment capabilities.

Sustainable Budgeting Practices

To guarantee the lifetime of digital assessment methods, institutions should use budgeting models that prioritise sustainability (see Table 11.9).

Table 11.9 Sustainable budgeting practices

Sustainable strategy	Description
Total cost of ownership (TCO)	Accounts for both upfront and ongoing costs of technology over its lifecycle
Participatory budgeting	Involves faculty and students in financial decision-making for tech investments
Grants and external funding	Leverages research funding for assessment innovation
Cross-departmental sharing	Pools resources for training and tech use across faculties
Phased implementation	Distributes costs over time for scalable deployment

11.1.9 Institutional Models and Practices

Organisations adapt and survive in different settings based on institutional models and practices. Institutional theory proposes that organisations obtain legitimacy and stability by following norms, regulations, and cultural expectations. Organisational structures, decision-making, and policy formulation are affected by these models. Additionally, frameworks like the Institutional Analysis and Development (IAD) model assist policymakers in developing successful interventions by incorporating institutional restrictions and incentives [15] (see Table 11.10).

Budgeting for Accessibility and Inclusion

Too frequently, accessibility is seen as an "add-on" rather than a primary financial goal. Budgeting for inclusion includes:

1. Captioning and transcribing services.
2. Screen reader compatibility testing
3. Multilingual platform development
4. Training on Universal Design for Learning (UDL).

Impact of Underfunding

When funding is insufficient or misplaced, digital assessment attempts might suffer from:

1. Tool mismatches: Use of generic instruments that do not meet the instructional aims.
2. Platform instability refers to frequent system interruptions during tests or submissions.
3. Faculty burnout: Lack of support results in excessive workloads for educators.
4. Reduced student confidence: Technical problems or uneven assessment styles weaken trust.

Table 11.10 Institutional models and practices

Institutional approach	Strengths	Challenges
Centralized budgeting	Ensures standardization and bargaining power with vendors	May overlook specific departmental needs
Decentralized budgeting	Allows customization to discipline-specific needs	Risks of duplication and inefficiency
Hybrid models	Combines core infrastructure with unit-specific innovation funding	Requires coordination and communication mechanisms

11.1.10 Institutional Resistance to Innovation

Despite the obvious pedagogical benefits of digital assessment, institutions often demonstrate tremendous resistance to innovation. This opposition stems from past habits, bureaucratic inertia, cultural conservatism, and fear of failure. As a consequence, even well-funded digital evaluation efforts may fail to establish momentum at all levels of an organisation. Resistance to digital innovation in assessment often stems from three interrelated domains [16] (see Table 11.11).

Common Manifestations of Resistance

Active resistance, such as protests, strikes, and demonstrations, challenges authority or policies. Passive resistance, like non-compliance, avoidance, or silent opposition, defies authority or policies. Cultural resistance, such as art, literature, and media that challenge dominant narratives, is symbolic resistance [17] (see Table 11.12).

Root Causes of Institutional Resistance

1. Universities operate in risk-averse cultures where failure might jeopardise reputation, financing, or compliance.
2. Change Fatigue: Frequent proposals without follow-through may breed indifference or scepticism of proposed improvements.
3. Disciplinary Silos: What succeeds in one faculty (for example, digital portfolios in Art) may be rejected in another (such as Law or Philosophy).
4. Misaligned Incentives: Promotion and tenure systems may prioritise research productivity above instructional innovation.

Table 11.11 Understanding resistance in institutional contexts

Domain	Source of resistance
Structural	Rigid administrative procedures, accreditation constraints
Cultural	Academic traditions, scepticism toward technology
Personal/interpersonal	Fear of change, lack of training, and generational divides in tech fluency

Table 11.12 Common manifestations of resistance

Manifestation	Impact on digital assessment
Delayed implementation	Stagnant pilot programs or incomplete system rollouts
Tokenistic adoption	Superficial use of tools without pedagogical integration
Sabotage or apathy	Faculty or staff disengage from or undercut innovation efforts
Policy misalignment	Institutional rules contradict digital practices (e.g., paper-based archiving requirements)
Preference for traditional exams	Departments resist replacing pen-and-paper tests with multimedia assessments

Table 11.13 Institutional theories on change and innovation

Theory	Relevance to digital assessment resistance
Diffusion of innovations	Adoption depends on perceived advantage, compatibility, and complexity
Organisational learning theory	Institutions must adapt, reflect, and incorporate feedback loops
Path dependency theory	Historical practices shape present decisions and resist deviations
Institutional isomorphism	Universities may mimic others, but still resist meaningful internal change

5. Technological scepticism refers to concerns about monitoring, depersonalisation, or "technologising" the learning process.

Institutional Theories on Change and Innovation

Institutional change and innovation theories examine how organisations and societies grow inside systems. These ideas emphasise how institutions formal rules, norms, and cultural expectations influence innovation and adaptability. Institutions may promote and hinder innovation by affecting new idea adoption, dissemination, and sustainability. Another method studies institutional change via gradual adjustments or external forces that challenge norms. Some theories also explain how cultural and historical legacies affect change throughout time via official and informal institutions [18, 19] (see Table 11.13).

11.1.11 Strategies for Overcoming Resistance

Addressing concerns, encouraging involvement, and building trust is necessary to overcome opposition. Leaders should communicate the advantages of change, listen to concerns, and teach and encourage people to adapt. Quick victories and organisational champions may boost change acceptability. Phased implementation and milestone celebrations may help ease implementation and build commitment (see Table 11.14).

Balancing Innovation with Academic Integrity and Autonomy

One of the most difficult tasks is reconciling innovation with respect for academic autonomy. Many instructors are concerned that digital evaluation systems impose external templates on their teaching philosophies. To address this:

1. Innovations must be adaptable and customised.
2. Institutional policies should express instructional ideals, not simply technical norms.

Table 11.14 Strategies for overcoming resistance

Strategy	Description
Stakeholder engagement	Involve faculty and students early in the planning and implementation
Pilot programs with evaluation	Run small, measurable trials with feedback mechanisms
Incentive structures	Recognize innovation in promotion, awards, or research grants
Change agents and champions	Appoint respected faculty to model and advocate for digital practices
Transparent communication	Regular updates, forums, and FAQs to reduce uncertainty and fear
Support infrastructure	Ensure technical and pedagogical support is readily accessible

3. Faculty growth should prioritise flexibility within structure, allowing for innovation while preserving discipline identity.

Monitoring and Evaluating Change

To preserve innovation while minimising backlash:

1. Set up evaluation innovation committees across faculties.
2. Use surveys, interviews, and analytics to track adoption trends and concerns.
3. Provide formative feedback on innovative initiatives without penalising them.

11.1.12 Cultural Shifts in Academic Workflows

The use of digital evaluation tools has transformed academic culture. Technological, social, cognitive, and procedural changes affect how educators create, deliver, and assess learning. Moving from static, static evaluation models to dynamic, continuous, multimedia-based systems changes academic work, cooperation, and communication. Digital assessment systems encourage collaborative learning, which challenges academic work's individualistic orientation (see Table 11.15).

Redefining the Role of the Educator

In digital assessment environments, educators increasingly serve as [20, 21]:

Table 11.15 From individualism to collaborative academic cultures

Traditional workflow	Digital-enabled shift
Siloed course design	Collaborative syllabus and assessment co-creation
Individual grading practices	Team-based or peer-reviewed evaluations
Independent teaching	Cross-disciplinary or co-teaching using shared digital resources

Table 11.16 Workload redistribution and temporal reorganisation

Aspect	Shift in workflow
Grading	Moves from batch processing at term's end to continuous or rolling evaluation
Feedback	Becomes more frequent, often multimodal (video, audio, screencasts)
Course preparation	Front-loaded with high initial effort in digital content and rubric design

1. Curators of material and tools, rather than just knowledge suppliers.
2. Facilitators of learning processes, assisting students using digital platforms.
3. Developers of interactive, student-centred assessment experiences.

This reframing necessitates new competencies such as digital pedagogy, learning design, and data interpretation that may not have been covered in conventional academic training.

Evolving Student Expectations and Behaviours
Digital assessment also affects pupils' cultural expectations.

1. Students want immediate and transparent grading and feedback.
2. There is an increasing desire for agency in tasks such as selecting assessment forms and co-creating rubrics.
3. Students are now expected to be digitally savvy producers rather than simply consumers of information.

Cultural Implications: Students may oppose or struggle with more autonomy, especially in settings where rote memorisation has been the primary teaching approach.

Workload Redistribution and Temporal Reorganisation

Digital evaluation alters the rhythms and intensities of academic work [22]: (see Table 11.16).

11.1.13 Pedagogical Paradigm Shifts

Technology, educational philosophies, and social requirements because pedagogical paradigm shifts, which alter teaching and learning methods. These changes might include switching from teacher-centered to student-centered learning, using digital technologies, and adopting inclusive and transformational pedagogies that value equality and involvement [23, 24] (see Table 11.17).

Resistance and Negotiation in Cultural Change
Cultural change is not universally accepted. Faculty may resist for the following reasons:

Table 11.17 Pedagogical paradigm shifts

Old paradigm	Emerging paradigm
Summative assessment at term end	Continuous and formative feedback loops
Text-based exams and essays	Multimodal, media-rich assessments
Uniformity across students	Personalized and adaptive assessment pathways
Assessment of memory and recall	Assessment of creativity, synthesis, and problem-solving

Table 11.18 Leadership and institutional culture

Leadership action	Cultural impact
Modeling innovation	Encourages experimentation and reduces stigma around failure
Recognizing digital teaching efforts	Shifts prestige structures toward valuing pedagogical innovation
Fostering safe spaces for learning	Supports risk-taking and community among faculty
Encouraging reflection	Embeds cultural change within a cycle of continuous improvement

1. Loss of control: Digital platforms may automate activities or enforce strict protocols.
2. Loss of identity: Teaching approaches based on lecture-and-exam traditions seem out of place.
3. Generational differences: Younger faculty members may adjust rapidly, but others struggle with digital standards.

Institutions must deal with these conflicts by providing training, maintaining pedagogical variety, and participating in reflective conversation about what constitutes "good teaching" and "valid assessment (see Table 11.18).

11.1.14 Time Constraints and Faculty Workload

One of the most persistent barriers to using digital evaluation systems is the problem of time restrictions and teacher workload. While digital technologies promise efficiency, the reality often includes a steep learning curve, greater planning needs, and continuous demands for technology maintenance and student assistance. Digital evaluation tools are often touted as time-saving, although they usually result in time-shifting rather than time-reduction (see Table 11.19).

Strategies to Mitigate Time Burdens

Effective productivity and efficiency methods are needed to decrease time constraints. Prioritisation methods like the Eisenhower Matrix assist prioritise work by urgency

Table 11.19 The myth of automation: time-saving versus time-shifting

Task	Pre-digital	Post-digital
Assignment creation	Simple template reuse	Designing interactive or multimedia assignments
Grading	Batch marking using paper	Continuous or rolling assessment with analytics review
Feedback	End-of-term comments	Frequent, multimodal (text, audio, video) feedback
System management	Minimal	Time spent navigating platforms, resolving tech issues

and priority. Time blocking reduces distractions by allocating specified times for concentrated work. The Pomodoro Technique, which uses short, intensive bursts with pauses, may boost focus and minimise burnout. Workflow automation may reduce manual effort by automating repetitive operations. Encourage delegation to spread tasks effectively and avoid burnout. Finally, promoting efficiency keeping meetings brief and eliminating extraneous tasks can considerably minimise time constraints.

Policy Recommendations and Structural Reforms

To effectively address faculty workload issues, schools should:

1. Incorporate digital teaching and evaluation practices into formal workload models.
2. Establish an annual innovation fund for curriculum revision.
3. Advocate for the inclusion of digital assessment work in promotion, tenure, and evaluation processes.
4. Facilitate collaborative teaching strategies that disperse effort across faculty teams.

Affective Dimensions of Time Pressure

Time limitations can have an emotional and cognitive impact:

1. Stress from frequent system alerts and student contacts.
2. Navigating several digital sites might lead to decision fatigue.
3. Task overload and fractured attention lead to less creativity.

Addressing faculty workload is more than just an issue of hours; it is also about well-being and professional dignity.

11.1.15 Policy Frameworks for Digital Integration

Effective policy frameworks are essential to the successful integration of digital technology in education in the ever changing field of digital pedagogy. These frameworks help organisations match digital technologies to inclusive practices, ethical standards, and educational goals. Policy frameworks are strategic tools that impact curriculum design, pedagogical practices, infrastructure development, capacity building, and student participation; they are not only administrative blueprints. A digital integration policy framework describes an institution's vision, tactics, laws, and standards for integrating digital tools and platforms within its educational environment. Infrastructure investment, digital literacy training, device and internet access, data privacy, intellectual property rights, and educator professional development are common policy topics [25, 26].

Rationale for Policy Frameworks in Digital Pedagogy
Several considerations highlight the need of clear policy frameworks.

1. Strategic Alignment: Policies guarantee that digital initiatives are consistent with the institution's overall strategic objectives and purpose.
2. Consistency and Quality Assurance: Frameworks encourage consistent implementation across departments and give criteria for assessing performance.
3. Equity and Access: Policies may require the distribution of digital resources to underserved communities, therefore bridging the digital gap.
4. Ethics and Privacy: Defined procedures safeguard students' data and intellectual rights in an increasingly surveillance-prone digital world.
5. Capacity Building Policies create criteria for digital proficiency among instructors and students, establishing a digitally literate academic environment.

Key Components of an Effective Digital Integration Policy

A well-rounded digital integration strategy includes many critical components to guarantee smooth adoption and successful use of digital technology. It starts with strategic alignment to ensure digital activities meet organisational or national objectives. Infrastructure and accessibility are essential to building inclusive digital frameworks, including internet connection and cybersecurity. Data management and governance ensure that data gathering and use are secure and ethical. Digital literacy and workforce development teach people how to use technology, while interoperability allows platforms to communicate. Strong regulatory and ethical considerations reduce hazards and ensure responsible digital use. User-centric design improves accessibility, inclusiveness, and engagement, and ongoing monitoring and improvement discover difficulties and refine tactics to innovate. These elements comprise a holistic digital transformation strategy that promotes efficiency, security, and sustainability [25] (see Table 11.20).

Developing a Policy Framework: Steps and Best Practices

The development of a successful digital integration policy should be collaborative, iterative, and context-sensitive. The steps below provide a roadmap [27, 28]:

1. Needs Assessment: Conduct audits to identify technical gaps, stakeholder digital preparedness, and infrastructure requirements.

Table 11.20 Core components of a digital integration policy framework

Component	Description
Vision and strategic goals	Outlines the long-term goals of digital integration in teaching, learning, and administration.
Infrastructure and access	Details plans for digital infrastructure, device accessibility, and internet connectivity.
Digital literacy and training	Provides guidelines for regular professional development and student training programs.
Curriculum integration	Describes how digital tools will be embedded in course design, assessment, and pedagogy.
Equity and inclusion	Ensures policies address the needs of disadvantaged or underrepresented student populations.
Data privacy and security	Sets standards for ethical use, data protection, and cybersecurity measures.
Monitoring and evaluation	Establishes KPIs and evaluation mechanisms to assess policy implementation and impact.
Stakeholder engagement	Involves faculty, students, administrators, and IT personnel in policy development and review.
Budget and resource allocation	Provides a financial blueprint for funding digital initiatives and maintaining infrastructure.
Legal and ethical compliance	Aligns institutional practices with national and international education and technology laws.

2. Stakeholder Consultation: Involve a wide range of stakeholders, including teachers, students, IT personnel, and policy experts, to obtain perspectives and create buy-in.
3. Draft the Policy: Create a policy that represents the institution's digital vision while being realistic and responsive to future developments.
4. Review and validation: Before implementation, conduct peer reviews and pilot stages to assess viability and gather input.
5. Implementation Planning: Develop an executable timetable that outlines roles, responsibilities, and resource allocations.
6. Training and Support: Implement comprehensive training modules to help staff and students connect with digital platforms.
7. Monitoring and input Loops: Implement frequent policy evaluations and revisions in response to changing digital trends, user input, and institutional objectives.

Policy Frameworks and Regulatory Contexts

National and international regulatory frameworks should inform educational policy on digital integration. Policy in Nigeria, for example, has to be in sync with the country's National Policy on Information and Communication Technology in Education, which advocates for expanding access, improving fairness, and enhancing the quality of education via the use of ICT. Institutions throughout the world may look to the ICT Competency Framework for Teachers put forth by UNESCO for guidance on how to educate students digitally and integrate technology into their lessons. When it comes to protecting students' personal information, regulations like the EU's General Data Protection Regulation (GDPR) and Nigeria's Data Protection Regulation (NDPR) are essential.

Case Example: University of Ibadan Digital Integration Strategy

One prominent example of integration driven by strategic strategy is the digital transformation path of the University of Ibadan. All departments were required to use blended learning models under the university's Digital Learning Policy, which was developed in response to the COVID-19 epidemic. Instructors might participate in LMS platform training as part of the strategy, and low-income students could take advantage of student access programs. Additionally, telecom companies could develop agreements to provide subsidised data plans. Academic continuity and digital engagement were both enhanced within two years at the institution. This achievement was made possible by the all-encompassing framework that dealt with the technical and human aspects of digital transformation.

Challenges in Policy Development and Implementation

Despite its promise, developing and implementing digital integration frameworks has various challenges:

1. Limited financing may limit the implementation of infrastructure and training initiatives.
2. Resistance to Change: Faculty or administrative reluctance to adopt new technology might stifle development.
3. Infrastructural disparities: Unequal access to the internet and digital devices may undermine policy efficacy in rural or underfunded institutions.
4. Policy incoherence: Fragmented or contradictory policies across departments may cause confusion and inefficiency.
5. Sustainability: Policies that lack long-term vision or fail to account for upkeep often fail after gaining early traction.

11.1.16 The Future of Digital Policy in Education

As technology advances, so do policy frameworks. Institutions must adopt adaptive governance models that enable real-time policy modifications, the incorporation of new technologies like artificial intelligence and virtual reality, and responsiveness to student-centred feedback systems. Policies must increasingly handle ethical quandaries, such as algorithmic bias in learning analytics or spying in remote proctoring, while encouraging innovation and experimentation [29].

Evaluating the Sustainability of Initiatives

Sustainability in digital pedagogy means long-term viability, relevance, and effectiveness. When investing in educational technologies, platforms, and training, institutions must evaluate their immediate effect and long-term viability, evolution, and value. Sustainability evaluation assures resource use, stakeholder demands are addressed, and the digital transformation meets institutional, environmental, and social objectives. Sustainability in education technology extends beyond environmental considerations [30, 31]. It includes:

1. Operational sustainability is the capacity to maintain and manage digital infrastructure, tools, and platforms without frequent external assistance.
2. Financial sustainability refers to the long-term affordability of digital systems, which includes licensing, upgrades, training, and hardware maintenance.
3. Social sustainability entails ensuring that digital efforts remain inclusive, accessible, and helpful to all stakeholders, especially marginalised communities.
4. Pedagogical sustainability is the ability of digital efforts to adapt to changing curricular demands, technology changes, and learner requirements.

Dimensions of Sustainability Evaluation

It is common practice to use a multi-dimensional framework to summarise the evaluation of sustainability, which entails evaluating several interrelated aspects (see Table 11.21).

Indicators and Metrics for Sustainability

Sustainability must be assessed using both quantitative and qualitative indicators. These measurements differ depending on the nature of the digital effort; nevertheless, popular indications include:

Table 11.21 Dimensions of sustainability in digital education initiatives

Dimension	Evaluation focus
Financial	Long-term funding, cost-benefit analysis, budget allocation, and funding diversification
Technical	Reliability of infrastructure, scalability, compatibility, ease of maintenance
Human capital	Staff digital competence, ongoing professional development, and leadership commitment
Institutional support	Policy backing, strategic alignment, and administrative engagement
Social and ethical	Equity, accessibility, user inclusion, data privacy, and cultural responsiveness
Environmental	Energy use, e-waste management, and sustainable procurement practices
Educational value	Learning outcomes, student engagement, teacher satisfaction, and adaptability

1. User Retention Rates: Students and faculty's continued usage of a platform or product throughout numerous academic sessions.
2. System Uptime and Reliability: The number of outages or disturbances that impact teaching and learning.
3. Cost-Effectiveness Ratios: A comparison between financial input and educational output (for example, cost per student engagement hour).
4. Digital Literacy Growth: Increased user proficiency as a consequence of training and use.
5. Feedback & Satisfaction Surveys: Students and staff provide regular evaluations of usefulness, relevance, and effect.
6. Policy Integration: The extent to which the initiative is included in institutional plans and evaluated policies.

Qualitative data, such as user narratives, case studies, and ethnographic observations, may supplement quantitative metrics by collecting contextual insights, impediments, and success stories.

Evaluating Long-Term Impact

Sustainability assessment is not a one-time occurrence. It must be incorporated throughout the initiative's lifespan. A longitudinal strategy, which involves recording data and feedback across semesters or academic years, may show use trends, efficacy, and obsolescence risk. The key methods include:

1. Formative evaluation is carried out throughout the development and early implementation stages to form the project.
2. Summative evaluations are conducted after a financing or academic cycle to review overall results.
3. Developmental evaluation is a flexible methodology that enables real-time adaptation and development, making it ideal for pilot or experimental digital initiatives.

Stakeholder Roles in Evaluating Sustainability

Sustainability assessment should be a collaborative and inclusive process. The key stakeholders include:

1. Educators analyse how tools help or impede learning goals.
2. Students provide input on accessibility, engagement, and results.
3. ICT departments examine system performance, maintenance requirements, and technological scalability.
4. Administrators are responsible for tracking return on investment, ensuring alignment with strategic goals, and providing policy assistance.
5. External partners, such as non-governmental organisations (NGOs) or educational technology companies, may provide new resources and viewpoints.

Institutions must build feedback channels and involve stakeholders on a regular basis in order to adjust digital efforts to real-world use.

11.1.17 Barriers to Sustainability

Despite the greatest efforts, various impediments might impede the viability of digital projects.

1. Overreliance on Donor financing: If internal capability is not established, projects may fail when external financing is no longer available.
2. Lack of Monitoring and Evaluation (M&E): Without data, it is difficult to detect shortcomings or opportunities for improvement.
3. Technological obsolescence occurs when tools and platforms are not updated or upgraded on a regular basis.
4. Digital Inequality: Longstanding differences in access may restrict the scope and effectiveness of programs.
5. Institutional silos: Fragmented methods across departments may result in duplication and inefficiency.

Overcoming these obstacles requires a proactive, systems-thinking approach that prioritises sustainability as a key design element rather than an afterthought.

Case Study: Sustaining Digital Learning in a Nigerian Public University

In response to COVID-19, a public institution in Kaduna State has established an e-learning platform to augment in-person instruction. The platform, which was first sponsored by donors, grew rapidly. However, after two years, sustainability concerns arose.

1. Financial: Subscription fees to the LMS provider become expensive after the donor period.
2. Technical issues, such as unstable internet and obsolete technology, hampered the user experience.
3. Human Resources: Only a small number of instructors have undergone digital pedagogy training.
4. Social Equity: Many rural pupils lacked equipment or internet access.

In response, the institution created a Sustainability Action Plan that included local hosting of the LMS to save costs, blended training programs for staff, and device lending schemes for low-income students. This change increased system availability, expanded access, and strengthened institutional ownership of the program.

Barriers to Sustainability

Despite the greatest efforts, various impediments might impede the viability of digital projects.

1. Overreliance on Donor financing: If internal capability is not established, projects may fail when external financing is no longer available.
2. Lack of Monitoring and Evaluation (M&E): Without data, it is difficult to detect shortcomings or opportunities for improvement.

3. Technological obsolescence occurs when tools and platforms are not updated or upgraded on a regular basis.
4. Digital Inequality: Longstanding differences in access may restrict the scope and effectiveness of programs.
5. Institutional silos: Fragmented methods across departments may result in duplication and inefficiency.

Overcoming these obstacles requires a proactive, systems-thinking approach that prioritises sustainability as a key design element rather than an afterthought.

Strategies to Enhance Sustainability
To increase the longevity of digital projects in education, institutions should:

1. Developing In-House Expertise: Train employees to administer and maintain platforms internally.
2. Budget for continuity: Include sustainability funds in your initial budgeting process.
3. Scale gradually: Start with test programs before spreading throughout the institution.
4. Encourage ownership: Involve people in design and decision-making to boost commitment and relevance.
5. Update regularly: Maintain platform and content consistency with current technology and educational requirements.

11.1.18 Challenges Encountered and Lessons Learned

Infrastructure constraints, change opposition, and cybersecurity issues make digital integration policy implementation difficult. Lack of internet access or obsolete equipment might slow growth, necessitating major expenditure in updates. Digital literacy training and change management methods are needed because workers and stakeholders may resist change owing to unfamiliarity with digital technologies or fear of job loss. Data protection and regulatory compliance are needed as systems grow increasingly linked, increasing cybersecurity threats. Integrating varied technologies might cause interoperability challenges, requiring standardised protocols for easy communication. Despite these hurdles, proactive stakeholder involvement, continual skill development, and adaptable policy for developing technology were essential lessons. Long-term sustainability and digital transformation depend on user-centric design, ethical governance, and innovation [32, 33] (see Table 11.22).

Framework for Building Adaptive Capacity

Drawing on these tales, a framework for fostering resilience in educational institutions could include: scenario planning for future disruptions; use of open-source, modular systems that can scale or shift; digital literacy ecosystem; peer groups, mentoring, and cross-institution collaboration; feedback loops.

Table 11.22 Challenges encountered and lessons learned

Challenge	Observed impact	Adaptation strategy
Poor internet connectivity	Interrupted live sessions, delayed submissions	Shift to asynchronous learning and offline apps
Inadequate teacher training	Low confidence and ineffective use of platforms	Cascade training model and peer coaching
Device unavailability for students	Learning interruptions, disengagement	Device loan programs and shared community access
Mental health issues	Burnout among staff and students	Counseling services and wellness campaigns
Resistance to change	Reluctance to adopt new methods	Incentives, recognition, and incremental change

Sustaining a Culture of Resilience

Resilience should not be a reactionary attribute, but rather a purposeful institutional approach. After the crisis, many institutions are incorporating digital readiness into strategic plans, developing emergency response protocols, and establishing blended learning as a standard rather than a contingency. This development represents a larger pedagogical shift: education is no more a static activity rooted on physical settings, but rather a dynamic process including networks, technology, and human adaptability.

References

1. Sikandar, A., Muhammad, A., Jazeb, N., Ahmad, F.: Multiagent Based System for Secondary Education Using Machine Learning. pp. 1310–1315 (2018). 10.1109/ICMLA.2018.00212
2. Mishra, P., Shrivastava, S.: Cloud AIoT based smart wheelchair using module for social distancing, temperature monitoring, and oximeter module. J. Inf. Technol. **7**(5), 29–35 (2021) http://ijitjournal.org/volume-7/issue-5/IJIT-V7I5P6.pdf
3. Povoda, L., Burget, R.: Genetic optimization of big data sentiment. Analysis. pp., 141–144 (2017)
4. Bozkus Kahyaoglu, S., Aksoy, T.: Survey on Blockchain based accounting and finance algorithms using bibliometric approach. 21st century approaches to Manag. Account. Res., 1–13 (2021). https://doi.org/10.5772/intechopen.98207
5. Omheni, N., Bouabidi, I., Gharsallah, A., Zarai, F., Obaidat, M.S.: Smart mobility management in 5G heterogeneous networks. IET Networks. **7**(3), 119–128 (2018). https://doi.org/10.1049/iet-net.2017.0208
6. Chew, E.Y., et al.: Standardization and clinical applications of retinal imaging biomarkers for cardiovascular disease: a roadmap from an NHLBI workshop. Nat. Rev. Cardiol. (2024). https://doi.org/10.1038/s41569-024-01060-8
7. Farhan, B.I., Jasim, A.D.: Performance analysis of intrusion detection for deep learning model based on CSE-CIC-IDS2018 dataset. Indones. J. Electr. Eng. Comput. Sci. **26**(2), 1165–1172 (2022). https://doi.org/10.11591/ijeecs.v26.i2.pp1165-1172
8. Hosamo, H.H., Imran, A., Cardenas-Cartagena, J., Svennevig, P.R., Svidt, K., Nielsen, H.K.: A review of the digital twin technology in the AEC-FM industry. Adv. Civ. Eng. **2022**, 1–17 (2022). https://doi.org/10.1155/2022/2185170

9. Olika, C.D.: Risk management practices among selected state governments in southwestern Nigeria. Eur. J. Sci. Innov. Technol. **1**(2), 24–31 (2024) https://www.researchgate.net/profile/Hendra-Manurung/publication/356633046_From_Connectivity_to_Digital_Improving_Employee_Readiness_toward_Organizational_Change_in_Digital_Transformation/links/61a5b00671a23a0084c9b061/From-Connectivity-to-Digital-Improvi
10. Agboola, O., Aremu, M.O., Eze, B.U., Wahab, N.O.: Effect of training and career development on Bank employees' performance: evidence from selected banks in Nigeria. EMAJ Emerg. Mark. J. **10**(2), 56–62 (2021). https://doi.org/10.5195/emaj.2020.204
11. Oyeronke, A., Fagbohun, M., Adebayo, O.: An Assessment of Computer and ICT Skills among Secondary School Teachers in Ota Ogun State. (2013). http://digitalcommons.unl.edu/libphilprac%5Cnhttp://digitalcommons.unl.edu/libphilprac/846%5Cnhttp://digitalcommons.unl.edu/libphilprac/
12. Osman, J., Khuzzan, S.M.S., Razaksapian, A.: Building information modelling: proposed adoption model for quantity surveying firms. Proc. Int. Conf. Inf. Technol. Soc. **8-9**, 151–165 (2015)
13. Brinken, J., Trojahn, S., Behrendt, F.: Sufficiency, consistency, and efficiency as a base for systemizing sustainability measures in food supply chains. Sustainability. **14**(11), 1–18 (2022). https://doi.org/10.3390/su14116742
14. Reja, V.K., Varghese, K.: Impact of 5G technology on IoT applications in construction project management. Proc. 36th Int. Symp. Autom. Robot. Constr. ISARC 2019. **36**(5), 209–217 (2019). https://doi.org/10.22260/isarc2019/0029
15. Hadullo, K., Oboko, R., Omwenga, E.: A model for evaluating E-learning systems quality in higher education in developing countries. Int. J. Educ. Dev. Using Inf. Commun. Technol. **13**(2), 185 (2017)
16. Arufe-Giráldez, V., Sanmiguel-Rodríguez, A., Ramos-Álvarez, O., Navarro-Patón, R.: News of the pedagogical models in physical education—a quick review. Int. J. Environ. Res. Public Health. **20**(3) (2023). https://doi.org/10.3390/ijerph20032586
17. Ibrahim, F.S., Esa, M., Rahman, R.A.: The adoption of iot in the Malaysian construction industry: Towards construction 4.0. Int. J. Sustain. Constr. Eng. Technol. **12**(1), 56–67 (2021). https://doi.org/10.30880/ijscet.2021.12.01.006
18. Tyler, C.: Power, alienation and performativity in capitalist societies. Eur. J. Soc. Theory. **14**(2), 161–179 (2011). https://doi.org/10.1177/1368431011403460
19. Martins, R., Oliveira, T., Thomas, M.A.: An empirical analysis to assess the determinants of SaaS diffusion in firms. Comput. Human Behav. **62**, 19–33 (2016). https://doi.org/10.1016/j.chb.2016.03.049
20. Anderson, M.D., Settee, P.: Knowledge and education for peoples' sovereignty. Globalizations. **17**(7), 1300–1309 (2020). https://doi.org/10.1080/14747731.2020.1783785
21. Purcell, C., et al.: A peer-led, school-based social network intervention for young people in the UK, promoting sexual health via social media and conversations with friends: intervention development and optimisation of STASH. BMC Public Health. **23**(1), 1–12 (2023). https://doi.org/10.1186/s12889-023-15541-x
22. Orunbon, N.O., Ibikunle, G.A., Badmus, A.A.: Principals' Digital Transformational Leadership, Teachers' Organisational Commitment and Job Satisfaction During Covid-19 in Lagos State Education District V, Nigeria. Int. J. Humanit. Technol. Civiliz. **8**(1), 68–73 (2023). https://doi.org/10.15282/ijhtc.v8i1.9424
23. Ghavifekr, S., Razak, A., Ghani, M., Ran, N., Meixi, Y., Tengyue, Z.: ICT integration in education: incorporation for Teaching & Learning Improvement. Malaysian Online J. Educ. Technol. **2**(2), 24–45 (2014)
24. Zou, D., Huang, X., Kohnke, L., Chen, X., Cheng, G., Xie, H.: A bibliometric analysis of the trends and research topics of empirical research on TPACK. Educ. Inf. Technol. **27**(8), 10585–10609 (2022). https://doi.org/10.1007/s10639-022-10991-z
25. Casini, M.: Extended reality for smart building operation and maintenance: a review. Energies. **15**(10), 3785 (2022). https://doi.org/10.3390/en15103785

26. Villa, V., Naticchia, B., Bruno, G., Aliev, K., Piantanida, P., Antonelli, D.: IoT open-source architecture for the maintenance of building facilities. Appl. Sci. **11**(12), 1–23 (2021). https://doi.org/10.3390/app11125374
27. Yaqoob, I., Salah, K., Jayaraman, R., Omar, M.: Metaverse applications in smart cities: enabling technologies, opportunities, challenges, and future directions. Internet of Things (Netherlands). **23**, 100884 (2023). https://doi.org/10.1016/j.iot.2023.100884
28. Shahinmoghadam, M., Motamedi, A.: Review of BIM-centred IoT deployment: state of the art, opportunities, and challenges. Proc. 36th Int. Symp. Autom. Robot. Constr. ISARC 2019. 1268–1275 (2019). 10.22260/isarc2019/0170
29. Ekechi CC, Chukwurah EG, Oyeniyi LD, Okeke CD: A Review of Small Business Growth Strategies in African Economies. Int. J. Adv. Econ. **6**(4), 76–94 (2024). 10.51594/ijae.v6i4.1071
30. Ghosh, A., Edwards, D.J., Hosseini, M.R.: Patterns and trends in internet of things (IoT) research: future applications in the construction industry. Eng. Constr. Archit. Manag. **28**(2), 457–481 (2021). https://doi.org/10.1108/ECAM-04-2020-0271
31. Khan, S.A.R., Piprani, A.Z., Yu, Z.: Digital technology and circular economy practices: future of supply chains. Oper. Manag. Res. **15**(3–4), 676–688 (2022). https://doi.org/10.1007/s12063-021-00247-3
32. Munawar, H.S., Ullah, F., Qayyum, S.: Big data in construction. Current Applications and Future Opportunities. (2022)
33. Fernandez-Vidal, J., Gonzalez, R., Gasco, J., Llopis, J.: Digitalization and corporate transformation: the case of European oil & gas firms. Technol. Forecast. Soc. Change. **174**, 1–31 (2022). https://doi.org/10.1016/j.techfore.2021.121293

Chapter 12
Future Directions and Global Perspectives

12.1 AI and Machine Learning in Education

The world of education is changing as a result of artificial intelligence (AI) and machine learning (ML), which provide new opportunities for data-driven decision-making, intelligent tutoring, administrative automation, and personalised learning. These technologies provide educators and politicians in the Global South a previously unheard-of chance to solve structural issues, such as curriculum delivery and assessment limitations, teacher shortages, and unequal access to education. But in order to guarantee their responsible and inclusive usage, its implementation also brings up significant pedagogical, ethical, and infrastructure issues that need to be resolved. Computer programs that simulate human intellect to carry out activities like thinking, problem-solving, language comprehension, and decision-making are referred to as AI. A form of artificial intelligence called ML uses data to teach computers to make judgments or predictions without needing to be explicitly programmed for each job. Tools that automate grading, provide real-time feedback, identify learning challenges, and customise instructional materials may all be powered by AI and ML in educational settings [1] (see Fig. 12.1).

Key Applications of AI and ML in Education

Improvements in learning outcomes, simplification of administrative tasks, and individualisation of lessons are just a few ways in which AI and ML are changing the face of education (see Table. 12.1).

Benefits of AI Integration in Education

1. AI can personalise educational experiences, leading to increased engagement and better learning results.
    (a) Efficient administrative tasks.

F. A. Alaba and I. G. Ajibola, *Digital Pedagogy: Bridging the Gap Between Software and the Humanities*, Information Systems Engineering and Management 85,
https://doi.org/10.1007/978-3-032-20335-9_12

Fig. 12.1 The role of AI and ML in Education [2]

Table 12.1 Key Applications of AI and ML in education

Application	Description	Example tools/Platforms
Adaptive learning systems	Customize content delivery based on learner pace, preferences, and performance	DreamBox, smart sparrow, Knewton
Intelligent tutoring systems	Simulate one-on-one tutoring by providing hints and feedback	Carnegie learning, squirrel AI
Automated assessment	Grade essays, quizzes, and provide instant feedback	Gradescope, Turnitin AI
Early warning systems	Predict student dropout or disengagement using behavioral data	Microsoft azure education insights
Learning analytics	Analyze learning behaviors for instructional planning	Google classroom analytics, canvas analytics
AI-powered Chatbots	Offer 24/7 student support and answer routine academic queries	Pounce, AdmitHub
Voice and language processing	Enhance accessibility via speech-to-text and translation tools	Google text-to-speech, Duolingo AI tutors

   (b) Automating duties like attendance monitoring, timetabling, and grading minimises educators' labour.

2. Real-time Feedback

   (a) Students get quick feedback on their performance, allowing for speedier correction of misconceptions.

Table 12.2 Challenges and ethical considerations

Challenge	Implication
Data privacy and security	Students' personal data must be protected from misuse or breaches
Bias in algorithms	AI trained on biased data can reinforce inequalities in assessment and feedback
Over-reliance on technology	Risks diminishing human-centered pedagogy and emotional intelligence in teaching
Infrastructure and connectivity gaps	Rural and low-income regions may lack necessary digital infrastructure
Teacher capacity	Teachers need training to understand, use, and critically evaluate AI tools

3. Enhanced Inclusion AI solutions for translating languages or converting voice to text may help students with impairments or linguistic obstacles.
4. Scalability

   AI solutions can assist a large number of students at once, which is very valuable in under-resourced educational settings.

Challenges and Ethical Considerations

While AI has the potential to transform education by allowing for personalised learning, automating administrative duties, and giving data-driven insights, its deployment is not without challenges (see Table. 12.2).

12.1.1 AI and the Role of the Teacher

Instead of replacing instructors, AI may enhance their talents by: identifying learning gaps using analytics, recommending diverse educational tactics, automating regular duties to make time for mentoring and creativity, and AI-curated materials support continuing professional growth [3].

AI in the Global South: Opportunities and Constraints

It is crucial to tailor the deployment of AI in the Global South to local contexts. Lightweight or open-source alternatives are becoming more popular as large-scale AI systems become more resource-intensive. Example: Using conversational AI in local languages, the "ConveGenius" EdTech platform in India reaches underprivileged students over WhatsApp with instructional information, ensuring that they do not need high-tech gadgets or internet connections.

Important Roles for AI in Developing Nations:

1. prioritising mobile users and optimising for low-bandwidth environments
2. Incentives and policy backing from the government
3. Collaborations with educational technology companies, both domestic and abroad
4. Datasets to train AI that are culturally relevant

Future Trends: What Lies Ahead?
New developments that will impact the future of AI in the classroom include:

1. Emotional intelligence refers to systems that are able to identify stress levels and learner emotions in order to provide more tailored answers.
2. Generative AI: ChatGPT and similar tools facilitate content generation, ideation, and essay writing; this trend necessitates fresh approaches to digital literacy.
3. Teaching students to think critically about AI and its effects on society is an important part of AI ethics education.
4. More open credentialing and academic records with blockchain and AI.

Policy Implications and Governance

To effectively employ AI in the classroom, governments and other organisations must put in place strong frameworks, such as:

1. Policies on the Moral use of AI: Simple rules for data use, permission, and algorithmic openness.
2. Initiatives for Strengthening Capacity: Providing teachers with the knowledge and abilities to harness and evaluate AI technologies.
3. Design Requirements for All Users: Making sure that low-income or disabled students can easily use the resources provided.

12.1.2 Immersive Technologies: AR/VR/XR

Immersive technologies, AR, VR, and XR are generating dynamic, engaging, and multimodal learning environments that are changing education. Learners may study abstract ideas, mimic real-world settings, and collaborate experientially using these tools. Immersive technology may replace conventional pedagogies in situations with limited physical resources, laboratory access, or field excursions [4] (see Table. 12.3).

Applications of Immersive Technologies in Education

Immersive technologies, such as VR, AR, and MR, are transforming education by enabling engaging, interactive, and immersive learning environments (see Table. 12.4).

Pedagogical Advantages

Table 12.3 Defining AR, VR, and XR in education

Technology	Definition	Educational use
Augmented reality (AR)	Overlays digital content (text, images, audio) onto the real world via devices like smartphones or AR glasses.	Interactive anatomy overlays, historical site visualizations
Virtual reality (VR)	Creates a fully immersive digital environment using headsets and sensors. Users are cut off from the physical world.	Virtual field trips, science labs, and language immersion
Extended reality (XR)	An umbrella term covering all immersive technologies, including AR, VR, and mixed reality (MR).	Combines real and virtual interactions across educational domains

Table 12.4 Applications of immersive technologies in education

Field/Subject	Immersive Use Case
Science and engineering	Simulated labs, molecule manipulation, physics experiments
Medicine and health	Surgical simulations, anatomy exploration, emergency response drills
History and geography	3D reconstructions of historical events and ancient cities
Vocational training	Safe simulations of machinery use, technical procedures
Special education	Social skill training, behaviour modelling, sensory regulation tools
Language learning	Virtual cultural immersion and conversational practice

1. Experiential Learning: Students learn by doing, which leads to increased retention rates and a better comprehension.
2. Access to Unavailable Experiences: Virtual field tours to museums, outer space, and disaster zones provide access to educational possibilities.
3. Increased Engagement: Immersive settings attract students' attention and limit distractions, particularly for digital natives.
4. Safe Learning Environments: Dangerous or high-risk training (such as electrical repairs, firefighting, or surgery) may be done safely.
5. Multisensory learning: integrates sound, vision, and motion to support many learning modalities, including visual, aural, and kinesthetic.

12.1.3 Barriers to Adoption

Despite their benefits, immersive technologies provide substantial practical and institutional problems, particularly when creating educational institutions. High equipment costs make VR headsets and AR-ready devices prohibitively costly, restricting availability. Infrastructure needs, such as high-speed internet, consistent electricity, and suitable devices, might further impede wider adoption. Additionally, content

Table 12.5 Comparison of immersive tools by accessibility and educational impact

Tool Type	Cost	Ease of Use	Infrastructure Need	Educational Impact
Smartphone AR apps	Low	High	Low to medium	Moderate to high
VR headsets (standalone)	Medium	Medium	Medium	High
Full XR labs	High	Low	High	Very high

localisation remains an issue due to the scarcity of region-specific, curriculum-aligned immersive material. Teacher training is also important, since many educators may lack the expertise or confidence to properly incorporate these technologies into their teaching approaches. Finally, health and safety problems, such as VR-induced motion sickness and eye strain from extended usage, are crucial considerations for student well-being. To address these obstacles, strategic investment, regulatory support, and continuing training are required to guarantee that immersive technologies are effectively and sustainably incorporated into education [5].

Designing for Inclusion and Relevance

To ensure that these technologies promote equitable education, design principles must include:

1. Low-bandwidth Compatibility: Use lightweight AR applications for cellphones.
2. Cultural sensitivity refers to visual and narrative information that represents local customs, history, and the environment.
3. Open-Source Tools: Platforms such as Mozilla Hubs and Unity enable instructors to create unique XR material.
4. Partnership Models: Collaborations between governments, EdTech developers, and non-governmental organisations (NGOs) may cut costs and increase scalability (see Table. 12.5).

12.1.4 Blockchain and Credentialing Systems

A safe and transparent way to authenticate academic accomplishments, abilities, and professional credentials is essential in a digital and globalised society. Blockchain, initially designed for cryptocurrency, is being used in education to establish tamper-proof, verifiable, and decentralised credentialing systems. These advances might change how learners are evaluated, accredited, and recognised, particularly across borders and institutions. Blockchain is a decentralised digital ledger that keeps transactions safe, immutable, and visible. Each record, or "block," is timestamped and connected to previous blocks, resulting in a chain that cannot be changed retroactively without network consent [6, 7].

Key Features of Education:

1. Immutability: Once granted, credentials cannot be changed or falsified.
2. Decentralisation: Records are not held in a single place, which reduces the likelihood of data loss or corruption.
3. Transparency and trust: Institutions, employers, and students may independently check credentials.
4. Interoperability: Credentials are recognised and accepted across borders and platforms.

Traditional Credentialing Versus Blockchain-Based Systems

Blockchain-based solutions and traditional credentialing evaluate and manage academic and professional certifications differently. Traditional credentialing is issued and validated by universities, accrediting agencies, and companies. Paper certificates, digital records in institutional databases, and third-party verification procedures may be time-consuming, fraudulent, and difficult to transfer across companies or regions. Blockchain-based credentialing stores and verifies credentials securely using decentralized, tamper-proof technology. By letting people communicate their verified credentials internationally without middlemen, blockchain technologies eliminate fraud, enhance accessibility, and validate instantly. With encryption and distributed ledgers, blockchain gives credential holders ownership of their data and improves transparency and control [8, 9] (see Table. 12.6).

Use Cases in Education

1. Digital Diplomas and Certificates: Universities and training institutes may use the blockchain to issue certificates. Learners are given a digital wallet in which they may save their accomplishments and access them at any time.
2. Micro-credentials and Badges: Short-term courses or skills training programs may give blockchain-verified badges, encouraging lifelong learning and skill development.
3. Cross-border Recognition: Blockchain provides the seamless verification of foreign degrees and certificates, which is particularly important for migration or employment in other countries.

Table 12.6 Traditional credentialing versus blockchain-based systems

Aspect	Traditional Credentialing	Blockchain-Based Credentialing
Storage	Centralised (institutional databases)	Decentralised ledger
Verification	Manual or institutional request	Instantaneous, peer-to-peer
Security	Vulnerable to fraud or loss	Cryptographically secure and immutable
Accessibility	Limited by geography or bureaucracy	Global, 24/7 access
Cost	High (reprinting, mailing, verification services)	Reduced or eliminated with digital access

4. Academic Record Portability: Students moving schools, particularly refugees or displaced learners, may keep their confirmed academic records regardless of institutional affiliation.

12.1.5 Blockchain Credentialing in the Global South: Opportunities

Several promising opportunities for blockchain-based credentialing could have far-reaching positive effects in many fields. By reducing the prevalence of fake certificates and unaccredited institutions, blockchain technology offers a significant advantage in combating credential fraud through secure, tamper-proof verification. Furthermore, it improves employability by facilitating rapid qualification verification for businesses, which streamlines recruiting procedures and guarantees correct talent representation. Another important advantage is that blockchain can help people in the informal sector advance in their careers by recognising non-traditional and skills-based learning, which is not possible with traditional educational systems. In addition, migrant workers and displaced people can utilise verified digital records to demonstrate their educational background internationally without having to rely on physical documents, which is a huge boon to their mobility. Last but not least, academic and professional record-keeping becomes more frictionless and sustainable with blockchain-driven credentialing, which increases cost-efficiency for institutions by lowering paperwork, storage needs, and the complexity of credential verification [10].

Blockchain Credentialing in the Global South: Opportunities

Several areas might be revolutionised by blockchain-based credentialing. It is vital in the fight against credential fraud because it reduces the incidence of fraudulent certificates and unaccredited institutions and ensures the security and verifiability of academic and professional credentials. Blockchain technology streamlines the hiring process by enhancing employability and enabling companies to instantaneously check a candidate's credentials. It also gives those working in the informal economy more agency by validating non-traditional and skills-based forms of education, which opens doors to jobs that would otherwise be out of reach. Additionally, this technology facilitates mobility by removing the need for displaced people and migratory workers to depend on tangible documents to verify their educational qualifications. These documents are sometimes misplaced or hard to confirm. Finally, blockchain improves institutions' cost-efficiency by doing away with unnecessary paperwork, cutting storage expenses, and streamlining the process of verifying credentials. Its potential to drive innovation, accessibility, and confidence in credentialing systems, as well as revolutionise education and workforce management, is highlighted by these advantages [11].

Challenges and Barriers of Blockchain Credentialing

1. Technological Infrastructure: Implementing blockchain systems requires a consistent internet connection, data security standards, and digital literacy factors that are not universally present in all countries.
2. Policy and regulatory frameworks: Legal recognition of digital credentials is continuously developing. Many national systems lack the necessary structures to implement blockchain-based credentials.
3. Interoperability between systems: Without established platforms and standards, blockchain solutions may stay fragmented and underappreciated.
4. Digital divide: Marginalized groups may be excluded owing to a lack of devices, connection, or comprehension of technology.
5. Cost of Implementation: Smaller institutions may face challenges due to initial setup expenses and the requirement for technical competence.

12.1.6 Designing Inclusive Blockchain Credentialing

To guarantee blockchain gains are evenly dispersed, particularly in the Global South, the following design principles are essential [12]:

1. Open Standards and Protocols: Use interoperable systems like Blockcerts and W3C Verifiable Credentials.
2. Mobile Compatibility: Ensure credentials may be accessed via low-cost mobile devices.
3. Decentralised Identity (DID): Allow people to manage their educational data and determine who to share it with.
4. Localisation: Support local languages and culturally suitable documentation.
5. Public-Private Partnerships: Collaborate with governments, NGOs, technology companies, and educational institutions to create solutions.

Role of Institutions and Governments on Blockchain Credentialing

Governments and institutions must take a proactive approach by [13]

1. Recognise digital credentials as legally valid.
2. Creating national repositories or blockchain consortia.
3. Staff and teachers will be trained in credential issuing and administration.
4. Adoption incentives may be provided via grants or policy frameworks.

Future Directions of Blockchain Credentialing

1. Smart Contracts for Learning Paths automate course progression and credential issuance based on performance.
2. Blockchain and AI: Intelligent analytics on learning outcomes and skill gaps.

Table 12.7 Comparative overview of credentialing approaches

Criteria	Paper-based	Digital (non-blockchain)	Blockchain-based
Tamper-proof	Low	Medium	High
Verification speed	Slow	Moderate	Instant
Accessibility	Limited	Moderate	High
Cost efficiency	Low	Medium	High (after setup)
User control of data	None	Low	High

3. Decentralised Education Marketplaces: Platforms where verifiable credentials enable worldwide job matching or academic mobility.
4. Sustainability in Credentialing: Compared to conventional credentialing, digital methods consume less paper, mail, and energy (see Table. 12.7).

12.1.7 Post-Pandemic Pedagogical Shifts

The COVID-19 epidemic caused a seismic upheaval in worldwide education systems, serving as both a disruptor and a motivator for change. Institutions were obliged to adjust quickly to remote teaching, instructors redesigned learning approaches, and students encountered new learning settings and problems. As civilisations emerge from the epidemic, the educational landscape evolves. When the epidemic began, educational institutions used emergency remote teaching, a temporary shift to online instruction in response to the crisis. Many people eventually switched to more sustainable hybrid or blended learning approaches, which included in-person and online training [14] (see Table. 12.8).

Key Pedagogical Shifts

1. Student-centered learning: The epidemic highlighted the need to tailor instruction to individual needs. Educators are increasingly turning to learner-driven models that provide for more flexibility in speed, style, and evaluation approaches.

Table 12.8 Post-pandemic pedagogical shifts

Mode	Definition	Post-pandemic application
Emergency remote teaching	Crisis-driven, rapid shift to online platforms	Temporary, often lacking in pedagogy
Online learning	Fully digital courses, often asynchronous	Permanent option for flexibility
Blended learning	Mix of face-to-face and online modalities	Common in universities and training centers
HyFlex learning	Students choose in-person or online attendance in real time	Emerging in higher education

2. Flipped classrooms: Instructors provide curriculum to students via videos or digital resources that they may examine at home, with in-person meetings for discussion and application.
3. Assessment reforms: Formative assessments, project-based learning, and e-portfolios are gaining popularity as alternatives to high-stakes tests that promote deeper comprehension and critical thinking.
4. Digital Literacy: A Core Competency: The pandemic transformed digital literacy from a desirable skill to a basic requirement for both educators and students.
5. Focus on Social-Emotional Learning (SEL): The integration of SEL frameworks into curricula was motivated by concerns about mental health and well-being.

Lessons Learned from the Pandemic

The epidemic demonstrated the value of resilience, flexibility, and creativity in many facets of life. One significant takeaway was the need for digital change, as remote employment, online education, and telemedicine became necessary, emphasising the necessity for solid technology infrastructure and accessibility. Furthermore, the crisis highlighted the necessity of public health preparation, emphasising the need for effective healthcare systems, fast response mechanisms, and worldwide collaboration in dealing with health catastrophes. The epidemic also exposed economic weaknesses, forcing businesses and governments to reconsider supply networks, employment models, and financial safety nets. Furthermore, societal dynamics changed, emphasising the need for mental health knowledge and robust social support networks. On a larger scale, it increased enthusiasm for scientific research and data-driven decision-making, as innovations in vaccinations and treatments demonstrated the value of teamwork. These lessons continue to influence legislation, companies, and everyday life, ensuring that societies are better prepared for future crises (see Table. 12.9).

Table 12.9 Lessons learned from the pandemic

Lesson	Implications for education
Connectivity is critical	Internet access must be considered a public utility
Teachers are innovators	Professional development in digital pedagogy is essential
Inequities are exacerbated by crises	Addressing access gaps is key to future preparedness
Parental and community roles are vital	Learning is a collaborative, multi-stakeholder effort
Adaptability is an educational competency	Flexibility in curriculum and delivery is essential

12.1.8 *Innovations Emerged*

1. Mobile learning tools, such as WhatsApp, Telegram, and local language apps, have proven effective in reaching learners without internet.
2. Community-based Learning Hubs: Safe zones were built in regions where pupils could not use the internet at home.
3. Radio and TV broadcasts, Kenyan, Nigerian, and Pakistani education ministries broadcast classes over national media.
4. EdTech Platforms LMS such as Google Classroom and Moodle have grown rapidly, as have localised systems customised to the country's curriculum.
5. Teacher Collaboration Networks Educators organised virtual communities of practice to exchange resources, techniques, and peer support.

Equity and Access in the Global South Due to inequalities in access to online schooling, many students were unable to adjust as quickly as others once the pandemic revealed large digital differences. Subsidised technology and offline learning materials are necessary since many students have difficulty due to limited digital infrastructure, such as inconsistent internet or a lack of gadgets. The significance of community engagement and safeguarding strategies to guarantee ongoing education was further highlighted when gender-based disparities manifested themselves via the withdrawal of girls from school in certain localities. Educational resources were often not accessible in local languages or did not provide enough assistance for students with disabilities; this highlighted the need for inclusive content development and translation initiatives. Another obstacle was the language barrier. Families were already struggling to pay for or devote enough time to their children's education before the economic downturn, so programs like school lunches and conditional cash transfers became even more important in closing the achievement gap. It will need long-term governmental initiatives, investments in technology, and a dedication to equal education access to fix these structural problems [15].

Rethinking Teacher Roles

Educators were tasked with creating material, troubleshooting technology, counselling students, and serving as community liaisons. Post-pandemic teaching requirements: digital pedagogical training, adaptive curriculum planning, multimodal content delivery, and trauma-informed teaching approaches. Investment in teacher capacity development has emerged as a worldwide strategic priority, especially in resource-constrained environments [16].

Policy Shifts and Strategic Planning

Governments and organisations are increasingly including remote learning strategies in national education policy. Create contingency plans for educational interruptions. Developing public-private partnerships for device distribution and connection. Implementing data collecting methods to monitor student involvement and performance [17].

12.1.9 Sustainability of Innovations

Although not all initiatives from the pandemic period will last, many have established important standards for future education resilience. Digital resource libraries are highly sustainable because they enable long-term access to learning resources for both instructors and students, promoting ongoing participation. Printed home learning kits, although beneficial in some circumstances, have a medium lifespan owing to logistical problems and high delivery costs. SMS-based learning reminders have a high potential for long-term viability, providing an inexpensive and accessible solution to keep students interested without needing complex technology. Temporary online platforms, on the other hand, have a limited likelihood of survival due to a lack of institutional or financial backing. These developments emphasise the necessity of adaptation, accessibility, and long-term investment in education systems for inclusive and sustainable learning solutions [18, 19].

A Post-Pandemic Vision for Education

An inclusive and resilient education system should:

1. Ensure ubiquitous access to gadgets and the internet.
2. Incorporate flexibility into the educational design.
3. In addition to academic accomplishment, place a premium on your health.
4. Micro credentials and alternate routes may help promote lifelong learning.
5. Promote local innovation rather than just adopting global patterns (see Table. 12.10).

Table 12.10 From pandemic response to pedagogical transformation

Area	Pre-pandemic	Pandemic response	Post-pandemic vision
Teaching model	In-person, teacher-centered	Online or hybrid, tech-reliant	Blended, learner-centered
Assessment	Exams, grades	Alternative, continuous	Competency-based, reflective
Equity strategies	Limited support	Emergency relief efforts	Systematic inclusion and equity
Curriculum delivery	Uniform, rigid	Flexible, modular	Personalised, adaptive
Teacher development	Infrequent training	Rapid digital upskilling	Ongoing, tech-integrated pedagogy

12.1.10 Digital Humanities in the Global South

The Digital Humanities (DH) discipline, which is frequently defined by the interface of computers and the humanities, has historically been formed by institutions and researchers in the Global North. However, researchers, archivists, and cultural practitioners from the Global South are rapidly adding to and redefining the area. DH has typically been characterised by approaches such as text analysis, digital archiving, data visualisation, and digital publication, which are often developed by Northern academic institutions. However, in the Global South, DH has evolved beyond methodological innovation to become a political and epistemological intervention that addresses historical injustices and amplifies marginalised perspectives.

Key Drivers of Digital Humanities in the Global South

1. The preservation of endangered languages and cultures: Many communities in the Global South are losing their cultural identity as a result of globalisation, conflict, and migration. DH offers instruments for digitising and sharing oral histories, endangered languages, indigenous knowledge, and religious practices.
2. Decolonising the Archives: There is a rising movement to recover historical narratives that have long been marginalised or misunderstood. Through storytelling, metadata design, and archival practice, projects often question Eurocentric historical narratives and strive to reclaim local agency.
3. Open Access and Knowledge Justice: Paywalls, bandwidth constraints, and language barriers are often used to restrict access to information. DH practitioners in the Global South are leading open-source projects, translation efforts, and alternative knowledge systems.
4. Youth-Driven Innovation and Mobile Technologies: With the increase in mobile phone prevalence and low-cost data models, young digital humanists are disseminating information and creating micro-archives using apps, WhatsApp, TikTok, and other mobile-friendly forms.

Notable Projects and Initiatives

Across different regions, several projects are actively working to digitize and preserve cultural narratives, educational resources, and marginalized experiences. In Kenya, the No One is Disposable Archive focusses on digitally preserving LGBTQ+ groups' experiences, ensuring their voices are preserved and accessible. Mukurtu CMS is a specialised content management system in the Australia/Pacific area that promotes cultural preservation via indigenous digital storytelling and community archiving. Latin America benefits from the Latin American Open Archives, which offers open-access collections from civil society and human rights organisations, making critical historical documents broadly accessible.

Infrastructure and Access Challenges

In spite of the fact that DH are seeing a continuous expansion in the Global South, there are still a number of obstacles that prevent its long-term viability and wider influence. There is still a significant challenge in the form of limited finance, since many DH programs are dependent on funds that are either short-term or multinational, which makes long-term planning challenging. Additionally, the dependency of technology on Northern infrastructure, which includes servers, software, and hosting, results in vulnerabilities and limits the ability of locals to exercise control over digital resources.

Community-Centred Approaches

Digital humanities in the Global South often use participatory and community-centered approaches rather than institutionally driven top-down frameworks. These techniques highlight:

1. Co-creating material with elders, youth, and local historians
2. Consent and access controls for culturally sensitive items.
3. Knowledge dissemination via community radio, storytelling events, and social media.
4. Orality and performance are legitimate digital outputs (not only text or graphics).

12.1.11 Towards a Southern Digital Humanities Epistemology

Scholars and practitioners are advocating for a Southern epistemology of DH, which is based on local knowledge systems, cultural practices, and social justice perspectives, rather than just reproducing Northern ones (see Table. 12.11).

Recommendations for Strengthening DH in the Global South

1. Invest in infrastructure and capacity building by supporting local data centres, low-bandwidth technologies, and digital skills training.
2. Foster regional networks and collaborations.
   Strengthen South-South DH collaborations to promote knowledge sharing and mutual assistance.
3. Promote Policy Support. Governments and organisations must acknowledge DH as a genuine subject of study and offer financing.

Table 12.11 Towards a southern digital humanities epistemology

Characteristic	Southern DH approach
Epistemic grounding	Local ontologies and ways of knowing
Tools and platforms	Adapted for low-resource, multilingual environments
Data and metadata	Community-defined taxonomies and protocols
Value system	Centering repair, remembrance, and reparation
Educational deployment	Integrated into local curriculums with decolonial framing

Table 12.12 Comparing digital humanities approaches

Aspect	Global north	Global south
Primary output	Peer-reviewed publications, datasets	Oral histories, storytelling, and local archives
Methodological emphasis	Computational tools, large-scale analysis	Community memory, ethics, and access
Funding source	Government and institutional grants	NGOs, small foundations, self-funding
Knowledge framework	Academic knowledge production	Epistemic justice, indigenous knowledge
Audience	Scholars, policymakers	Communities, students, and civil society

4. Engage Diasporic Communities. Utilise the global South's diasporas to co-create archives and narratives that cross continents.
5. Support Multilingual DH. Encourage efforts that translate material into and out of indigenous languages to reach a larger audience (see Table. 12.12).

12.1.12 Decolonizing Pedagogy and Knowledge

The worldwide push for decolonisation in education poses a threat to established power structures, knowledge systems, and educational paradigms that have their origins in colonial times. Pedagogy and knowledge decolonisation necessitates a fundamental reevaluation of whose knowledge is valued, how it is taught, and for what purpose; it goes beyond just including varied viewpoints. Decolonisation in education is the critical breakdown of colonial conceptions of knowledge production and transmission. It challenges the universalisation of Western epistemologies, aiming to recognise and legitimise indigenous, local, and non-Western methods of knowing (see Table. 12.13).

Table 12.13 Pedagogy

Colonial pedagogy	Decolonial psedagogy
Hierarchical, lecture-based	Dialogic, participatory
Eurocentric curriculum	Multivocal, culturally situated content
Standardized assessments	Diverse modes of learning and evaluation
Text-centric, theory-heavy approaches	Praxis-oriented, experiential, and community-rooted
Focus on the "discovery" of knowledge	Recognition of pre-existing knowledge systems

12.1.13 Historical Context and Legacy

Colonial education systems were meant to support imperial aims, such as teaching local elites to work for colonial bureaucracies and eradicating indigenous knowledge systems. These regimes enforced Eurocentric ideals while undermining native languages, customs, and worldviews. Many Global South countries' post-independence reforms preserved the structural DNA of these systems, frequently influenced by foreign donors or standardisation initiatives.Decolonising pedagogy entails tackling these residual power relations and developing a curriculum and classroom practice that: Reflects the local history and priorities. Encourages linguistic and cultural inclusivity. Recognises the coloniality ingrained in academic fields [20].

Key Dimensions of Decolonizing Pedagogy

1. Curriculum reform: Decolonising the curriculum entails more than just including a few books by non-Western writers. It requires: Rethinking fundamental canons in literature, history, science, and philosophy. Contextualising knowledge within local histories and ecosystems includes critical theory from African, Asian, Indigenous, and Latin American intellectuals. Promoting trans-epistemological discussions (between Western and non-Western cultures) [21].
2. Pedagogical approaches: The manner of instruction is as essential as the topic presented. Decolonial Pedagogy: Respects oral traditions, storytelling, and communal knowledge. Encourages collaborative learning rather than top-down teaching. Respects learners as knowledge bearers. Uses local analogies, examples, and languages to promote comprehension.
3. Language and Power: Language is a key component of colonial heritage. English, French, Spanish, and Portuguese are the dominant languages in many previously colonised countries' educational institutions. Decolonisation advocated for: multilingual education and mother tongue teaching. The revitalisation and modernisation of indigenous terminology creates academic space for non-Western grammars of thinking. Encourage code-switching as an acceptable linguistic strategy [22].

Key Contrasts Between Colonial and Decolonial Educational Models

The comparison between colonial and decolonial education approaches reveals basic disparities in knowledge generation and distribution. The colonial model views Western elite institutions as the dominant source of knowledge, based on universalist and positivist epistemologies that emphasise standardised facts. Curriculum under this framework focuses on Western-centric narratives, sometimes overlooking various views, and education is offered in colonial or foreign languages, restricting accessibility for local groups. Classroom authority is teacher-centred, reinforcing hierarchical learning structures, and evaluation procedures are based on standardised testing, which prioritises uniformity above individual comprehension. In contrast, the decolonial paradigm values varied, local, indigenous, and diasporic knowledge sources, encouraging pluralist, contextual, and situated epistemologies that acknowledge many viewpoints. Its courses are multicultural, transnational, and critically engaged, assuring inclusion and diverse representation. Instruction is delivered in local and indigenous languages, which improves accessibility and cultural relevance. In the classroom, authority is shared, with a focus on community-informed learning, and evaluations emphasise reflexive, dialogical, and performance-based techniques that suit individual learning styles and life experiences [23].

The Role of Technology in Decolonizing Knowledge

Digital platforms provide a double-edged sword. On the one hand, they make it easier to access a variety of knowledge systems; on the other, they often reproduce hegemonic structures via algorithms, content regulation, and linguistic supremacy [23]. Promising developments include:

1. Open Educational Resources (OERs) created in local languages
2. Community-led digital archives chronicle oral traditions.
3. Using mobile learning platforms for distant rural education.
4. Decolonial podcasts and YouTube channels strengthen marginalised voices.

Decolonizing Research Methodologies

To prevent extractive and exploitative behaviours, both research and education must be decolonised. The key concepts include:

1. Community-Based Participatory Research (CBPR)
2. Ethical frameworks informed by local rules, consent
3. Respecting Collective Intellectual Property
4. Allowing people to co-author and own their data.
5. Transitioning from community research to community-based research

Recommendations for Educators and Institutions

For education to be decolonised, deliberate steps must be taken to promote diversity and confront colonial preconceptions. In order to create a more inclusive and fair learning environment, it is helpful to conduct curriculum audits in order to detect biases and under-representation. By providing resources for faculty training in decolonial pedagogy, we may encourage the adoption of alternative pedagogical practices and epistemologies that value diverse forms of knowledge. Cultural relevance and accessibility are enhanced when educational resources prioritise linguistic variety, enabling students to interact with learning in their home languages. Learning becomes more meaningful and relevant when communities are involved in the creation of curricula, since this guarantees that education is based on real-life experiences. Also, a more comprehensive and flexible way to measure academic achievement may be achieved by rethinking assessment procedures to include more than just standardised tests [24–26].

12.1.14 Global Collaborative Learning Models

The need for cooperative, international educational models has never been higher than it is now, in a globally linked world. To foster cross-cultural understanding, knowledge co-creation, and global competency development, Global Collaborative Learning (GCL) is a great tool. GCL refers to organised educational activities that bring learners, instructors, and institutions together across geographical and cultural borders. It uses technology and pedagogical practices to promote collaborative learning, conversation, and problem-solving. These models are grounded on:

1. Constructivist learning theories (Vygotsky and Piaget)
2. Social learning theory (Bandura)
3. Intercultural competency frameworks (Deardorff and Bennett)
4. Critical pedagogy (Freire), particularly when focused on power and inclusiveness.

Core Components of Global Collaborative Learning

A number of essential elements that promote participation, teamwork, and diversity are included in effective teaching methods. Intercultural discussion promotes mutual respect and cultural understanding among learners, allowing for meaningful interactions across varied backgrounds. Co-creation of knowledge enables instructors and students to collaborate on content development and real-world problem solving, ensuring that learning is participatory and immediately relevant. The utilisation of ICT applications, such as Zoom, Padlet, Moodle, and WhatsApp, allows for synchronous and asynchronous interactions, making education more accessible and dynamic. Shared curricular goals assist institutions in aligning learning objectives to preserve relevance and coherence, ensuring that educational programs successfully meet student requirements. Finally, reflective practice promotes ongoing self-assessment, enabling both students and teachers to critically review their methods and positionality, resulting in more meaningful and powerful learning experiences [27].

Models of Global Collaborative Learning

1. SUNY and other worldwide institutions developed the Virtual Exchange (VE) and Collaborative Online International Learning (COIL). connects classrooms in different countries through project-based learning. Promotes cross-institutional faculty cooperation Can be added into current courses without needing to travel.
2. Virtual classrooms: Multi-institutional courses offered by professors from two or more countries. May include cooperative evaluations, shared readings, and peer-reviewed tasks. For example, colleges in the United States, Ghana, and Indonesia together teach Global Environmental Politics.

3. MOOCs and Global Forums: MOOCs featuring regulated discussion boards, including students from all around the globe. Platforms like as edX, FutureLearn, and Coursera provide worldwide case studies and multilingual assistance. Some provide regional breakout groups and cross-border capstone projects.
4. Collaborative research and community projects: Student or faculty-led initiatives addressing the UN SDGs (e.g., water security, gender equity, climate action). Often associated with service-learning or community-based research. Can be funded by platforms such as UNESCO UNITWIN, Erasmus+, or AUF.

Benefits of Global Collaborative Learning

Engaging in inclusive and multidisciplinary learning contexts provides numerous significant advantages for students. Individuals' cultural intelligence (CQ) grows when they are exposed to other worldviews and value systems, which fosters mutual respect and flexibility. The utilisation of different platforms improves digital literacy, allowing students to gain technical fluency and efficiently manage digital technologies. Global citizenship is developed by increasing empathy, civic duty, and ethical awareness, allowing students to interact critically with global challenges. Collaborative problem-solving abilities are honed as students work on real-world situations that need multidisciplinary and intercultural cooperation, preparing them for complex global interactions. Furthermore, equity via access guarantees that students from underserved backgrounds may engage in educational opportunities, therefore levelling the playing field and promoting greater inclusion in learning [28].

Challenges and Limitations

Despite its revolutionary potential, GCL faces substantial challenges to its efficacy and accessibility. The digital gap remains a serious issue since uneven access to dependable internet and gadgets inhibits fair participation, especially in underserved areas. To accommodate various worldwide participants, scheduling becomes more problematic when several time zones are involved, necessitating flexible techniques and, in many cases, asynchronous learning models. Language hurdles exacerbate involvement, with English dominance marginalising non-native speakers and restricting inclusion. Furthermore, curriculum misalignment complicates standardising shared assessment since varied academic calendars and educational standards produce discrepancies. Power disparities also raise issues, since Northern institutions usually dominate partnerships, increasing the possibility of neocolonial processes and restricting the agency of institutions in the Global South. Addressing these issues is critical to ensuring that GCL maintains a genuinely egalitarian and effective educational paradigm [29].

Strategies for Inclusive and Equitable GCL

To promote inclusive and fair practices in global cooperation, the following techniques are necessary:

1. Co-design Projects with Equal Stakeholders: All institutions should participate equally in the design, delivery, and evaluation stages.
2. Use Multilingual and Multimodal Communication: Provide translation tools, closed captioning, and support for local language integration.
3. Plan for digital accessibility by providing offline choices, mobile-first platforms, and low-bandwidth solutions.
4. Culturally Responsive Facilitation: Educators should be trained in intercultural sensitivity and anti-bias techniques.
5. Contextualised Content: Make sure that themes represent both local reality and global storylines.

Tools and Platforms for GCL

In the classroom, students may work together and participate in meaningful ways with the help of several digital resources. Padlet promotes collaborative brainstorming and asynchronous reflection, enabling students to exchange ideas and insights in a visually appealing way. Moodle and Canvas are systems for hosting shared course materials, discussion forums, and peer feedback, which promotes organised learning and engagement. Zoom and Microsoft Teams allow for live conversations, lectures, and breakout group activities, making distant learning more interactive and engaging. Flipgrid promotes linguistic and cultural interchange by allowing learners to express themselves and engage with other points of view via video reflections. Google Docs enables real-time collaborative writing and peer editing, facilitating seamless collaboration on academic and professional projects. Together, these technologies provide an adaptable ecology for participatory and inclusive learning [22].

Evaluation and Assessment in GCL Contexts

Evaluation of global collaborative learning should be formative, reflective, and collaborative:

1. Peer and self-assessment: Encourage students to consider group dynamics and contributions.
2. Process Journals: Track individual learning trajectories throughout the project lifetime.
3. Intercultural Competence Rubrics: Evaluate improvement in empathy, communication, and openness
4. Digital Portfolios: Archive contributions across platforms for an overall evaluation.

GCL aims not just to extend students' perspectives but also to strengthen their empathy, justice, and co-creation skills. As the education industry faces crises ranging from pandemics to climate change, global alliances will be critical in redesigning learning environments as collaborative, democratic, and inclusive. Institutions must go beyond symbolic relationships and engage in long-term, reciprocal, equity-driven cooperation. The future of education is not limited by boundaries, but rather transcends them [30].

12.1.15 Lifelong Learning and Micro-Credentials

In an era of fast technological advancement, changing labour markets, and global crises, the need for flexible, personalised, and ongoing education is increasing. Traditional degrees, although still useful, are no longer sufficient for a workforce that must constantly adapt.

Conceptual Frameworks

Lifelong learning is the continuous, voluntary, and self-motivated pursuit of information for personal or professional purposes. It includes official, non-formal, and informal learning paths. Lifelong learning is defined by UNESCO [31] as "a holistic and inclusive approach that integrates learning across life stages, contexts, and purposes." Micro-credentials are brief, competency-based certifications that demonstrate mastery of a certain skill or topic. They may be stackable, enabling students to work towards entire degrees, or stand alone for specific instruction (see Table 12.14).

Drivers of Lifelong Learning and Micro-Credential Adoption

Several significant variables are influencing the changing landscape of education and professional development. Changing labour demands, driven by automation and AI developments, need ongoing reskilling and upskilling to stay up with technological innovation. The globalisation of work has increased the need for digital and intercultural skills. Continuous learning has become more accessible because to digital learning platforms such as MOOCs, LMSs, and smartphone applications, which allow users to quickly acquire new skills. Furthermore, credential transparency has become critical, as companies increasingly want verifiable, thorough verification of abilities over conventional degrees alone. Government and agency policy changes contribute to this transformation by supporting the acknowledgement of alternative credentials, ensuring that varied learning routes are recognised and appreciated in the contemporary workforce. Collectively, these elements help to create a more adaptive and inclusive learning environment [32].

Table 12.14 Definition of terms

Term	Definition
Digital badge	Visual representation of a micro-credential; often embedded with metadata
Stackable credential	A sequence of modular achievements that can lead to formal qualifications
Nano-degree	Industry-designed program focused on job-ready skills (e.g., data analytics)
MOOC certificate	Credential earned via massive open online courses (Coursera, edX, etc.)

Global Examples of Implementation

Several nations and regions are establishing micro-credential systems to improve education accessibility, transparency, and alignment with employment requirements. The European Micro-Credentials Framework in the European Union aims to standardise credentials across institutions and borders, assuring uniformity in learning recognition. Australia's National Micro-Credentials Marketplace functions as a digital catalogue for stackable learning, enabling students to gradually develop abilities. In the United States, Credential Engine serves as a registry to increase credential transparency, allowing people and companies to quickly check qualifications. Kenya and Rwanda have launched MOOC-based TVET and Digital Skills Programs to increase access and fairness in technology education, hence increasing participation in digital learning. Meanwhile, Canada's eCampus Ontario Micro-Credentials Initiative promotes collaboration between companies and educators, successfully connecting education and employment needs. These projects show the worldwide trend of flexible, competency-based learning paradigms [33, 34].

Technological Infrastructure for Delivery

Micro-credentials depend significantly on digital platforms that enable: Learning Management Systems (LMS) (such as Moodle and Canvas). Credentialing platforms (such as Credly and Badgr). Blockchain-based verification (tamper-proof digital certificates). AI-driven Learning Analytics (tracking and personalisation), and Mobile-first design is extremely relevant in the global south [35]. While micro-credentials provide flexibility and accessibility, they also represent issues that may exacerbate existing disparities if not correctly managed. Limited digital connectivity continues to be a significant obstacle, since students in disadvantaged places may not have access to dependable internet or devices that are suited for their needs, which limits their ability to participate.

Micro-credential programs use rigorous standards to assure consistency, transparency, and confidence in skill validation. Specified learning outcomes provide quantifiable capabilities for learners to exhibit. Assessment standards, which include valid and reliable assessments such as projects, peer reviews, and tests, guarantee that skills are verified. Credit portability enables credentials to be recognised across institutions and businesses, promoting career and educational mobility. Credential metadata improves openness by including important facts like the issuer, date, skills covered, and degree of expertise. Furthermore, accrediting alignment connects micro-credentials to national or worldwide qualification frameworks, such as the European Qualifications Framework (EQF), increasing their credibility and acceptability in global education and workforce systems [36].

12.1.16 Pedagogy for Digital Citizenship

In an age where digital technologies pervade every part of life, promoting digital citizenship has arisen as a vital need for educators, politicians, and students. Digital citizenship is the responsible, ethical, and educated use of digital technology to participate in society, politics, education, and the workplace. As digital platforms transform civic life, pedagogy must adapt to provide students with not just technical skills, but also the key competences required to traverse digital environments securely, inclusively, and ethically [37].

Digital citizenship refers to a wide set of behaviours, rights, duties, and literacy skills required for successful and ethical interaction in digital contexts. It includes:

1. Access and understanding of technology
2. Critical appraisal of digital information.
3. Civic involvement on digital platforms
4. Online Etiquette and Empathy
5. Cybersecurity Awareness
6. Digital Rights and Responsibilities

Digital competence comprises many critical qualities that allow people to navigate the online world efficiently and ethically. Digital literacy refers to the capacity to access, assess, and generate digital material, allowing individuals to interact critically with digital platforms. Digital ethics focuses on courteous and ethical online behaviour, encouraging beneficial interactions while avoiding harmful activities. Understanding digital rights is critical for protecting freedom of speech, data privacy, and giving people control over their personal information. Digital law focuses on understanding intellectual property rights and online legal duties in order to ensure regulatory compliance and ethical usage of digital content. Digital health and wellness emphasise the significance of regulating screen time, emotional well-being, and ergonomic practices in order to maintain a healthy digital lifestyle. Digital commerce allows people to securely engage in e-commerce and financial technology, guaranteeing informed decisions in online transactions. Finally, digital security is critical for safeguarding identity, data, and devices from cyberattacks while also increasing cybersecurity awareness and risk mitigation measures. Together, these aspects provide a complete framework for ethical and productive digital involvement [38].

Role of Educators in Fostering Digital Citizenship

Educators have a critical role in incorporating digital citizenship into teaching and learning. Pedagogical techniques must evolve away from passive technology usage and towards active, critical, and interactive engagement [39, 40]. Effective digital citizenship education involves:

1. Inquiry-based learning examines ethical concerns in digital life.
2. Combating disinformation with media literacy education
3. Collaborative initiatives using digital technologies for civic involvement
4. Simulation games about online rights and responsibilities.
5. Case studies of digital activism and online dispute resolution.

One of the most prominent models is Ribble's Nine Elements of Digital Citizenship, which is often included into school curriculum globally (see Table 12.15).

Table 12.15 Ribble element

Ribble's Elements	Examples in Education
Digital access	Ensuring all students can use online resources
Digital communication	Teaching appropriate online communication norms
Digital literacy	Assigning projects that assess credibility of digital sources
Digital etiquette	Discussing cyberbullying and respectful discourse
Digital law	Lessons on plagiarism and digital copyright laws
Digital Rights & Responsibilities	Debates on censorship versus. freedom of speech online
Digital Health & Wellness	Strategies for balanced tech use and screen breaks
Digital security	Practicing safe password management and data protection
Digital commerce	Simulated online purchasing and digital financial literacy

Case Studies from Global Contexts

There have been campaigns to encourage ethical internet use and digital literacy in a number of nations and areas. To help its citizens learn to critically analyse and assess news stories, Finland has implemented a new national media literacy curriculum. To ensure that young learners acquire fundamental competencies for navigating the internet environment, Kenya's DLP focuses on basic digital skills and digital responsibility. In Canada, kids are encouraged to interact responsibly with technology via the Media Smarts project, which offers materials for grades K-12 that focus on digital citizenship and online ethics. The goal of the Cybercafés program in public schools in the Philippines is to protect students from cyberbullying and make the internet a safer place for them to use. In the meantime, the Digital Education Action Plan of the European Union promotes online democracy and digital competency by encouraging educated and engaged involvement in the digital sphere. Collectively, these efforts demonstrate a worldwide drive to establish a society that is more cognisant of and accountable for its digital footprint.

Addressing Challenges in Teaching Digital Citizenship

Even if more and more people are becoming aware of the need for digital citizenship education, there are still several obstacles (see Table 12.16).

Tools and Resources for Digital Citizenship Education

At all grade levels, students may benefit from the many resources available to them to help them become digital citizens and practise appropriate online behaviour. Common Sense Education offers a digital citizenship program and materials for children in grades K-12, guiding them towards responsible and secure online behaviour. Middle school students may learn important digital decision-making situations via the interactive game Digital Compass. Be Internet Awesome is a Google initiative that aims to educate kids about digital citizenship and cybersecurity via fun and engaging activities. The UNESCO Media and Information Literacy Curriculum

Table 12.16 Challenges

Challenge	Implication
Digital inequity	Uneven access limits inclusion in digital citizenship education
Overloaded curricula	Educators struggle to integrate digital literacy within tight schedules
Cultural and contextual variability	Norms of digital behavior differ across regions and communities
Limited educator training	Teachers may lack preparation in digital ethics and critical pedagogy
Youth engagement issues	Learners may be passive consumers rather than critical digital participants

promotes ethical use of digital resources and analytical thinking on a global scale. Educators and students may participate in positive digital interactions on collaborative platforms like Edmodo and ClassDojo, which encourage safe online behaviour. Together, these resources promote responsible, well-informed, and ethical use of technology and boost digital competence [41].

12.1.17 Visions for Post-Digital Education

Post-digital education goes beyond integrating digital technologies into teaching and learning. It imagines a society where digital is integrated into educational culture, infrastructure, and philosophy. Digital technologies are normalised, their limits critically appraised, and their usage directed by educational, ethical, and humanistic concepts in post-digitality.

The term "post-digital" was first used in arts and cultural theory to characterise a situation in which digital technologies are firmly ingrained yet critically scrutinised (Cramer, 2013). In education, this paradigm: recognises the widespread integration of digital technologies, emphasises human agency above technical determinism, promotes transdisciplinary and multimodal literacy, resists both technological utopianism and technophobia, and advocate for ethical, egalitarian, and critical engagement with technology. "Post-digital education is less about devices and platforms, and more about values, relationships, and systems of knowing" [42].

Modern education has numerous key characteristics that improve learning experiences and accessibility. Hybrid learning has become the norm, featuring mixed settings that incorporate online and offline training in both synchronous and asynchronous modes, giving learners more freedom. Data-conscious pedagogy promotes critical interaction with algorithms, learning analytics, and surveillance technologies, resulting in ethical and informed use of digital tools. Technological humility emphasises the importance of simplicity, low-tech solutions, and thoughtful digital behaviours, putting accessibility ahead of unnecessary complexity. Open knowledge ecosystems emphasise the value of open access, Open Educational Resources (OER), and collaborative knowledge development, resulting in a more diverse learning environment. Multisensory and embodied learning incorporates physical, emotional, and cognitive experiences into digital settings, resulting in a more comprehensive and immersive approach to education. Furthermore, contextual and localised methods acknowledge cultural variety and local epistemologies, ensuring that educational frameworks are relevant and responsive to various populations. Together, these elements provide a more adaptive and meaningful learning environment [43].

The Human-Centred Turn

Post-digital education prioritises the human in technical systems. It reclaims education as a social, emotional, and moral activity. Teachers operate as designers of learning environments, customising resources to the requirements of their students rather than the other way around. Learners as Co-Creators: Students actively shape content, conversation, and even the tools themselves. Ethics and Care: Emphasis on empathy, community, mental health, and wellness in digital environments [44].

Challenges and Contradictions

While the idea for post-digital education is enticing, some difficulties remain. Digital colonialism strengthens reliance on platforms and epistemologies from the Global North while weakening local sovereignty. Meanwhile, the emergence of platform capitalism poses a danger of commodifying education by prioritising profit above teaching. Data privacy issues are growing as learning analytics and AI present surveillance hazards, sometimes without meaningful permission. Furthermore, infrastructural disparities make ubiquitous access to a perceived standard in post-digital models difficult for marginalised areas. Finally, the rapid rate of technological change has left many educators puzzled and trying to adapt. To accomplish its promise, post-digital education must remain flexible, inclusive, and critically self-aware while actively tackling these systemic issues [45].

Post-Digital Pedagogies in Practice

Recognising that technology is intrinsic to learning settings rather than an independent entity, post-digital pedagogies place an emphasis on digital tool integration within larger educational contexts. Instead of technology dictating pedagogy, these methods put the emphasis on critical interaction with digital platforms. Teachers who embrace post-digital pedagogies often combine online and offline approaches, using interactive technologies to enhance learning while preserving a focus on human-centered approaches. To further guarantee that technology progress promotes egalitarian and inclusive education, these pedagogies promote contemplation of data protection, accessibility, and digital ethics (see Table 12.17).

Table 12.17 Approaches

Approach	Application example
Critical digital pedagogy	Dialogues on algorithmic bias, digital labour, and platform ethics
Slow tech integration	Using fewer, deeper tools intentionally—e.g., one collaborative platform used meaningfully
Analog-digital fusion	Using physical artifacts in tandem with digital tools (e.g., zines, hand-drawn concept maps shared online)
Community-embedded learning	Locally relevant, socially embedded projects supported by digital means
AI-augmented human teaching	Teachers using AI for scaffolding and feedback while preserving empathy and intuition

References

1. Shahbazi, Z., Nowaczyk, S.: Towards personalized cardiometabolic risk prediction: a fusion of exposome and AI. Heliyon. **11**(1), e40859 (2025). https://doi.org/10.1016/j.heliyon.2024.e40859
2. Baratzadeh, F., Hasheminejad, S.M.H.: Customer behavior analysis to improve detection of fraudulent transactions using deep learning. J. AI Data Min. **10**(1), 87–101 (2022). https://doi.org/10.22044/JADM.2022.10124.2151
3. Sikandar, A., Muhammad, A., Jazeb, N., Ahmad, F.: Multiagent based system for secondary education using machine learning., 1310–1315 (2018). https://doi.org/10.1109/ICMLA.2018.00212
4. Casini, M.: Extended reality for smart building operation and maintenance: a review. Energies. **15**(10), 3785 (2022). https://doi.org/10.3390/en15103785
5. De Cicco, L., Mascolo, S., Palmisano, V., Ribezzo, G.: Reducing the network bandwidth requirements for 360 ∘ immersive video streaming. Internet Technol. Lett. **2**(4), e118 (2019). https://doi.org/10.1002/itl2.118
6. Beillahi, S.M., Ciocarlie, G., Emmi, M., Enea, C.: Behavioral simulation for smart contracts. Proc. ACM SIGPLAN Conf. Program. Lang. Des. Implement. **20**(4), 470–486 (2020). https://doi.org/10.1145/3385412.3386022
7. Steffen, S., Bichsel, B., Vechev, M.: Zapper: Smart Contracts with Data and Identity Privacy. (2022). https://doi.org/10.1145/3548606.3560622
8. Qian, P., Liu, Z.G., He, Q.M., Huang, B.T., Tian, D.Z., Wang, X.: Smart contract vulnerability detection technique: a survey. Ruan Jian Xue Bao/Journal Softw. **33**(8), 3059–3085 (2022). https://doi.org/10.13328/j.cnki.jos.006375
9. Samreen, N.F., Alalfi, M.H.: VOLCANO: Detecting Vulnerabilities of Ethereum Smart Contracts Using Code Clone Analysis. 1(1), 1–18 (2022). http://arxiv.org/abs/2203.00769
10. Gilani, K., Ghaffari, F., Bertin, E., Crespi, N.: Self-sovereign identity management framework using smart contracts. In: Proceedings. IEEE/IFIP Network Operations and Management Symposium: Network and Service Management in the Era of Cloudification, Softwarization and Artificial Intelligence NOMS. 2022 (2022). https://doi.org/10.1109/NOMS54207.2022.9789831
11. Masenya, T.M.: Breakthrough barriers to knowledge sharing using modern technologies in academic libraries in South Africa. Int. J. Sociotechnology Knowl. Dev. **16**(1), 1–19 (2024). https://doi.org/10.4018/IJSKD.347380
12. Williams, P., Kaylan, I., Daoud, H., Bayoumi, M.: Internet of things a survey on security in internet of things with a focus on the impact of emerging technologies. Internet Things. **19**(7), 10–23 (2022). https://doi.org/10.1016/j.iot.2022.100564

13. Tsabary, I., Manuskin, A., Eyal, I.: LedgerHedger: Gas Reservation for Smart-Contract Security. vol. 1(1). Association for Computing Machinery, (2022). https://eprint.iacr.org/2022/056
14. König, J., Jäger-biela, D.J., Glutsch, N.: Adapting to online teaching during COVID-19 school closure: teacher education and teacher competence effects among early career teachers in Germany. Eur. J. Teach. Educ. **43**(4), 608–622 (2020). https://doi.org/10.1080/02619768.2020.1809650
15. Neitzert, F., Petras, M.: Corporate Social Responsibility and Bank Risk, vol. 92, no. 3. Springer Berlin Heidelberg (2022). https://doi.org/10.1007/s11573-021-01069-2
16. Aswir, H.M.: Smartphones as personal digital archives? Recentering migrant authority as curating and storytelling subjects. Photosynthetica. **2**(1), 1–13 (2018) http://link.springer.com/10.1007/978-3-319-76887-8
17. Herweijer, C., Combes, B., Ramchandani, P., Sidhu, J.: Fourth Industrial Revolution for the Earth: Harnessing Artificial Intelligence for the Earth (2018). https://www.pwc.com/gx/en/sustainability/assets/ai-for-the-earth-jan-2018.pdf
18. Knutsen, W.L.: Value as a self-sustaining mechanism: why some nonprofit organizations are different from and similar to private and public organizations. Nonprofit Volunt. Sect. Q. **42**(5), 985–1005 (2013). https://doi.org/10.1177/0899764012457244
19. Gupta, S., Nawaz, N., Alfalah, A.A., Naveed, R.T., Muneer, S., Ahmad, N.: The relationship of csr communication on social media with consumer purchase intention and brand admiration. J. Theor. Appl. Electron. Commer. Res. **16**(5), 1217–1230 (2021). https://doi.org/10.3390/jtaer16050068
20. Silva, C., Oliveira, D., Silva, C., Oliveira, D., Sanin, C., Szczerbicki, E.: Sciencedirect sciencedirect human feedback and knowledge discovery: towards cognitive human feedback and knowledge discovery: towards cognitive systems optimization systems optimization. Procedia Comput. Sci. **176**, 3093–3102 (2020). https://doi.org/10.1016/j.procs.2020.09.179
21. López Rupérez, F., García García, I., Expósito-Casas, E.: School leadership in Spain. Evidence from PISA 2015 assessment and recommendations. Leadersh. Policy Sch. **21**(2), 378–397 (2022). https://doi.org/10.1080/15700763.2020.1770806
22. Al Hamad, N.M., Adewusi, O.E., Unachukwu, C.C., Osawaru, B., Chisom, O.N.: Bridging the gap: Using robotics to enhance emotional and social learning in K-12 education. Int. J. Sci. Res. Arch. **11**(1), 231–243 (2024). https://doi.org/10.30574/ijsra.2024.11.1.0025
23. Ekechi, C.C, Chukwurah, E.G., Oyeniyi,L.D., Okeke, C.D.: A review of small business growth strategies in African economies. Int. J. Adv. Econ. **6**(4), 76–94, 2024. 10.51594/ijae.v6i4.1071
24. Trépanier, S.-G., Fernet, C., Austin, S.: Social and motivational antecedents of perceptions of transformational leadership: a self-determination theory perspective. Can. J. Behav. Sci. Can. des Sci. du Comport. **44**(4), 272 (2012)
25. Alharbi, S., Drew, S.: Using the technology acceptance model in understanding academics' behavioural intention to use learning management systems. Int. J. Adv. Comput. Sci. Appl. **5**(1), 143–155 (2014). https://doi.org/10.14569/ijacsa.2014.050120
26. Roberts, C.: Blockchain Security and Innovations for Colleges and Universities (2022). https://doi.org/10.13140/RG.2.2.33348.91528
27. Zhao, S., Pan, Q., Gao, D., Cheng, J.: Integrating internet of things and mixed reality to teach performance-based architectural design: a case study of shading devices. Educ. Inf. Technol. (2022). https://doi.org/10.1007/s10639-022-10998-6
28. Heydari, J., et al.: Towards a circular economy for packaging waste by using new technologies: the case of large multinationals in emerging economies. J. Clean. Prod. **11**(1), 438–450 (2019). https://doi.org/10.1016/j.resconrec.2017.08.017
29. Unachukwu, C., Ijiga, A.C., Okunade, B.A., Osawaru, B., Akinwolemiwa, D.I.: A comprehensive review of multilingual leadership: the role of French language skills in global business dynamics. World J. Adv. Res. Rev. **20**(3), 941–951 (2023). https://doi.org/10.30574/wjarr.2023.20.3.2493

30. Olika, C.D.: Risk management practices among selected state governments in southwestern Nigeria. Eur. J. Sci. Innov. Technol. **1**(2), 24–31 (2024) https://www.researchgate.net/profile/Hendra-Manurung/publication/356633046_From_Connectivity_to_Digital_Improving_Employee_Readiness_toward_Organizational_Change_in_Digital_Transformation/links/61a5b00671a23a0084c9b061/From-Connectivity-to-Digital-Improvi
31. Obi, K.C., Iwuanyanwu, O.: Innovation in the assessment of primary school mathematics: an approach to revamping Nigeria education. J. Teach. Perspect. **18**(1), 51–66 (2023) http://repositorio.unan.edu.ni/2986/1/5624.pdf%0Ahttp://fiskal.kemenkeu.go.id/ejournal%0Ahttp://dx.doi.org/10.1016/j.cirp.2016.06.001%0Ahttp://dx.doi.org/10.1016/j.powtec.2016.12.055%0Ahttps://doi.org/10.1016/j.ijfatigue.2019.02.006%0Ahttps://doi.org/10.1
32. Sousa, D., Lamurias, A., Couto, F.M.: Using neural networks for relation extraction from biomedical literature. Methods Mol. Biol. **2190**(23), 289–305 (2021). https://doi.org/10.1007/978-1-0716-0826-5_14
33. Ribeiro, J.M., et al.: Artificial intelligence and transcatheter interventions for structural heart disease: a glance at the (near) future. Trends Cardiovasc. Med. **2**(4), 1–7 ((202)1). https://doi.org/10.1016/j.tcm.2021.02.002
34. Kumar, N.M., et al.: Distributed energy resources and the application of AI, IoT, and blockchain in smart grids. Energies. **13**(21) (2020). https://doi.org/10.3390/en13215739
35. Mediavilla, M.A., Dietrich, F., Palm, D.: Review and analysis of artificial intelligence methods for demand forecasting in supply chain management. Procedia CIRP. **107**, 1126–1131 (2022). https://doi.org/10.1016/j.procir.2022.05.119
36. Furrer, L., Jancso, A., Colic, N., Rinaldi, F.: OGER++: hybrid multi-type entity recognition. J. Cheminform. **11**(1), 1–10 (2019). https://doi.org/10.1186/s13321-018-0326-3
37. Isa, A., Nte, N.D., Baba-Ahmadu, A.: Critical thinking for intelligence analysis for Nigerian students: evaluating the pedagogical challenges in real life scenarios. Indones. Educ. Adm. Leadersh. J. **5**(2), 46–60 (2023). https://doi.org/10.22437/ideal.v5i2.28639
38. Ghavifekr, S., Kunjappan, T., Ramasamy, L.: Teaching and learning with ICT tools: issues and challenges from teachers' perceptions. Malaysian Online J. Educ. Technol. **4**(2), 38–57 (2016)
39. Luhamya, A., Bakkabulindi, F.E.K., Muyinda, P.B.: Integration of ICT in teaching and learning: a review of theories. Makerere J. High. Educ. **9**(1), 21 (2017). https://doi.org/10.4314/majohe.v9i1.2
40. Naveh, G., Tubin, D., Pliskin, N.: Information technology diffusion in academic teaching: an institutional perspective. In: International Federation for Information Processing. **287-235**, 563–567 (2008). https://doi.org/10.1007/978-0-387-87503-3_35
41. Musungwini, S., Mugoniwa, B., Furusa, S.S., Rebanowako, T.G.: An analysis of the use of cloud computing among university lecturers: a case study in Zimbabwe. Int. J. Educ. Dev. Using Inf. Commun. Technol. **12**(1), 53–70 (2016) http://files.eric.ed.gov/fulltext/EJ1099577.pdf
42. Sarkingobir, Y.: Assessment of food security, living condition, personal hygiene health determinants and relations among Almajiri students in Sokoto metropolis, Nigeria. J. Thu Dau Mot Univ. **63–76**, 10.37550/tdmu.ejs/2023.01.372 (2023)
43. Alshurideh, M.T., et al.: Fuzzy assisted human resource management for supply chain management issues. Ann. Oper. Res. (2022). https://doi.org/10.1007/s10479-021-04472-8
44. Ibrahim, F.S.B., Esa, M.B., Kamal, E.B.M.: Towards construction 4.0: empowering bim skilled talents in Malaysia. Int. J. Sci. Technol. Res. **8**(10), 1694–1700 (2019)
45. Rüth, R., Netzer, T.: The key elements of cultural intelligence as a driver for digital leadership success. Leadership, Educ. Personal. An Interdiscip. J. **2**(1), 3–8 (2020). https://doi.org/10.1365/s42681-019-00005-x

The manufacturer's authorised representative in the EU is Springer Nature Customer Service Centre GmbH, Europaplatz 3, 69115 Heidelberg, Germany. If you have any concerns regarding our products, please contact ProductSafety@springernature.com

Printed and bound by CPI Group (UK) Ltd, Croydon, CR0 4YY
16/09/2026
02217815-0015